CONTENTS

List of Figures

FIGURE   **PAGE**

FIGURE   **PAGE**

FIGURE   **PAGE**

FIGURE   **PAGE**

List of Tables

Preface

The Galileo Project: Commitment, Struggle, and Ultimate Success

I am glad to see that someone has followed through and done a really comprehensive and workman-like job capturing the history of the Galileo project. This book goes all the way back to the initiation of the project, when it was more or less just a thought in a few peoples' minds, and traces the whole evolution from there. I got a great deal out of reading it.

I have a vested interest in the Galileo project—I was deeply involved in it for over a decade. A lot of people know about the mission and its terrific science return, but they don't know about the struggle putting the project together, getting it started, and keeping it going through all of the reprogramming and restructuring. One of the arguments that we used with people on Capitol Hill to keep the program alive through delays in the congressional budgeting process was that Galileo would be a nonthreatening manifestation of our country's technological capabilities, and this would send a powerful message to the rest of the world. It did just that.

Galileo meant a lot to the United States, but it also meant a lot to our space science community, because at the time that we were going through the development of Galileo, it was the only major deep space project. There were Earth satellite launches going on, but nothing to the planets. It was Galileo that really helped NASA and the U.S. space science community maintain viability during a period of extreme drought in program development. A lot of capability would have disappeared over the course of the 10 years that Galileo was in development.

The commitment that individuals made to Galileo was extraordinary. Many individuals committed a third or more of their professional lifetimes to executing this project. Over the years, situations developed so many times where it looked like there was just no way out for the project, but we always managed to come up with a solution. The number of times we managed to pull the fat out of the fire was truly remarkable.

The Galileo project was complex in that it required funding for science instrument and spacecraft development from numerous sources, including NASA and its Centers, the Department of Energy, U.S. universities, and the Europeans (especially the Germans). Just how the pieces of the fabric were woven together into what turned out to be a very successful program—one that required an investment of almost two decades of preparatory and execution work to bring about—is an interesting story. Revisiting the program from a historical point of view is what motivated me to read this book. I think that people who are interested in the space program, its science achievements, and its contribution to technology in general will really appreciate this history. It's comprehensive, it's complete, and it seems to me to be pretty even-handed. I'm very appreciative of what Michael has done.

—John Casani, First Galileo Project Manager

Foreword

We Are All Standing on the Bridge of Starship *Enterprise*

This book details the history of the Galileo mission. Galileo had political ups and downs, technical challenges and hurdles, and was a multigenerational task. We had people starting the mission on advisory committees and senior management jobs who have now passed on. Many people went through parts of Galileo in stages of their careers. I have friends who still mark their anniversaries and the birthdates of their children in terms of, "That was when we were on the beginning stages of prelaunch preparation," or, "It was during our first Europa encounter when that happened." The Galileo team felt much more as a family than a pure professional enterprise. People worked together for long periods of time, through good times and bad, to accomplish this thing not just for them, but for everybody.

The Galileo mission to Jupiter was part of the grand sweep of solar system exploration. You can view planetary exploration as a human endeavor—a wave sweeping outward from Earth. We went to the close-in places first—the Moon, Venus, and Mars. The outer solar system was the next big frontier, and it was an order of magnitude more difficult to explore. The distances are truly staggering, and the problems of developing spacecraft that could survive on their own for long periods of time were major challenges.

The early explorations of the outer solar system were performed by relatively fast-trajectory spaceships, like the Pioneers and Voyagers. They and Galileo were major steps forward in being able to develop reliable craft that would operate for decades, continue to send data back without failing, and be smart enough to take care of themselves out of communication with Earth. At Jupiter, the time available to send a radio signal is typically 45 minutes, and another 45 minutes before you get an answer back. You are well beyond being able to do the types of things that can be done in Earth orbit.

NASA's outer solar system missions helped turn that region into a known place, rather than just the realm of astronomers. The outer solar system became someplace that can be talked about and thought about in geological and geophysical terms. The person on the street and kids in school can say, "Hey, I saw a picture of a moon of Jupiter the other day and it had volcanoes on it."

We planned the Galileo mission in that context. The scientific advisory committees to NASA and the U.S. government laid out an exploration strategy of fast reconnaissance missions followed by missions such as Galileo, in which we orbited planets and did more detailed studies. There was a leapfrogging characteristic to this type of exploration. The missions take so long to plan and execute that the next wave of exploration must be prepared even before the current one can be launched. We began work on Galileo in 1972 in its infant form, and we really got to work on it in 1974 with detailed studies, even before Voyager was launched in 1977.

On Galileo, we combined orbital and in situ exploration strategies. We believed that if we were going through so much effort and so many resources to get there, we ought to not only study the planet, its satellites, and magnetic fields, but that we also should understand the chemistry of the atmosphere in detail with an entry probe that actually went into the atmosphere, grabbed a sample, analyzed it, and sent the data back to the mother ship before burning up in the atmosphere. This was very ambitious.

We had a strategic plan that said we were going to go orbit Jupiter, get into its atmosphere, and then inform that plan with tactics derived from Voyager results. It is a good example of the way exploration progresses. You learn things that lead to new questions you want to answer, and so forth. We were preparing to follow up and understand Jupiter's miniature planetary system at a very detailed level, even as Voyager was continuing on from Jupiter to Saturn, opening up the rest of the outer solar system.

People frequently ask, "Why should the average person be interested in what's going on in the Jupiter system?" There are answers to that on all levels, ranging from the visibility of high technology to developing new things that have spinoffs to enhancing national prestige to satisfying pure curiosity. But really, it is all about changing the way we look at the universe and the world. We want to know how planets tick and understand the processes that control us here on Earth—everything from geophysics to climatology to global warming. The universe is effectively a laboratory waiting for us to study these things.

Most of the people who have worked on Galileo over the years probably regard their biggest contribution as having changed the textbooks. Kids today learn about such things as the moons of Jupiter, and they know what they're talking about. To them, the planets become not just dots in the sky that you can barely see with a telescope. The planets become real places, and kids know their characteristics.

There is always a tension in the national debate about how much robotic exploration (such as Galileo) we should do versus so-called human exploration (such as Apollo). This misses the point! What we call robotic exploration is in fact human exploration. The crews sitting in the control room at Jet Propulsion Laboratory as well as everyone out there who can log on to the Internet can take a look at what's going on. So, in effect, we are all standing on the bridge of Starship *Enterprise*.

It is important to note that Galileo was an international mission. Science is, by its nature, international. We had people contribute from countries all over the European community and other countries as well. This was one of the first missions in which the analysis of data more or less continued around the clock, around the globe. We'd wake up in the morning and hear that one of our colleagues in Berlin had processed some images overnight and brought new data to the table—and we could look at it immediately, while they had a chance to sleep.

The intellectual children, grandchildren, and great-grandchildren of the people who worked on Galileo have now spread through the crews operating the Cassini spacecraft at Saturn and the MER Rovers, Spirit and Opportunity, on the surface of Mars. It is an ongoing spirit of exploration and, to me, that's really the bottom line of what Galileo was all about. It is important to record the history of these types of things, both because of the intrinsic interest that the public has and because there are always lessons to be learned. I sure hope people will read this book and get that feeling from it.

—Torrence V. Johnson, Galileo Chief Scientist

Acknowledgments

So many people gave me valuable input for this book. I want to thank them for their time and interest in this project. Some of these people include:

Torrence V. Johnson, who related with eloquence how decisions were made and operations were carried out on the mission.

John Casani, whose project leader perspectives on the mission's importance and on the political battles that had to be fought were hugely important for my book. Also, thank you to the mission's other project managers: Dick Spehalski, Bill O'Neil, Bob Mitchell, Jim Erickson, Eilene Theilig, and Claudia Alexander, all of whom gave me valuable insights on what Galileo was all about.

Thanks go out to the mission engineers and technicians at Jet Propulsion Laboratory (JPL), especially Nagin Cox, who is indeed a poet when she talks about Galileo, Gerry Snyder, Duane Bindschadler, Brad Compton, Gregory C. Levanas, Theodore Iskendarian, and various members of the Galileo mission support crew; experts from Ames on the Probe, including Charlie Sobeck, Ed Tischler, and Joel Sperans; and Krishan K. Khurana of UCLA. Also, thanks to Dick Malow, a former congressional committee staff director; Allan J. McDonald of Thiokol Propulsion; Carl Fromm for his help and encouragement; Julie Cooper for patient help in finding graphics for the book; and Craig Waff for sharing his extensive research on the Galileo mission. In addition, a special thanks goes to the staff of JPL Archives, especially Russell Castonguay, who was incredibly helpful in locating long-buried letters and documents.

At NASA Headquarters, the History Division staff, in particular, former NASA Chief Historian Roger Launius and the current NASA Chief Historian Steven Dick, deserve much credit for their support and oversight. Archivists Colin Fries and Jane Odom helped a great deal by finding and organizing essential archival material. Steve Garber oversaw the production process. Interns Liz Suckow, Jennifer Chu, Giny Cheong, and Gabriel Okolski all helped tremendously in obtaining and organizing images.

Also at NASA Headquarters, the talented professionals in the Office of Printing and Design deserve much credit. Lisa Jirousek and Dyana Weis carefully copyedited the manuscript, designers Cathy Wilson and Smahan Upson laid out the manuscript in a very attractive manner, printing specialists Jeffrey McLean and Henry Spencer handled this crucial last step, and supervisors Steven Johnson and Gregory Treese managed the whole process.

And finally, a thank you to my wife, Naisa Kaufman, for her talented and tough manuscript editing.

"All truths are easy to understand once they are discovered. The point is to discover them."

—Galileo Galilei

Introduction
MEETING THE GRAND CHALLENGE

IN AN

ADDRESS TO THE U.S. SENATE SUBCOMMITTEE ON Science, Technology, and Space, author James Michener asserted that "it is extremely difficult to keep a human life or the life of a nation moving forward with enough energy and commitment to lift it into the next cycle of experience . . . . There are moments in history when challenges occur of such a compelling nature that to miss them is to miss the whole meaning of an epoch. Space is such a challenge."[1]

The Galileo mission to Jupiter successfully explored a vast new frontier, had a major impact on planetary science, and provided invaluable lessons for the design of subsequent space vehicles. In accomplishing these things, Galileo met the challenge of "such a compelling nature" that Michener envisioned. The impact of the mission was felt by those who worked on it, the country that supported it, and the people from other parts of the world who were deeply impressed by it. In the words of John Casani, the original Project Manager of the mission, "Galileo was a way of demonstrating . . . just what U.S. technology was capable of doing."[2] An engineer on the Galileo team expressed more personal sentiments when she said, "I had never been a part of something with such great scope . . . . To know that the whole world was watching and hoping with us that this would work. We were doing something for all mankind . . . I'd walk outside at night and look up at Jupiter, and think, my ship's up there."[3]

---

[1] James A. Michener, "Space Exploration: Military and Non-Military Advantages" (speech delivered before the U.S. Senate Subcommittee on Science, Technology, and Space, Washington, DC, 1 February 1979). Published in Vital Speeches of the Day (Southold, NY: City News Publishing Company, 15 July 1979).

[2] John Casani interview, tape-recorded telephone conversation, 29 May 2001.

[3] Nagin Cox interview, tape-recorded telephone conversation, 15 May 2001.

Like other grand voyages of discovery, Galileo altered the way we view our surroundings (in this case, our planetary surroundings). It is thus fitting that this mission to the Jovian system was named after a man whose own astronomical observations radically challenged the way that people of his time viewed their universe. The discoveries of both Galileo the man and Galileo the spacecraft brought us new perceptions of our planetary system, made our lives richer and more interesting, and breathed new vitality into our quest to understand ourselves and our universe.

It's continuing to be a marvelous voyage of discovery.[1]

—Torrence V. Johnson, Galileo Chief Scientist

Chapter 1
THE IMPORTANCE OF THE GALILEO PROJECT

WHEN THE GALILEO SPACECRAFT LAUNCHED ON 18 October 1989, it began an interplanetary voyage that took it to Venus, two asteroids, back to Earth (twice), and finally on to Jupiter. It studied the planet's intense magnetic fields, belts of radiation, and high-energy particles, and it sent off a planetary probe that accomplished the most difficult atmospheric entry ever attempted. The Galileo spacecraft repeatedly swooped by Jupiter's Galilean moons, using a suite of scientific instruments to delve into their structures and properties. The mission amassed so many scientific "firsts" and key discoveries that it can truly be called one of the most impressive feats of exploration of the 20th century. The Galileo mission added dramatically to our understanding of the Jovian system and our entire solar system. Moreover, the mission was a triumph of teamwork and ingenuity under exceedingly difficult conditions.

The Lure of Jupiter for People on Earth

For centuries, the Jovian system has stirred our imaginations and been the frequent subject of our observations. The astronomer Galileo Galilei, using an early telescope that he himself made, discovered that Jupiter was not only a planet, but also the center of its own system of moons. This discovery called into question the Aristotelian-Ptolemaic view

---

[1] Statement by Torrence V. Johnson, Galileo chief scientist. Reported in "Galileo's Bounty," *San Francisco Chronicle* (31 July 2000): A4.

of the universe, which was accepted at the time by the Catholic Church, that all things revolve around Earth and, by extension, us. Thus, there could be only one center of revolution, and yet, Jupiter did not appear to obey that law.[2]

The Aristotelian-Ptolemaic model held that Earth was fundamentally different from the other planets in that it was "corruptible," "alterable," and "naturally dark and devoid of light."[3] Earth, it was thought, was also the only planet accompanied by a moon. Galileo's discovery showed that this was not so. Jupiter was more like Earth than had been previously imagined. Using this new knowledge, Galileo argued that Earth indeed might not be fundamentally different from the other planets in additional ways. In his 1632 treatise entitled *A Dialogue Concerning the Two Chief World Systems,* Galileo used his discovery of the Jovian moons, as well as many other observations of Jupiter, the other planets, and Earth's Moon, to defend the Copernican model of our planetary system. In Copernicus's view, Earth was not immovable. All of the planets, including Earth, revolved around the Sun. Such a concept challenged the anthropocentric ideas of the time and flew in the face of official Church doctrine.[4]

Jupiter's sheer size and energy have added to our interest in the planet. It is one of the brightest objects in the night sky. Unlike other planets, it emits more thermal radiation than it absorbs from the Sun. It is also a radio-wave source. A storm bigger than Earth (the Great Red Spot) has been raging on its surface for centuries.

Jupiter contains 75 percent of the total nonsolar matter in the solar system. The planet is so big that it constitutes a transitional object between a terrestrial-type planet and a star. Its composition—mainly hydrogen and helium—is closer to the Sun's composition than to that of Earth.

Another important aspect of Jupiter is the effect that its enormous gravity has on its satellites. Nowhere else in the solar system do planets interact so strongly with their moons. In particular, Jovian gravity causes mammoth tides and extensive volcanism on the satellite Io, the most tectonically active body in the solar system. Geological changes occur so rapidly on Io that in the space of only five months, Galileo was able to observe significant alterations in a large section of the moon's surface (see the Io section in chapter 9 for a fuller discussion of these observations).

The Galileo mission was the first to provide extended observations of Jupiter's dynamic magnetic fields. The Jovian magnetosphere is the largest of any planet's, so expansive that it could envelop the Sun and much of its corona. If our eyes could see magnetic field lines, Jupiter's magnetosphere would appear larger than our Moon, in spite of its distance from us.

Finally, Jupiter is a window into our own past. Many scientists believe that the composition of the planet is little altered from that of the original solar nebula. As such, the Galileo spacecraft's observations provided a look back in time to the early stages of

---

[2] Office of the Vice President of Computing, Rice University, "Jupiter and Her Moons: One Planet's Quest to Defy Aristotle," The Galileo Project, *http://es.rice.edu/ES/humsoc/Galileo/Student_Work/Astronomy95/jupiter.html* (accessed 25 July 2003).

[3] Galileo Galilei, A Dialogue Concerning the Two Chief World Systems, 1632, available online at *http://webexhibits. org/calendars/year-text-Galileo.html* (accessed 24 July 2003).

[4] Roger D. Launius and Steve Garber, "Galileo Galilei," *Biographies of Aerospace Officials and Policymakers, E–J,* National Aeronautics and Space Administration (NASA) Office of Policy and Plans, *http://www.hq.nasa.gov/office/ pao/History/biose-j.html* (accessed 20 November 2001); "Nicolaus Copernicus," JOC/EFR, University of St. Andrews, Scotland, November 2002, *http://www-history.mcs.st-andrews.ac.uk/history/Mathematicians/Copernicus.html*; A. Mark Smith, "Galileo," *World Book Online Americas Edition, http://www.worldbookonline.com/ar?/na/ar/co/ ar215300.htm* (accessed 20 March 2003); Galileo Galilei, *A Dialogue.*

our solar system. Being able to perform such an examination was important because it helped untangle the "bewildering array of processes and phenomena which have affected the evolution of the planets and which control their environments and futures."[5]

Galileo: A New Phase in the Study of the Outer Planets

The two Voyager spacecraft, which launched in 1977, completed visits to all of the planets known to ancient astronomers. The Galileo mission began a different phase in the study of the outer planets: an era of more careful, systematic study, featuring close flybys and in-depth analyses of planetary system characteristics.[6]

Galileo was not a Jupiter mission in the traditional sense of the term, which typically refers to a project focusing on a single target. Galileo's objectives were much broader, encompassing a holistic analysis of the entire Jovian system of satellites, primary planet, magnetic fields, and particle distributions.

Galileo's Impact on Future Deep Space Missions

The Galileo project team's approach to exploration, spacecraft technical design, and operational problems will all strongly affect the manner in which other projects are conducted in the decades to come. Among the Galileo mission's memorable achievements were the repeated successes of its staff in solving serious technical problems, even though the spacecraft was hundreds of millions of miles from home. Examples include the responses of the team to the high-gain antenna deployment failure and to the jamming of the spacecraft's data tape recorder, both of which will be discussed in detail in later chapters. In each instance, the team attacked potentially mission-ending problems and found ingenious ways to keep the spacecraft operational and productive. Results of these efforts, such as enhancement of the Deep Space Network (DSN) and development of new data-compression software, constituted technological improvements that will benefit future missions as well.

The Many Firsts of Galileo

Galileo was the first deep space mission designed to launch on an expendable launch vehicle from the Space Shuttle's payload bay rather than from Earth. This was not a trivial achievement. Designing enough thrust into the craft to reach Jupiter while carrying its large, sophisticated payload took a great deal of creativity, especially after the *Challenger* disaster precluded the use of liquid fuel due to safety considerations.

The Galileo mission was also the first to make a close flyby of an asteroid (Gaspra), discover an asteroid moon (Ida's satellite Dactyl), orbit a gas giant planet

---

[5] T. V. Johnson, C. M. Yeates, and R. Young, "Space Science Reviews Volume on Galileo Mission Overview," *Space Science Reviews* 60.1–4 (1992): 5–6.

[6] Eric G. Chipman, Donald L. De Vincenzi, Bevan M. French, David Gilman, Stephen P. Maran, and Paul C. Rambaut, "The Worlds That Wait," chapter 7-2 in A Meeting with the Universe: Science Discoveries from the Space Program (Washington, DC: NASA Publication EP-177, 1981), *http://www.hq.nasa.gov/office/pao/History/EP-177/ch7-2.html* (accessed 20 March 2000).

(Jupiter), send a probe down into the Jovian atmosphere, and make direct observations of a comet (Shoemaker-Levy) smashing into a planet.[7]

Other Milestone Accomplishments

The Galileo spacecraft made two flybys of Earth before leaving the inner solar system, using our planet's gravity to alter its trajectory in critical ways. During these flybys, the spacecraft made important observations of our Earth-Moon system. For instance, the spacecraft confirmed the existence of a huge impact basin on our Moon's far side that had been inferred, but never observed, from Apollo data. Galileo also provided evidence of more extensive volcanism on the Moon than previously thought.[8]

To reach Jupiter, Galileo traveled 2.4 billion miles at an average speed of 44,000 miles per hour (mph). After leaving Earth, it required only 67 gallons of fuel to control its flightpath and orientation. So accurate was Galileo's trajectory that the craft missed its target point near the Jovian moon Io by a mere 67 miles. This feat has been likened to shooting an arrow from Los Angeles to New York and missing a bull's-eye by 6 inches.[9]

On its way to Jupiter, Galileo discovered the most intense interplanetary dust storm ever observed. It found a very strong radiation belt above Jupiter's cloud tops and Jovian wind speeds exceeding 720 kilometers per hour (450 miles per hour, or mph). These speeds remained fairly constant to depths far below the clouds, unlike the jetstreams on Earth. Probe measurements within the Jovian atmosphere revealed a surprisingly small amount of water compared to what had been expected from Voyager observations, as well as far less frequent lightning than predicted. The Probe observed only one-tenth the lightning activity per unit area of that found on Earth, although the strength of individual events exceeded those on Earth by an order of magnitude.[10]

The spacecraft observed that Jovian helium abundance (24 percent) is very nearly that of the Sun (25 percent)—one reason why Jupiter can be considered a transitional object between a planet and a star. From magnetic data collected by Galileo, scientists suspect that Ganymede, the largest Jovian satellite, creates its own magnetic field by means of an Earth-like dynamo mechanism within its core. This was the first moon found to exhibit such magnetic field characteristics. But the most dramatic and important discovery, according to many on the Galileo team, was the strong evidence of a deep saltwater ocean beneath the icy crust of the moon Europa. Such an ocean could conceivably harbor simple forms of life.[11]

---

[7] National Space Science Data Center, "Galileo Project Information," *http://nssdc.gsfc.nasa.gov/planetary/galileo.html*, p. 2 (accessed 11 March 2000).

[8] "Galileo Project Information," pp. 2–3.

[9] Everett Booth, "Galileo: The Jupiter Orbiter/Space Probe Mission Report," December 1999, p. 19, JPL internal document, Galileo—Meltzer Sources, folder 18522, NASA Historical Reference Collection, Washington, DC.

[10] "Galileo Project Information," p. 3; Donald J. Williams, "Jupiter—At Last!" *Johns Hopkins APL Technical Digest* 17:4 (1996): 347, 354; "The Probe Story: Secrets and Surprises from Jupiter," *Galileo Messenger* (April 1996): 3.

[11] "Galileo Project Information," p.3; Booth, p.9.

The Multiteam Cooperation Required To Develop the Galileo Spacecraft

The success of the Galileo mission was partly due to the efficient management of an incredibly complex and diverse network of specialist teams. Such a network was required to develop the sophisticated spacecraft that could perform the many mission functions required.

The Galileo spacecraft was characterized by its highly integrated design. Its diverse subsystems, developed by different project teams, needed to interface flawlessly with each other over many years. Subsystem dependencies needed to be fully understood and controlled because the environment experienced by a typical subsystem was strongly affected by the behavior of other connected subsystems. For example, the limited electric power that was available aboard Galileo had to be shared by temperature control and many other spacecraft subsystems in a manner that minimized energy use. Large temperature swings inside the spacecraft could result from this situation, and this placed stringent design requirements on the subsystems that had to withstand those swings.[12]

The task that had been laid before Galileo project management[13] was to design, fabricate, test, modify as needed, and ultimately fly the highly integrated, complex spacecraft. To accomplish this, communications among a large number of different organizations had to be smooth and effective. The lead NASA Field Center for the project was the Jet Propulsion Laboratory (JPL) of the California Institute of Technology, working for NASA's Solar System Exploration Division. Galileo's Orbiter, which conducted extensive flybys of Jupiter and its moons, was developed by JPL, operating in close coordination with a range of university, industry, and government laboratory groups.

Galileo's Jupiter Probe, which entered the planet's atmosphere, was designed by NASA's Ames Research Center and manufactured by Hughes Aircraft Company under contract to Ames. The U.S. Department of Energy provided the plutonium-powered electrical generators that ran the various spacecraft functions. NASA's Lewis Research Center (now known as Glenn Research Center) helped develop the propulsion system and integration of the payload with the Space Shuttle. The Shuttle itself was supplied by Johnson Space Center, while Kennedy Space Center made available launch and landing facilities.

The Galileo Project also involved international cooperation with the Federal Republic of Germany, which provided instrumentation for both the Orbiter and the Probe, a propulsion system for the Orbiter, and telemetry and command support from the German Space Operations Center.[14]

The agreements forged in this project established a context for future space missions and the activities and complexities that surround them. Important management lessons were learned from the international nature of the mission. For instance, when mission-critical equipment is being designed, manufactured, and tested by other countries, NASA needs to create parallel organizations in both the managing country and "subcontractor" countries to handle the day-to-day coordination issues that arise.

During the Galileo Project, parallel organizations in the United States and Germany developed a contractual instrument called an "Interface Requirements Document" (IRD) that contained all the technical specifications for both the propulsion system being built

---

[12] R. J. Spehalski, "Galileo Spacecraft Integration: International Cooperation on a Planetary Mission in the Shuttle Era," *Earth-Orient. Applic. Space Technol.* 4.3 (1984): 139.

[13] John R. Casani of JPL was the Project Manager during most of Galileo's development.

[14] Spehalski, "Galileo Spacecraft Integration," pp. 139–140.

in Germany and its interface with the U.S.-built Orbiter. The IRD gave the United States a vital measure of control over the work performed by Germany. Contractual arrangements, however, necessitated the development of the IRD early in the project, while the spacecraft design was still in a period of flux. A great deal of time, more than originally planned, was expended by both countries in negotiating acceptable designs and verification procedures for the propulsion system development. The unplanned time and effort, which were deemed necessary due to the mission-critical nature of the propulsion system, resulted in unforeseen costs and delays.

NASA might have avoided the above cost overruns and schedule delays if it had not placed responsibility for mission-critical equipment in one country and project management in another. Lessons learned from the Galileo experience were implemented in planning the International Solar Polar Mission, in which NASA and the European Space Agency each took on the responsibility of designing and fabricating functionally independent spacecraft. A similar management model could be applied to an international Orbiter/Probe effort by assigning one nation the job of developing the Orbiter and another nation the job of delivering the Probe.[15]

Importance of Galileo for the U.S. Space Program

John Casani, Galileo's first Project Manager, believed that our country's space exploration program has had strong benefits. It has elicited worldwide respect and admiration for the United States and has effectively demonstrated, in a nonthreatening way, our country's technological capabilities. Moreover, our space program has been a source of pride and inspiration for millions of Americans, as well as a motivation for young people to pursue careers in science and technology. Our ventures into space have demonstrated, loud and clear, that the U.S. holds a position of unparalleled preeminence in the fields of space science and space technology.[16]

The Galileo mission was critical for preserving NASA's planetary exploration capability and for advancing the technology and survivability of robotic spacecraft. Galileo was an international cooperative program that strongly demonstrated to the European community the dependability of U.S. space exploration commitments. Such a demonstration was especially pertinent because it came in the wake of the International Solar Polar Mission's termination.[17]

Galileo was a cornerstone of NASA's planetary exploration program and achieved the highest scientific objectives for the outer solar system. The project maintained high visibility in a very positive manner for the United States at what can be considered a modest cost. During Galileo's development, the entire planetary exploration program constituted only 5 percent of NASA's budget, which was in turn only 1 percent of the federal budget.[18]

---

[15] Ibid., pp. 148–150.

[16] John R. Casani to Thomas G. Pownall, President, Martin Marietta Corporation, "Why Galileo?" 8 December 1981, John Casani Collection, Galileo Correspondence 11/81–12/81, folder 24, box 3 of 6, JPL 14, JPL Archives.

[17] Casani, "Why Galileo?" p. 2.

[18] John R. Casani to Fred Osborn, 4 December 1981, John Casani Collection, Galileo Correspondence 11/81–12/81, folder 24, box 3 of 6, JPL 14, JPL Archives.

In its 1981 budget authorization, the U.S. Senate's Committee on Commerce, Science and Transportation recognized the importance of the Galileo mission, as well as that of the rest of the planetary exploration program. The Committee noted the "spectacular accomplishments" of the exploration program, particularly with respect to the investigation of Saturn, Jupiter, Venus, and Mars. The Senate Committee was concerned, however, about the four-year gap that would occur between Voyager 2's 1981 encounter with Saturn and its encounter with Uranus, as well as the dearth of planetary imaging data that would result during this time. The Committee expressed its strong support for our country's world leadership position in planetary exploration but was quite apprehensive about other nations' aggressive programs in this area. In particular, the Committee regretted the lack of planned U.S. representation in missions to Halley's Comet when it made its closest approach to the Sun in 1986. The Soviet Union, France, and other European nations, as well as Japan, were all developing missions to investigate the comet.[19]

The U.S. House of Representatives Committee on Science and Technology concurred with the Senate's concerns, commenting that the House Committee would "not preside silently over this abandonment of U.S. leadership in planetary exploration." The House Committee went on to point out the need to "nurture planetary science with new data," expressing the fear that "where there is no new information to work with, a vital science becomes arid academic speculation."[20]

The Committee also commented that it considered NASA's civilian programs a "national resource" that has contributed substantially to the U.S. economy and our scientific preeminence. Results from these programs have, in the Committee's view, added to our understanding of the universe and to key questions regarding life, matter, and energy.[21]

How Important Is Our Need To Continually Pioneer New Frontiers?

Throughout human history, we have repeatedly wondered about the nature of our universe and have sought to understand it better. Ancient Greek, Roman, Chinese, Arabic, Egyptian, Mayan, and many other cultures expended considerable resources trying to fathom what lay beyond their immediate lines of sight. These early scientific quests resulted in bodies of knowledge that continue to inspire today's space scientists.[22]

Plato sensed the significance of these types of inquiries. In *Republic VII*, Socrates asked, "Shall we make astronomy the next study?"[23] Glaucon replied that certainly they should, listing the practical benefits that a knowledge of seasons and lunar and solar cycles have for everyone, from military commanders to sailors and farmers. But Socrates

---

[19] Senate Authorization Report No. 97-100 (p. 37), Committee on Commerce, Science and Transportation, John Casani Collection, Galileo Correspondence 11/81–12/81, folder 24, box 3 of 6, JPL 14, JPL Archives.

[20] House of Representative Report No. 97-32 (pp. 11-8 and -9, 12-10, 16-20), Committee on Science and Technology, John Casani Collection, Galileo Correspondence 11/81–12/81, folder 24, box 3 of 6, JPL 14, JPL Archives.

[21] Senate Authorization.

[22] Roger Launius, ed., *The U.S. Space Program and American Society* (Carlisle, MA: Discovery Enterprises, Ltd., 1998), p. 4.

[23] Martin J. Collins and Sylvia D. Fries, eds., *A Spacefaring Nation: Perspectives on American Space History and Policy* (Washington, DC: Smithsonian Institution Press, 1991), p. 29, quoting from Plato, *The Republic: Book VII*, circa 370 to 375 BCE.

chided Glaucon for his eminently sensible view, hinting that astronomy may be important to study even if it is not immediately profitable. Former NASA Administrator James C. Fletcher spoke much more directly to this point when he linked the study of space to the highest of human endeavors—the "salvation of the world."[24]

The benefits of space exploration, as well as scientific research in general, have never been easy to quantify in terms of dollars and cents, although we have identified various ways in which such activities enrich us. Reaching out into space and developing the technology to make this possible have added to our understanding of Earth, our solar system, and the universe that surrounds us. Increasing our knowledge of these things is satisfying in its own right—we feel wiser and better knowing more about our universe. But our quest is also highly practical. The technology we develop can help improve our lives. Furthermore, we must make sure that control of space by other nations is not used to endanger our national security. Our explorations help prepare us to use our access to space to defend ourselves if necessary.

We enhance our country's prestige among the peoples of the world by successful endeavors beyond our atmosphere, while at the same time adding to our scientific, technological, industrial, and military strength. Last but not least, we satisfy the compelling urge of our species to explore and discover, to set foot where no one else has. Our unstoppable curiosity distinguishes us from other species on our planet. Since Earth's surface has largely been explored, we must now turn our gaze upward and outward.[25]

[24] Collins and Fries, *Spacefaring Nation*, p. 29.

[25] Launius, *The U.S. Space Program*, p. 18.

Chapter 2

FROM CONCEPTION TO CONGRESSIONAL APPROVAL

NASA STRUGGLED FOR MANY YEARS TO DEVELOP A MISSION that would intensively study Jupiter and its satellites. Two decades of analyses, debates, and political maneuvering elapsed before Congress finally gave its stamp of approval to such a mission in 1977. The tortuous path to a congressional "yes" reflected not only the technical difficulties and uncertainties of outer planet missions, but also the dramatic social upheaval of the 1960s and 1970s. Space exploration expenditures had to be balanced against the country's new commitment to social reforms, as well as to the tremendous cost of the Vietnam War. Also debated were the types of space exploration most appropriate to the country's goals and needs. Table 2.1 gives an overview of the many steps involved in creating the Galileo project.

Early Concepts of Outer Planet Missions

Four Types of Missions

Although serious planning for the Galileo project did not begin until the 1970s, seeds were planted as early as 1959, when JPL scientist Ray Newburn, Jr., developed several types of mission concepts for exploring the outer solar system. *Deep space flights* would pass only through interplanetary space, making measurements of the regions encountered. *Flyby* missions would briefly encounter planets on their way to other parts of the solar system. The data collected would be limited by the encounter time, but flyby missions had the advantage that several planets could be studied during one trip. Flyby missions might be compared to "windshield tours" of a city's major landmarks. *Orbiter* missions would place spacecraft in long-term trajectories around planets, providing opportunities for intensive study and data collection. *Planetary entry and lander* missions would involve penetration of the planetary atmosphere, allowing observations that spacecraft flying above the atmosphere could not obtain.[1]

In a 1960 planning report, JPL staff envisioned a planetary exploration program that incorporated some of Newburn's concepts and that, by 1970, would have "demonstrated and exploited" spacecraft capable of planetary orbiter and lander missions.[2]

Scientific Analyses Proposed

Ray Newburn's section in Albert Hibbs's 1959 report on the feasibility of solar system exploration envisioned particular types of experiments that could be performed by spacecraft visiting other planets. For Jupiter flyby and orbiter missions, he identified photographic, radar, and spectral investigations that would yield data on the planet's atmosphere, meteorology, and thermal structure. He imagined ways to study Jupiter's gravitational and magnetic fields and outlined what was known about the planet's radio emissions. He also mentioned pressure, density, temperature, and opacity analyses that could be performed during missions that penetrated the Jovian atmosphere. Newburn sensed the importance of outer planet exploration and understood the impact that new knowledge about Jupiter and its satellites would have on theories of the solar system's origin and evolution.[3]

Table 2.1. **Milestones leading to Galileo mission approval.**

1959–1961	JPL identifies the importance of outer planet exploration and envisions experiments for Jupiter flybys and orbiters.
1960	The Army Ballistic Mission Agency (ABMA) investigates propulsion system requirements for a Jupiter mission.
Early 1960s	Nuclear-electric propulsion concepts are analyzed by JPL's Advanced Propulsion Engineering Section. Problems are identified, and NASA focuses on chemical propulsion.
1964	Lockheed studies payload design concepts and launch vehicle requirements for asteroid belt and Jupiter missions.
Mid-1960s	Work on outer planet programs moves from conceptual studies to initial mission planning. JPL, Goddard Space Flight Center, and Marshall Space Flight Center all agree that the first outer planet missions should be Jupiter flybys.
Summer 1965	The National Academy of Sciences panel lays out plan for outer solar system exploration. A progression of Earth-based studies, flybys, orbiters, atmospheric probes, and landers is suggested.
Late Summer 1965	The possibility of a single spacecraft's Grand Tour of Jupiter, Saturn, Uranus, and Neptune (made possible by a rare configuration of the planets) is identified.

---

[1] R. L. Newburn, "Scientific Considerations," in *Exploration of the Moon, the Planets, and Interplanetary Space*, ed. Albert R. Hibbs, JPL Report No. 30-1 (Pasadena, CA: JPL, 30 April 1959); Craig B. Waff, "A History of Project Galileo. Part 1: The Evolution of NASA's Early Outer-Planet Exploration Strategy, 1959–1972," *Quest* 5.1 (1996): 4.

[2] JPL, *Ten-Year Plan*, JPL Planning Report No. 35-1 (Pasadena, CA: JPL, 1 September 1960), p. 11.

[3] Newburn, "Scientific Considerations," pp. 60, 76, 89; Waff, "A History of Project Galileo," p. 5.

Table 2.1. **Milestones leading to Galileo mission approval.** (continued)

1966–1970s	Popular and government support for post-Apollo space exploration wanes. Budgets decline. Nevertheless, outer planet mission conceptual studies continue.
February 1968	NASA initiates serious planning for two Pioneer missions to Jupiter and beyond.
December 1968	NASA authorizes its Lunar and Planetary Missions Board to suggest space exploration strategies. The board does not form a clear strategy on outer planet exploration but recommends more study.
1969	NASA forms an Outer Planets Working Group of scientists and planners from its laboratories. The group develops an outer planet trip schedule for 1970s and early 1980s that includes flybys, orbiters, and probes. The recommended flybys collectively constitute a Grand Tour.
June 1969	NASA convenes a "summer-study panel" of experts from the U.S. space science committee. The panel recommends five missions, all of which include atmospheric entry probes. Like the Working Group, the panel recommends outer planet flybys that collectively constitute a Grand Tour.
Summer 1969	NASA begins formulating a budget request to Congress for starting up a Grand Tour program.
Summer 1969	The Nixon administration issues budget guidelines that discourage NASA from seeking Grand Tour funding for FY 1971. The budget request is delayed for a year.
1969	The ASC-IITRI (Illinois) study recommends the use of two types of spacecraft (spin-stabilized and three-axis-stabilized) to meet different outer planet exploration objectives.
July–August 1970	The National Academy of Sciences convenes an Academy Review Board to recommend how NASA's reduced budget should be apportioned among its seven scientific disciplines. No clear consensus is achieved on outer planet exploration.
1970	The Office of Management and Budget (OMB) tries to delay startup of Grand Tour program from FY 1972 to FY 1973 but fails.
January 1971	NASA announces that it will choose an outer planet program direction by November. JPL and Ames Research Center are authorized to evaluate costs and benefits of different mission concepts.
Spring 1971	Outer Planet Missions funding is reduced by the Senate. OMB requests that NASA conduct simpler, less costly endeavors. The House and Senate agree.
July 1971	To build unified support in the U.S. scientific community for outer planet objectives, NASA appoints members of the community to a Science Advisory Group (SAG). JPL and Ames report their findings to SAG. But SAG cannot reach consensus on the best mix of Grand Tour flybys versus orbiter/probe missions.
August 1971	The National Academy of Sciences Space Science Board also reviews the JPL and Ames findings but, like SAG, fails to establish a unified position on outer planet mission strategies.
Summer 1971	NASA Administrator James Fletcher asks Congress for Grand Tour funding for FY 1973, but his position is weakened by the scientific community's divided opinions. The Grand Tour startup is funded, but with a greatly reduced budget.

Table 2.1. **Milestones leading to Galileo mission approval.** (continued)

December 1971	President Nixon strongly supports the startup of the Space Shuttle program but does not want also to fund a Grand Tour program that relies on developing the expensive new Thermoelectric Outer Planet Spacecraft (TOPS).
January 1972	NASA proposes two less ambitious Jupiter-Saturn flybys using already-developed Mariner spacecraft.
Spring 1972	Congress approves the Mariner missions (eventually called Voyagers 1 and 2).
March 1972	Pioneer 10 launches, bound for a Jupiter flyby.
1972	NASA's SAG, concerned about the lack of adequate heatshield technology and radiation exposure protection, recommends that Jupiter orbiter/probe missions not launch until the early 1980s.
1973	NASA studies the possibility of launching planetary missions from the Space Shuttle. The Jovian "satellite tour" concept is born.
April 1973	Pioneer 11 launches, bound for a Jupiter flyby.
December 1973	Pioneer 10 discovers that Jupiter's radiation is less severe than was feared. This finding, coupled with advances in heatshields, makes a Jovian orbiter/probe mission more feasible.
May 1974	NASA considers teaming with the European Space Research Organization (ESRO) to launch a Jupiter orbiter/probe in 1980.
Summer 1974	The National Academy of Sciences' Committee on Planetary and Lunar Exploration (COMPLEX) recommends Jupiter orbiter missions in 1980 and 1985, then changes its position, urging more study and, due to the budget situation, a decision between flyby and orbiter missions.
February 1975	NASA conducts further studies and recommends that a Pioneer Jupiter orbiter/probe mission be conducted, as well as outer planet flybys. The National Academy of Sciences also backs a Pioneer Jupiter orbiter/probe mission. NASA authorizes Ames to initiate "Phase B" development efforts for such a mission.
Fall 1975	Management of the Jupiter orbiter/probe development is transferred to JPL from Ames, which will only develop the atmospheric probe. The mission is officially named the "Jupiter Orbiter Probe," or JOP.
Fall 1975	The Air Force chooses a solid-fueled rocket design to launch JOP from the Space Shuttle. A Centaur liquid-fueled rocket was rejected because of fire danger during transport in the Space Shuttle's cargo bay.
Late 1975	NASA develops a marketing plan for convincing Congress to fund an FY 1978 start for JOP.
January 1977	A proposed project budget of $270 million is approved by NASA's top management and OMB and is submitted to Congress.
May 1977	The House Housing and Urban Development (HUD)–Independent Agencies Appropriations Subcommittee votes to eliminate all funding for JOP from the budget. The subcommittee is opposed to funding both JOP and the Hubble Space Telescope.

Table 2.1. **Milestones leading to Galileo mission approval.** (continued)

May–June 1977	The U.S. space science community organizes a massive lobbying effort for JOP. The community is successful in obtaining the Senate HUD–Independent Agencies Appropriations Subcommittee's approval for JOP (as well as the Hubble Space Telescope).
Summer 1977	A vote on the House floor supports JOP two to one. The JOP project officially begins on 1 October 1977.

Engineering Challenges

Newburn foresaw the types of scientific data that would be vital to collect on outer planet missions, as well as the technologies that would be needed to make the necessary observations. During this same time, other visionaries were establishing the engineering requirements necessary to transport a spacecraft. Warren H. Straly and Robert G. Voss of the Army Ballistic Missile Agency (ABMA) examined questions of trajectory; flight environment; and spacecraft design, guidance, control, and data transmission. They also examined the energy required versus payload for an initial mission to Jupiter and its satellites. They anticipated possible hazards in the flight environment that could end the mission, such as the following:

- Collisions with large chunks of rock in the asteroid belt between Mars and Jupiter. Such collisions were thought to be probable, although this later proved not to be the case.

- Effects from Jupiter's radiation belts, which were feared to be intense enough to damage spacecraft systems. Years later, a Pioneer mission showed these belts to be less hazardous than suspected.

- The difficulty of keeping electronic and other equipment operational through years of transfer time from Earth to Jupiter, during which the equipment would be subjected to dramatic extremes of temperature.

- The great distances involved in a Jupiter mission, which would make accurate navigation and dependable spacecraft-to-Earth communication and data transmission very difficult.[4]

Propulsion Systems

Straly and Voss identified many of the hurdles that would have to be overcome in order for a scientific payload to reach Jupiter in operative condition. It was not at all clear in these early planning stages, however, what propulsion system would best transport the payload to its destination. While Straly and Voss imagined that it would be a large chemical

---

[4] Warren M. Straly and Robert G. Voss, *Basic Requirements for the Exploration of Jupiter and Its Moons*, preprint no. 60-57 (American Astronautical Society, Sixth Annual Meeting, 18–21 January 1960, obtained from JPL Archives); Waff, "A History of Project Galileo," p. 6.

rocket, their colleague at ABMA, Ernst Stuhlinger, advocated nuclear-electric drives for a range of space programs, including sending a craft to the outskirts of the solar system. He suggested that nuclear-electric propulsion was a good technology to deliver robotic probes into Jupiter and Saturn orbits, for it could reduce the travel from about three years (for a chemically driven craft) to a little over one year. The propulsion system he envisioned would be capable of delivering continual thrust from Earth to Jupiter. Once Jupiter orbit was attained, the nuclear-electric system would also supply abundant power for maintaining spacecraft operating temperatures, communications, antenna orientation, attitude control, and data transmission functions.[5]

In-depth analyses of nuclear-electric and other propulsion designs were performed at JPL in the early 1960s by the newly formed Advanced Propulsion Engineering Section, headed by John J. Paulson. At first, nuclear-electric systems looked like an excellent choice. A detailed conceptual design of a Jupiter mission envisioned that a 45,000-pound spacecraft would be boosted to Earth orbit by conventional rocket stages and then propelled by a 1-megawatt nuclear-electric system. One-third of the spacecraft's weight (15,000 pounds of propellant, possibly cesium) would be consumed by the time the spacecraft reached Jupiter. The trip to Jupiter would include 70 days of powered flight to escape Earth orbit, an additional 145 days of powered flight to adjust the trajectory, 400 days of coasting, and 60 days of reverse thrust to slow down during the final approach.

Spacecraft propulsion appeared to constitute an ideal use of nuclear power. In 1963, the JPL group expressed its strong conviction that it was necessary to develop such propulsion systems for exploring the solar system. The group considered this technology as the logical next step to chemical rockets, which, in its view, would have to be exceedingly large to deliver only modest payloads.

Although nuclear-electric systems initially held tremendous promise, many practical difficulties were identified in more thorough analyses. Belief in the inevitability of nuclear-electric propulsion decreased severely when these difficulties, and the inherent costs of working through them, were examined. Nuclear reactors required large amounts of heavy shielding materials, and this was problematic for a spacecraft with an extremely high cost for transporting each pound of its weight out of Earth orbit. The necessity of physically separating the reactor from spacecraft electronics and payload also presented difficulties, as did the requirement for a large radiator area to control reactor temperatures. It was thought in the early 1960s that such a radiator, which might have to be fairly flimsy due to spacecraft weight restrictions, would be quite vulnerable to micrometeorite threats. Furthermore, the long duration of outer planet missions required that nuclear-electric system lifetimes approach or even exceed 20,000 hours. In the early 1960s, most engineers considered that to be a major difficulty.[6]

As a result of the perceived difficulties with nuclear-powered space vehicles, NASA focused its efforts on chemical means of propulsion. The Agency continued to "redirect" ballistic missile weapon systems into launch vehicles for the space program.[7]

---

[5] Straly and Voss, *Basic Requirements for the Exploration of Jupiter and Its Moons*, pp. 31–35; Waff, "A History of Project Galileo," p. 7.

[6] "Appendix E: Launch Vehicles for Mars Missions," *On Mars: Exploration of the Red Planet, 1958–1978, http://www. hq.nasa.gov/office/pao/History/SP-4212/app-e.html* (accessed 29 May 2000).

[7] Lieutenant Colonel C. L. Gandy, Jr., and Major I. B. Hanson, "Mercury-Atlas Launch-Vehicle Development and Performance," chap. 5 in *Mercury Project Summary, Including Results of the Fourth Manned Orbital Flight* (Washington, DC: NASA SP-45, October 1963), online at *http://www.hq.nasa.gov/office/pao/History/SP-45/ch5.htm* (accessed 29 May 2000).

The missions of the 1960s were driven by such launch vehicles as Atlas-Agena and Atlas-Centaur combinations, Saturns, and Titans. These used such chemical fuels as liquid hydrogen and oxygen (Centaur D-1A), modified kerosene and liquid oxygen (Atlas SLV series), fuming nitric acid and dimethylhydrazine (Agena B), or powdered aluminum and ammonium perchlorate (solid-fueled rocket motors).[8]

The Lockheed Study

In July 1964, NASA contracted with Lockheed Missiles & Space Company for a study that would help the Agency move forward on missions exploring the asteroid belt and Jupiter. The Lockheed team reported various findings regarding appropriate propulsion systems. The team noted that using Atlas-Agena or Atlas-Centaur vehicles for launching outer solar system missions had certain advantages in that they were smaller and more economical than, for instance, a Saturn-series vehicle. But it would take two to three years to reach Jupiter, depending on the particular launch time selected, and the team felt that this would tax the reliability of spacecraft systems. This difficulty could be avoided by switching to more powerful launch vehicles such as the Saturn 1B or Titan IIIC. Such a change would decrease trip time to as low as 450 days and could also increase the maximum payload size.

It was the Lockheed team's opinion that before detailed outer planet missions could be conducted, the asteroid belt between Mars and Jupiter had to be properly characterized. Asteroids were of considerable scientific interest in themselves, but in addition, the team thought that the asteroid belt might pose a threat to spacecraft passing through. Existing sensing equipment was not adequate for studying this region of the solar system, and the team thus stressed the need to develop meteoroid detection equipment.

The Lockheed team had much to say about the most appropriate type of power source for keeping spacecraft systems operational during a Jupiter mission. Solar panels were ruled out for generating power, due in part to the decreasing insolation (incident solar radiation) as the craft journeyed away from the Sun. Large areas of photovoltaic panels would also present a considerable weight penalty and might be damaged by the high particle fluxes that were anticipated. Because of the distributed mass, it would be more complex to control spacecraft orientation, and the panel array might at times block some of the sensing equipment. Many engineers recommended using radioisotope thermal generators (RTGs) instead, with the hope that the steep costs of such systems would decrease and availability of their isotope fuel would increase by the time construction of the spacecraft commenced. One side benefit of the RTG systems was that the heat they generated could be used to help maintain spacecraft operating temperatures.

Adequate spacecraft-to-Earth communications could be maintained, according to Lockheed, using a 7-foot-diameter antenna dish operating at 10 watts of power. Two possible payload scenarios were envisioned. One involved video observations and required fairly complex data handling. The other employed an x-ray detector, radio receiver, sounding system, and visual spectrometer. The data from this latter system would have been simpler to handle, but a heavier instrument payload would have resulted.[9]

---

[8] "Appendix E: Launch Vehicles for Mars Missions."

[9] Waff, "A History of Project Galileo," pp. 8–10.

The various conceptual studies of the late 1950s and first half of the 1960s identified a range of possible design scenarios and tradeoffs for the spacecraft, such as minimizing propulsion system costs at the risk of compromising spacecraft reliability, or minimizing payload weight while having to deal with difficult data-handling requirements. In the following years, selections would have to be made from the spectrum of what was possible, narrowing down system options to the best choice, given the budgetary and technical constraints that were being imposed on the space program. Such choices were to be made more difficult because the fiscal policies and social environment under which NASA operated were about to undergo a period of rapid change.

Jupiter Mission Planning During the Apollo and Mariner Years

The 1960s were golden years for NASA. Its momentous Apollo lunar program and the Mariner investigations of Venus and Mars conveyed the image of an agency that could accomplish great things and send spacecraft anywhere in the solar system. Never was morale higher. NASA Administrators radiated confidence and a can-do attitude toward the future. Apollo had the strong support of Presidents John F. Kennedy and Lyndon B. Johnson. President Johnson, in fact, saw the quest to capture outer space from the evil hands of the Soviets as a 20th-century equivalent to the Roman Empire's road system or the British Empire's mighty 19th-century navy.[10]

It was during the heady time of the mid-1960s that work on outer planet programs moved from conceptual studies to initial mission planning activities. The staff at JPL, Goddard Space Flight Center, and Marshall Space Flight Center agreed that the first outer planet efforts should be flyby missions of Jupiter. The spacecraft would conduct particle and field research and also evaluate dangers that other missions would encounter, such as those posed by asteroids and Jupiter's intense radiation.

NASA staff recognized the necessity of flyby Jovian reconnaissance missions, but they were not in consensus regarding subsequent programs. This question was addressed in some detail when the Space Science Board of the National Academy of Sciences convened a study panel to analyze key areas of space research. The Space Science Board was established in 1958 and served as the scientific community's chief agency for advising NASA on space programs. The study panel that the Board organized met in Woods Hole, Massachusetts, during the summer of 1965. A working group of 33 scientists examined planetary and lunar exploration issues, and three of the panelists, G. B. Field of the University of California at Berkeley, Raymond Hide of the Massachusetts Institute of Technology, and J. C. Jamieson of the University of Chicago, served on a Panel on Jupiter and Other Major Planets.[11]

The working group identified planetary exploration as "the most reward-ing scientific objective for the 1970–1985 period," and, among other accomplishments, the group laid out a plan for exploring the outer solar system. The group suggested

---

[10] Joan Hoff, "The Presidency, Congress, and the Deceleration of the U.S. Space Program in the 1970s," in *Spaceflight and the Myth of Presidential Leadership*, ed. Roger D. Launius and Howard E. McCurdy (Urbana and Chicago: University of Illinois Press, 1997), p. 93.

[11] Space Science Board, *Space Research: Directions for the Future, Part One* (Washington, DC: National Academy of Sciences, December 1965), pp. iii, 142; Andrew J. Butrica, "Voyager: The Grand Tour of Big Science," chap. 11 in *From Engineering Science to Big Science: The NACA and NASA Collier Trophy Research Project Winners*, ed. Pamela E. Mack (Washington DC: NASA SP-4219, 1998), p. 253, online at *http://history.nasa.gov/SP-4219/Chapter11.html*; Waff, "A History of Project Galileo," p. 11.

that each outer planet first be analyzed in depth with studies performed by telescopes on Earth, as well as with observations taken from Earth-based balloons, rockets, and satellites. Once these baseline data were obtained, each outer planet should then be explored sequentially by flybys, orbiters, atmospheric probes, and eventually landers.[12]

Flybys, the panel noted, would provide valuable data upon which subsequent missions could build. In particular, flybys could assess magnetic fields, charged particle densities, particulate matter quantities, and infrared emissions. Subsequent planetary orbiters would then carry out many more measurements, including high-resolution photography; infrared surveys as functions of local time and latitude; laser reflection determination of cloud heights; ionospheric sounding; and detailed mapping of fields, particles, and radio sources over space and time.

Specialized satellites ejected from the planetary orbiter were envisioned. Low-altitude satellites would seek to make precise measurements of a planet's surface features; satellites with eccentric orbits would encounter a planet's moons; and close, inclined satellites would study areas of high interest such as Jupiter's Great Red Spot. Data collected by the specialized satellites could be relayed to Earth by the planetary orbiter.

The panel also underlined the importance of an entry probe (or, in the panel's terminology, a "drop sonde"), ejected from the planetary orbiter, that would plunge into Jupiter's atmosphere, radioing back temperature, pressure, and other data as it descended. Such a probe could obtain decisive information about pressure-density relationships deep in the atmosphere that would be available in no other way. In addition, the point at which the probe was destroyed could possibly identify the location of a fluid-solid interface. Acoustic sounding might also be able to identify interfaces.[13]

The panel recognized that an outer planet lander would be the most ambitious type of mission, requiring formidable reserves of fuel to land successfully once it reached its target planet. Also noted was the difficulty that would be encountered in transmitting data to Earth from vehicles of any kind among the outer planets. The panel observed that transmitting data to Earth from Jupiter, Saturn, or Uranus would be 12, 40, or 160 times more difficult, respectively, than transmitting from Mars. To attain a sufficiently high rate of information transfer, space vehicle power supplies would need to be improved and Earth-based receiving antennas would need greater sensitivity.[14]

The efforts of the National Academy of Sciences panel were pivotal in laying out the U.S. space program's course of further development. This was the first major analysis by the U.S. scientific community of how outer planet exploration might be conducted. From the panel sprang a basic vision for outer planet mission structures that provided a foundation for Galileo and other programs.

The Grand Tour Concept

Shortly after the National Academy of Sciences panel met, there was a discovery that greatly enhanced both public and government interest in an outer planet mission. In

---

[12] *Space Research*, p. 50; Waff, "A History of Project Galileo," p. 11; Butrica, "Voyager: The Grand Tour of Big Science," p. 253.

[13] *Space Research*, pp. 47–51; Waff, "A History of Project Galileo," p. 11.

[14] *Space Research*, pp. 51–52, 59–60.

late summer of 1965, a Caltech graduate student demonstrated that Jupiter, Saturn, Uranus, and Neptune would soon be positioned just right for a single spacecraft to visit all of them in a relatively short amount of time. The student, Gary Flandro, was working at JPL at the time and used the work of another graduate student, Michael Minovitch, in his demonstration.

Minovitch, a University of California (UCLA) student also working at JPL, had been examining a spacecraft orbital strategy called "gravity-assist," in which a planet's gravitational field is used to modify a craft's trajectory. Gravity-assists could, Minovitch believed, increase a spacecraft's velocity and alter its trajectory without violating the physics principle of conservation of energy. The planet giving the gravity-assist would lose precisely the amount of energy that it gave to the spacecraft. If the gravity-assist concept could be successfully applied to an interplanetary flight, the amount of fuel and flight time that a spacecraft would require to reach its destination might be significantly reduced. A spacecraft could take advantage of the gravitational influence of a planet, Minovitch wrote in a 1963 JPL report, if it passed the planet "on a precisely calculated trajectory that will place it on an intercept trajectory with another planet."[15]

Gary Flandro called a potential mission to four outer planets of the solar system the "Grand Tour." Michael Minovitch, in his 1963 JPL report, had recognized the value of finding a trajectory that would guide one spacecraft to flybys of several planets by using multiple gravity-assists. Minovitch also noted in a 1965 JPL report that such windows of opportunity do not occur that often, but mentioned that a 1976 or 1977 mission could employ planetary gravity-assists to reach all of the outer planets.[16] Flandro demonstrated that a gravity-assisted Grand Tour of Jupiter, Saturn, Uranus, and Neptune would optimally launch at the beginning of November 1979 and would take less than eight years to complete if the spacecraft was given a launch energy of 140 kilometers per second squared. By means of comparison, a direct, non-gravity-assisted flight with the same launch energy would take nearly as much time to reach only Jupiter and Neptune.[17]

Public attention was drawn to the Grand Tour idea after it was described in a 1966 *Astronomy and Aeronautics* article, written by JPL's Homer Joe Stewart, that was then referenced in a range of national newspaper articles. Rousing such public interest was rare indeed for a robotic mission idea. The name given to the mission definitely helped—it harkened back to 1920s ocean liners setting sail for a "grand tour" of the wonders of Europe. Capitalizing on this interest, JPL included in its advertisements for new employees this promise: "We're going to give Jupiter, Saturn, Uranus and Neptune the run-around. Join us on Project Grand Tour."[18]

The Grand Tour concept also elicited excitement at NASA Headquarters. Donald Hearth, the program manager for advanced programs and technology in the Planetary

---

[15] Michael A. Minovitch, *The Determination and Characteristics of Ballistic Interplanetary Trajectories Under the Influence of Multiple Planetary Attractions*, Technical Report No. 32-464 (California: JPL, 31 October 1963), p. viii; Craig B. Waff, "The Struggle for the Outer Planets," *Astronomy* 17.9 (1989): 44–45; Bob Mitchell, former Galileo Project Manager, interview, tape-recorded telephone conversation, 10 May 2001.

[16] Minovitch, *Utilizing Large Planetary Perturbations for the Design of Deep-Space, Solar-Probe, and Out-of-Ecliptic Trajectories*, JPL Technical Report No. 32-849 (California: JPL, 15 December 1965), p. 67.

[17] Flandro, "Utilization of Energy Derived from the Gravitational Field of Jupiter for Reducing Flight Time to the Outer Solar System," *JPL Space Programs Summary* 4, no. 37-35 (31 October 1965): 22.

[18] Minovitch, *Determination and Characteristics of Ballistic Interplanetary Trajectories*, p. viii; Butrica, "Voyager: The Grand Tour of Big Science," pp. 254–255; Waff, "The Struggle for the Outer Planets," pp. 44–45.

Programs Office, viewed the Grand Tour as a mission with "incredible sex appeal," as well as a wonderful way to carry out an ambitious multiplanet reconnaissance project. A Grand Tour was alluring not only because of what it could accomplish, but also because of how rare this opportunity truly was. Such a planetary configuration as was about to occur only happened approximately once every 175 years. In marketing the Grand Tour concept, Hearth noted that the last time the planets had been so configured was during Thomas Jefferson's presidency. Federal budget personnel always asked NASA why a mission had to be done at this time rather than later. The rarity of the Grand Tour opportunity provided a compelling answer to this question.

In 1968, Hearth funded JPL and the Illinois Institute of Technology Research Institute (IITRI) to carry out studies of the Grand Tour mission concept. While recognizing its potential as an initial reconnaissance mission, Hearth did not lose sight of its limitations. Other mission concepts might be more appropriate for providing the type of in-depth planetary data needed in order to more fully understand the natures of the outer planets. With this in mind, Hearth also allocated funds for advanced studies of orbiter and atmospheric-entry probe missions to Jupiter.[19]

Developing New Missions During a Time of Decreasing Budgets

Planning for post-Apollo missions took place in a context of declining support, both fiscal and political, for NASA's programs. The country's euphoric excitement over space exploration began to lessen after 1965, along with NASA's budget. In the words of Bruce Murray, who became the Director of JPL in 1976, there was "less willingness to gamble on the future as did John Kennedy when he created the Apollo program in 1961." The U.S. became "a divided country in terms of presidency versus Congress, since that period of time."[20]

Although President Johnson continued to recognize the space program's importance, political pressures forced reductions in its budget. The escalating Vietnam War and Johnson's "Great Society" social reform programs were both experiencing spiraling costs.[21] While Johnson never reneged on supporting the Apollo Program's goal of landing a man on the Moon by the end of the decade, he resisted making firm commitments to post-Apollo missions.[22]

Johnson's position was no doubt made easier by eroding public and congressional support for the space program. By the summer of 1965, a third of the nation favored cuts in the space budget, while only 16 percent wanted to increase it. A White House survey of congressional leaders at the end of 1966 revealed pronounced sentiment for "skimping on post-Apollo outlays." Confidence in NASA was further damaged by the

---

[19] Butrica, "Voyager: The Grand Tour of Big Science," pp. 254–255; Flandro, "Utilization of Energy," p. 22; Waff, "The Struggle for the Outer Planets," p. 45.

[20] Robert Dallek, "Johnson, Project Apollo, and the Politics of Space Program Planning," in *Spaceflight and the Myth of Presidential Leadership*, pp. 80, 83–84; "Biography: Dr. Bruce C. Murray," 13 July 1976, Bruce C. Murray Collection, 1975–1982, folder 1, box 1, JPL 216, JPL Archives; Bruce Murray, "Planetary Exploration and the U.S.: What's the Future? (Some Private Views for the JPL Advisory Council)," 15 August 1976, Bruce C. Murray Collection, 1975–1982, folder 3, box 1, JPL 216, JPL Archives.

[21] Hoff, "The Presidency, Congress, and the Deceleration of the U.S. Space Program in the 1970s," p. 92.

[22] Dallek, "Johnson, Project Apollo, and the Politics of Space Program Planning," pp. 80, 83–84.

tragic Apollo command module fire in January 1967, which killed three astronauts and portrayed the Agency as careless in its attempts to achieve a Moon landing too quickly.[23]

A federal budget crisis brought on in the summer of 1967 by the Vietnam War motivated Johnson to ask Congress for a 10-percent increase in income taxes. In order to persuade Congress, Johnson offered to make cuts in his programs, starting in October 1967. As a result, NASA's budget was targeted for a half-billion-dollar reduction. This was followed in September 1968 by a White House–proposed quarter-billion-dollar cut. After congressional appropriations committees recommended only a $3.99-billion NASA budget for 1969 (down from a high of $5.25 billion in 1965), James Webb resigned as the head of the Agency. Webb was a Washington insider who had been very successful since the early 1960s in garnering political and economic support for NASA's programs. He was considered a "master at bureaucratic politics" with "a seamless web of political liaisons."[24] Nevertheless, Webb was not able to change the reality of declining budgets and a lack of government enthusiasm for NASA's post-Apollo space exploration plans.[25]

Although government budget cuts made it difficult to start ambitious new exploration programs, such programs were being studied in depth during the latter years of the decade. IITRI and JPL were analyzing Jupiter orbiter mission concepts. IITRI's J. E. Gilligan also considered various designs for Jupiter atmospheric entry probes. Such probes, he believed, could provide data on the Jovian atmosphere's composition and structure that could not be obtained from Earth-based measurements, flybys, or orbiters. But there were severe technical obstacles to overcome. The most challenging would be to protect the probe from the high temperatures it would experience due to its high entry velocity. Gilligan estimated the entry velocity to be between 30 and 37 miles (48 to 60 kilometers) per second. Gas surrounding the probe would soon heat to temperatures 10 times greater than those that were currently being experienced by spacecraft reentering Earth's atmosphere. The danger was that a substantial part of the probe would burn up before much information could be obtained.[26]

During 1968–1969, an important study was conducted by the Astro Sciences Center (ASC) of IITRI on the types of spacecraft that would best accomplish various mission objectives. An ASC team led by John Niehoff indicated that scientific returns could be maximized if two differently designed spacecraft were employed. A spin-stabilized craft such as Ames Research Center's Pioneer, which revolved at about 60 revolutions per minute (rpm), had certain advantages for making particles-and-fields measurements. If placed in a highly elliptical orbit around Jupiter with an apoapsis, or most distant point, of 100 Jupiter radii, it would have the chance to measure various field regions surrounding the planet, including its magnetosphere and bow shock. A "three-axis-stabilized" craft such as JPL's Mariner, which used a series of jets and gyroscopes for precise attitude control, offered an excellent platform for planetary and satellite photography and remote sensing measurements.[27] The ASC study had a significant effect on outer planet mission planning, which often included the two types of spacecraft in various program scenarios.

Although a range of outer planet mission conceptual studies took place throughout the 1960s, NASA did not move from such efforts to serious mission planning until late

---

[23] Ibid.

[24] James E. Webb," *NASA History Office Biographical and Other Personnel Information*, http://www.hq.nasa.gov/office/pao/History/Biographies/webb.html (accessed 9 June 2003).

[25] Dallek, "Johnson, Project Apollo, and the Politics of Space Program Planning," p. 85.

[26] Waff, "The Struggle for the Outer Planets," p. 46.

[27] Waff, "A History of Project Galileo," p. 14.

in the decade. The Agency's reduced budget was largely focused on achieving piloted lunar landings and other Apollo objectives. The staff resources and funding committed to Project Apollo made a focused effort to develop outer planet missions very difficult.[28]

The Pioneer Missions

In January 1968, an important step toward outer planet exploration was taken when NASA initiated planning for two missions using Pioneer-series spacecraft that would eventually reach Jupiter and points beyond. The spacecraft were to function in some ways as scouts for future, more intensive outer planet exploration missions (such as Galileo). One objective of the Pioneer missions was to study the asteroid belt between Mars and Jupiter. This region was considered a potential hazard for spacecraft traveling to the outer solar system. Also of interest was "the gradient of the Sun's influence on interplanetary space,"[29] as well as the changes in cosmic radiation intensity as the craft moved out of the solar system and into galactic space.[30]

Ames Research Center was given responsibility for preproject planning (prior to congressional approval) and sought a spacecraft design derived from Pioneers 6 through 9 (also known as Pioneers A through D). This subseries of the spacecraft was intended to obtain continuous measurements of interplanetary phenomena from widely separated points in space.[31] These four spacecraft had been launched during the period from 1965 through 1968, with missions to investigate the solar wind; cosmic rays; and interplanetary electron density, dust, and magnetic fields. The spacecraft were all spin-stabilized, with the spin axis perpendicular to the ecliptic plane.

The Pioneer series originated in 1958, when the first three spacecraft were launched to attempt lunar orbits or flybys. None reached their targets due to failures of the launch vehicles, although Pioneer 4, which took off the following year, did make a successful lunar flyby. It was the first U.S. spacecraft to escape Earth's gravitational field.[32]

For a mission to Jupiter, the basic design of Pioneers 6 through 9 needed significant modifications, one of which was in the mode of on-board power. The photovoltaic panels that generated power for inner solar system trips could not perform nearly as well at Jupiter because of the low solar energy density at that distance from the Sun. RTGs, which employed nuclear power to supply electricity, were considered as possible alternatives. They had a distinct weight advantage over the large solar panel array that would be needed for a Jupiter mission.

The Pioneer missions to Jupiter were officially approved in 1969, and Ames was given management responsibilities for the project. The initial spacecraft designs still called for solar power, due to an apparent reluctance to commit to RTGs. This position

---

[28] Waff, "The Struggle For the Outer Planets," p. 1; Butrica, "Voyager: The Grand Tour of Big Science," p. 256.

[29] Waff, "A History of Project Galileo," p. 12.

[30] "Pioneer Flights Revised/Two More Missions Added," *Space Daily* (9 February 1968): 226.

[31] "Pioneer 6," *National Space Science Data Center Master Catalog*, Spacecraft ID: 65-105A, 20 January 1999, *http://nssdc.gsfc.nasa.gov/cgi-bin/database/www-nmc?65-105A*.

[32] "Pioneer 1," *http://nssdc.gsfc.nasa.gov/database/MasterCatalog?sc=1958-007A*; "Pioneer 2," *http://nssdc.gsfc.nasa.gov/database/MasterCatalog?sc=PION2*; "Pioneer 3," *http://nssdc.gsfc.nasa.gov/database/MasterCatalog?sc=1958-008A*; "Pioneer 4," *http://nssdc.gsfc.nasa.gov/cgi-bin/database/www-nmc?59-013A* (sites were reached via *National Space Science Data Center Master Catalog, http://nssdc.gsfc.nasa.gov/nmc/sc-query.html*, accessed 3 June 2003).

was reversed after RTG designs were studied in more detail. Pioneer 10 was scheduled to launch in February or March of 1972; Pioneer 11, 13 months later.[33]

An Ambitious Plan for the Future

In December of 1968, NASA became more serious about post-Apollo planning when its Associate Administrator for Space Science and Applications, John Naugle, was tasked with examining which outer solar system missions were the most appropriate to focus on. Naugle asked a NASA-appointed advisory group, the Lunar and Planetary Missions Board, to suggest strategies and scientific objectives for outer planet exploration.

The Board was chaired by James Van Allen, of the University of Iowa, and included Thomas Gold of Cornell and Von R. Eshleman of Stanford. These men had already spent considerable effort in conceptualizing a Pioneer mission to Jupiter and returned a report within two months. In its early 1969 report, the Board said much that was complimentary about a Grand Tour mission, mentioning that it would constitute "a technological and scientific tour de force of heroic dimensions." The amount of outer planet data that could come out of such a flyby mission for a relatively modest cost particularly impressed the Board.

The principal alternative to the Grand Tour that the Board studied was a series of intensive orbiter missions, one to each of the four major outer planets (Jupiter, Saturn, Uranus, and Neptune). A Jupiter orbiter would probably be the first objective. The Board recommended that such orbiter missions be included as a major NASA goal for the 1970s.

Although the Board identified many positive aspects of both a Grand Tour and a series of orbiter missions, it did not clearly recommend one over the other. Instead, it suggested that NASA conduct a more exhaustive examination of outer planet mission options.

Responding to the Board's suggestion, NASA formed an Outer Planets Working Group that began meeting almost immediately, drawing scientists and advanced mission planners from JPL, Ames Research Center, Goddard Space Flight Center, and Marshall Space Flight Center. In addition, NASA planned an intensive study that would take place in June 1969.

Although the Outer Planets Working Group recognized that a Jupiter orbiter project would have "greater scientific potential" than a series of multiplanet flybys such as the Grand Tour, it favored the latter because of the chance to accomplish many objectives at a relatively low cost. But instead of one four-planet Grand Tour, the Working Group envisioned two truncated flyby missions—one to Jupiter, Saturn, and Pluto, and the other to Jupiter, Uranus, and Neptune—which collectively would constitute a more ambitious Grand Tour than was envisioned by Gary Flandro and would allow more to be accomplished because all five outer planets would be visited.[34]

---

[33] "Pioneer 10," *http://nssdc.gsfc.nasa.gov/database/MasterCatalog?sc=1972-012A*; "Pioneer 11," *http://nssdc.gsfc.nasa. gov/database/MasterCatalog?sc=1973-019A* (sites were reached via *National Space Science Data Center Master Catalog, http://nssdc.gsfc.nasa.gov/nmc/sc-query.html*, accessed 3 June 2003); Waff, "A History of Project Galileo," pp. 12–13.

[34] Waff, "The Struggle for the Outer Planets," pp. 47–48; Butrica, "Voyager: The Grand Tour of Big Science," pp. 255–256.

By May 1969, the Working Group had folded its ideas into an ambitious outer planet trip schedule with launch dates throughout the 1970s and early 1980s. All three mission types—flybys, orbiters, and probes—were included in the schedule:

- 1972–73: Two Jupiter flybys. These missions were already approved by this time and would become Pioneers 10 and 11.

- 1974: Jupiter flyby/solar system escape. The mission trajectory would allow study of the unexplored region between the solar system and the surrounding interstellar neighborhood.

- 1976: Jupiter orbiter.

- 1977: Jupiter-Saturn-Pluto flybys.

- 1978: Multiple Jupiter atmospheric entry probes.

- 1978: Saturn orbiter and probes.

- 1979: Jupiter-Uranus-Neptune flybys.

NASA's summer-study panel convened in June 1969, shortly after the Working Group issued its recommendations. The summer-study panel consisted of 23 members who were asked to make recommendations for outer planet exploration from 1972 through 1980. Five missions with possible launch dates were selected by the summer-study panel and prioritized in order of scientific importance:

- 1974 or 1975: Jupiter deep-entry probe. A trajectory out of the solar system's ecliptic plane would be used.

- 1976: Jupiter orbiter combined with deep-entry probe.

- 1977: Jupiter-Saturn-Pluto mission combined with an atmospheric-entry probe for one of the planets.

- 1979: Jupiter-Uranus-Neptune mission that also would include a Neptune atmospheric probe.

- Early 1980s: Jupiter-Uranus mission with a Uranus probe.

As shown above, there was considerable overlap between the summer-study panel's recommendations and those of the Working Group. The summer-study panel, however, included atmospheric-entry probes in *all* of its suggested missions. The panel believed that data obtained from probes actually entering a planetary atmosphere would help greatly in interpreting subsequent atmospheric data from flybys and orbiters.[35]

---

[35] Waff, "The Struggle for the Outer Planets," pp. 47–48; Butrica, "Voyager: The Grand Tour of Big Science," pp. 255–256.

In a later meeting, the Working Group expressed doubt that entry probes could be designed and built in time for the mission schedule that the summer-study panel envisioned. That schedule required a probe to be ready for launch by 1974. The Working Group recommended instead that orbiter missions be sent off first and that NASA wait until 1978 to launch missions involving probes.

The area in which the Working Group and summer-study panel most closely agreed was on schedules for Grand Tour missions. Both recommended a 1977 Jupiter-Saturn-Pluto flyby and a 1979 Jupiter-Uranus-Neptune flyby (although the summer-study panel also wanted to add probes to both of these missions). Perhaps because of the two groups' agreement on Grand Tour schedules, and certainly because of the narrow window for launching such missions (1976 through 1980), NASA decided to focus on Grand Tour missions throughout the 1970s. During the summer of 1969, the Agency began formulating Fiscal Year 1971 funding requests for starting up a Grand Tour program. Initial funding was sought for a 1974 flyby test flight to Jupiter, followed by the 1977 and 1979 multiplanet missions that would collectively form a Grand Tour.

NASA planned to submit the funding request to the federal Bureau of the Budget in September 1969. But before it could, the administration of newly inaugurated President Richard Nixon issued budget guidelines that compelled NASA to reexamine its priorities.[36]

Changing of the Presidential Guard and Its Effect on Outer Planet Exploration

The year 1969 was not an easy time to nurture new space projects. The costs of the Vietnam War were spiraling upward, as were those of the Great Society social reform programs that Nixon had inherited from Johnson. American society was becoming disillusioned with and mistrustful of everything that the federal government did. Government priorities and funding allocations were frequently perceived as inappropriate, if not criminal. This lack of confidence in all things federal extended to the space program. In addition, Richard Nixon, who became president in January 1969, instituted very conservative fiscal policies that took their toll on NASA and its grand plans for the future. Today's outer planet exploration program grew from the tensions of these times.

NASA, touting its impressive success record, battled continuously against ever-growing budget constraints, an unsympathetic administration, and a mistrustful populace. There were many in the country who perceived NASA as arrogant and elitist. This negative image was furthered by astronauts promoting personal business endeavors and marketing space trinkets. Project Apollo launches were becoming chic media events, complete with high-visibility celebrities and the country's "beautiful people" in attendance. The lunar landings in 1969 elicited tremendous excitement, but they were also viewed as part of NASA's "three ring circus" and may have added to the public's disillusionment with Project Apollo. In short, the U.S. government and NASA in particular were perceived as doing a poor job of responding to the American people's needs and desires.[37]

President Nixon's emphasis on frugality led to a decelerated rate of space exploration in the 1970s and a philosophy of maximum scientific return for minimum cost.

---

[36] Waff, "The Struggle for the Outer Planets," pp. 46–48; Hoff, "The Presidency, Congress, and the Deceleration of the U.S. Space Program in the 1970s," pp. 92–94; Butrica, "Voyager: The Grand Tour of Big Science," pp. 256–257.

[37] Hoff, "The Presidency, Congress, and the Deceleration of the U.S. Space Program in the 1970s," pp.92–94.

Though Nixon portrayed astronauts as American heroes and used them to further his political aims, he never displayed "the personal enthusiasm for or expansive commitment to the space program that [Lyndon] Johnson and John F. Kennedy had shown." Nixon did not need the space program to prove himself able to deal with the Soviets, and this may have contributed to his attitude. Or perhaps he had so many staggering economic issues to deal with that he was forced to retrench and cut costs wherever he could.[38]

Several other factors contributed to the space program's downsizing in the 1970s. First, none of Nixon's close advisors promoted the space program. Neither of NASA's Administrators during that time, Thomas O. Paine and James C. Fletcher, had the ear of the President or of his inner circle of advisors. Second, NASA's highly visible budget overruns hurt its reputation. This situation was exacerbated because NASA upper management did not develop as close a working relationship with the Bureau of the Budget and, later, the Office of Management and Budget as it might have. Such a relationship was necessary in order for NASA to receive the highest priority consideration for its space exploration and other projects.

Third, since NASA's healthy budget of the early 1960s had been created partly because of Cold War pressures and perceptions, that same budget suffered when U.S.-Soviet relations improved. Beating the Soviets to outer space seemed less important than it had under Johnson and Kennedy. Finally, by 1969, the political tide had turned against increased NASA funding. Liberals and conservatives in both parties were loathe to add more dollars to the space program when severe domestic problems such as poverty, crime, urban renewal, racism, and the deteriorating environment were staring them in the face.[39]

Living with Reduced Means

The budget guidelines that President Nixon issued in the summer of 1969 were of great concern to NASA. The Agency had been planning to ask for $33 million in startup funding for its Grand Tour missions and $3 million for advanced studies of outer planet orbiter and probe missions. However, it became clear that Nixon was planning to cut $75 million, or almost 20 percent, from NASA's planetary programs budget. With the ongoing Viking Mars program costs escalating, NASA did not think that it would receive sufficient funds for initiating a Grand Tour program. The Agency delayed its request for startup funding for another year, which meant that the Grand Tour test flight would not take place until 1975 at the earliest.[40]

Seeing the severe budget cuts that NASA would soon be forced to endure, National Academy of Sciences president Philip Handler suggested that a Space Science Board review panel be convened to develop priorities for spending. The review panel would evaluate how to balance resources among the seven different scientific disciplines that NASA supported: planetary exploration, lunar exploration, astronomy, gravitational physics, solar-terrestrial physics, Earth environmental sciences, and life sciences.

NASA Administrator Thomas Paine agreed to the convening of a Space Science Board review panel. Before the panel met, however, Paine submitted his resignation to

---

[38] Ibid.

[39] Ibid., pp. 94-95.

[40] Butrica, "Voyager: The Grand Tour of Big Science," pp. 256–258; Waff, "A History of Project Galileo," p. 15.

President Nixon. His recent NASA budget request of about $3.3 billion was "down more than $500 million from the previous fiscal year," and Congress appeared to be "in a mood to cut space funding even more." But according to one history of the Apollo program, Paine probably quit not just because of the erosion of NASA's budget, but also because he saw no prospect of the situation's improving.[41]

The Space Science Board review panel met at Woods Hole, Massachusetts, from 26 July to 15 August 1970. Nearly 90 scientists participated, divided among the seven scientific disciplines and an executive committee that was to integrate the various proposals and recommendations.[42] The scientists agreed on some issues, for instance, that a large program such as the Grand Tour should not displace smaller but nonetheless important missions such as the exploration of Venus. But in certain key areas, consensus was not achieved, and this would eventually have consequences for the space program.

In particular, the participants did not reach agreement on what the highest priority for outer planet exploration should be. One of the scientific discipline groups, the Planetary Exploration Working Group, held that the Grand Tour opportunity was so important and rare that it should not be missed. The executive committee was more interested in the scientific data that could be generated from orbiter and probe missions to Jupiter. It thought that such intensive investigations would generate more useful information of the type needed to launch other successful outer planet missions. Furthermore, the executive committee viewed Grand Tour projects as quite risky. During such long flights, the various spacecraft systems might not continue to operate reliably. Also, the need for gravity-assists would constrain the mission trajectory so much, the committee believed, that studies of the outer planets' many satellites would be hindered.[43]

OMB, which the Nixon administration had recently formed to replace the Bureau of the Budget, tried to postpone the startup of Grand Tour development from fiscal year (FY) 1972 to FY 1973, which would have further delayed launch dates. NASA fought successfully to keep the startup in FY 1972 but began considering the executive committee's negative views on the mission. NASA was worried that once Congress learned of the executive committee's position, Congress would wonder why a space exploration program was moving forward when heavy hitters in the U.S. scientific community opposed it.

After discussions with the Space Science Board, NASA decided to keep several outer planet options open for the time being. At its budget briefing in January 1971, the Agency announced that by November, it would choose to develop a Jupiter-Saturn-Pluto mission, a Jupiter orbiter program, or a combination of both. The Agency also decided to investigate the costs and benefits of using two different spacecraft in a Jupiter orbiter program—a spin-stabilized Pioneer and a sophisticated three-axis-stabilized vehicle that NASA wanted to build, the Thermoelectric Outer Planet Spacecraft (TOPS).[44]

During the spring of 1971, JPL and Ames studied Jupiter orbiter concepts based on TOPS and Pioneer spacecraft. The study results were to have been presented to NASA and National Academy of Sciences panels during the summer. But before the panels even met, OMB requested NASA to study "simpler, less costly alternatives to the TOPS

---

[41] W. David Compton, "First Phase of Lunar Exploration Completed: Cutbacks and Program Changes," chap. 12-2 in *Where No Man Has Gone Before: A History of Apollo Lunar Exploration Missions* (Washington, DC: NASA SP-4214, 1989), *http://www.hq.nasa.gov/office/pao/History/SP-4214/ch12-2.html*.

[42] Waff, "A History of Project Galileo," p. 15.

[43] Waff, "A History of Project Galileo," p. 15; Butrica, "Voyager: The Grand Tour of Big Science," p. 258.

[44] Waff, "A History of Project Galileo," p. 16; Butrica, "Voyager: The Grand Tour of Big Science," p. 260.

spacecraft for the Outer Planet Missions."[45] (The name "Outer Planet Missions" had been given by NASA to a program that encompassed both Grand Tour flybys and Jupiter orbiters.) In June 1971, a Senate budget committee voted to slash NASA's Outer Planets Missions funding from $30 million to $10 million. Although this amount was raised to $20 million by a combined House and Senate conference committee, the conferees maintained the need for NASA to develop cheaper, less sophisticated spacecraft and recommended that NASA reexamine whether developing the expensive TOPS was appropriate.

NASA strongly wanted to build an outer planet program that would have the unified support of the scientific community, for it recognized the importance of this in securing congressional funding. Toward this end, NASA organized a Science Advisory Group (SAG) in July 1971 and appointed members of the U.S. space science community. JPL and Ames presented their findings on various mission concepts to SAG in July 1971. JPL cautioned that substituting orbiter missions for Grand Tour flybys would add considerably to the cost, especially if TOPS was the spacecraft used. By means of comparison, four TOPS multiplanet flyby missions would cost approximately $750 million, while two such missions and two TOPS Jupiter orbiter trips would raise that to $925 million. The orbiter missions were more expensive because, among other reasons, they required a new propulsion system to be developed for inserting the space vehicles into Jovian orbit.

JPL presented a fallback plan that was cheaper. The above mission concepts assumed that sophisticated TOPS spacecraft would be designed and constructed for the endeavors. If this was not possible due to budget constraints, then already-available Mariner-class spacecraft could be employed instead, at a savings of several hundred million dollars. Ames had a similar backup plan but suggested using the even less expensive Pioneer-class craft for orbiting Jupiter.

Unfortunately, NASA's SAG did not reach consensus on which outer planet plan should be pursued. Four Grand Tour flyby missions would have "broad exploratory appeal," it said, but two Grand Tours combined with two Jupiter orbiters would better serve those who thought that "more intensive investigation of Jupiter and its satellites will yield greater physical insight."[46]

The National Academy of Sciences Space Science Board reviewed the JPL/Ames findings in August 1971, but like NASA's SAG, the Board could not establish a unified position on the flyby versus orbiter debate. Shortly thereafter, NASA Administrator James Fletcher argued with Congress for inclusion of the Grand Tour in NASA's FY 1973 budget, claiming that it would have "great popular interest as mankind's farthest reach out into space,"[47] but his position was weakened by the lack of clear support from the scientific community. Fletcher only managed to get a reduced-budget Grand Tour project approved, with its funding level slashed from the planned $100 million to $29 million.

Prospects for outer planet missions continued to worsen. The Space Shuttle Program was then under consideration for funding and had the strong support of President Nixon. In December 1971, NASA learned that he was ready to approve it. Unfortunately, the President was of the mind that both the Shuttle and a Grand Tour program relying on the development of an expensive TOPS spacecraft could not simultaneously be funded.

---

[45] Waff, "The Struggle for the Outer Planets," p. 50.

[46] Waff, "The Struggle for the Outer Planets," pp. 49–51; Butrica, "Voyager: The Grand Tour of Big Science," pp. 260–261. Note: Quotations are from the Waff article.

[47] Waff, "The Struggle for the Outer Planets," p. 51.

The chances of getting the Grand Tour program funded with the development of a TOPS spacecraft looked dismal indeed.

As a result, in late December 1971 and early January 1972, NASA management decided to propose that less sophisticated Mariner-class spacecraft perform two less ambitious Jupiter-Saturn flyby missions. These reduced missions were approved by Congress in 1972.

At a NASA budget briefing and news conference, Fletcher pointed to the "less than enthusiastic response from certain elements of the scientific community" as pivotal in foiling a more extensive outer planet effort. His deputy, George Low, added that "the simple truth was that there wasn't unity among scientists regarding the value of the Grand Tour, and this made it an easy target for cancellation."[48]

A few months later, Bruce Murray, a scientist who later became JPL's Director, expressed the disappointment of those who'd believed in the value of a Grand Tour. Its cancellation, he felt, would have a serious impact on the U.S. space program. In a testimony to Congress, he stated:

> The Grand Tour cancellation was a self-inflicted setback of unprecedented magnitude. Those are strong words, but that's how I feel. Our prospects for future achievements suddenly were narrowed. We did it to ourselves. What had been established as a major commitment with widespread popular support and unique scientific promise was allowed to tremble and collapse. As a consequence, I feel the credibility of our commitment to outer planet exploration is brought into question.[49]

It is interesting that one factor that made the Grand Tour opportunity so marketable, its rarity (occurring only once every 175 years), also worked against the mission's implementation. The inflexible constraints on mission schedules and on the research and development that had to take place to meet those schedules ultimately presented a formidable barrier to carrying out the Grand Tour.[50]

A Reduced but Still Fruitful Outer Planet Program

NASA did not get its sophisticated TOPS spacecraft built, and it looked doubtful in 1972 that a Grand Tour would be accomplished. Nevertheless, outer planet efforts did go on during the 1970s using tried and true Pioneer- and Mariner-series spacecraft, and the information from these was vital for designing a Jupiter orbiter mission that would eventually become Galileo.

The Pioneer Missions

Pioneer 10, the first outer solar system mission, launched in March 1972 and passed within 120,000 miles (200,000 kilometers) of Jupiter on 3 December 1973. It conducted 15

---

[48] Butrica, "Voyager: The Grand Tour of Big Science," pp. 261–262; Waff, "The Struggle for the Outer Planets," pp. 51–52.

[49] Waff, "The Struggle for the Outer Planets," p. 52.

[50] Ibid., p. 51.

experiments studying magnetic fields, solar wind characteristics, cosmic rays, the helio-sphere, hydrogen abundance, dust particles, the Jovian atmosphere, aurorae and radio waves, and the planet's satellites, especially Io. Pioneer 11 launched in April 1973 and passed within 21,000 miles (34,000 kilometers) of Jupiter's cloud tops in December 1974. It went on to Saturn, reaching the planet in 1979.[51]

Pioneers 10 and 11 were important scouts for future, more intensive investiga-tions. In particular, they charted Jupiter's radiation belts. This was critical for determining design requirements that future spacecraft would need in order to survive in the planet's radiation environment for visits that might last years. One of Pioneer 10's most useful finds was that Jovian radiation levels were not as severe as previously feared.

The Mariner Voyager Missions

The two Mariner Jupiter-Saturn flyby missions, which would eventually become Voyagers 1 and 2, launched in 1977, and both flew by Jupiter in 1979. JPL engineers designed Voyager 1's path so that the spacecraft would also pass near the Saturnian satellite Titan, which at the time was the only moon known to have an atmosphere. Voyager 2's trajectory was designed so the spacecraft could potentially undertake the original "Grand Tour," that is, a Jupiter-Saturn-Uranus-Neptune mission, should the spacecraft continue to function that long and should additional mission funding become available. Congress eventually became more receptive to this idea and provided funds for extending the mission to both Uranus and Neptune.[52]

Voyagers 1 and 2, the last of the Mariner series spacecraft, discovered three new Jovian satellites, as well as a set of rings around the planet. Voyager 1 went on to Saturn, arriving the following year; Voyager 2 flew by Saturn a year after that and then turned toward Uranus and Neptune, reaching those gas giants in 1986 and 1989, respectively. Gravity-assists were used to modify the spacecraft's trajectories. When Voyager 2 reached Neptune, it had covered all four of the original Grand Tour planets envisioned by Gary Flandro 24 years earlier.[53]

Planning for the Jupiter Probe and Orbiter

Following the approval in the spring of 1972 of what would become Voyagers 1 and 2, the NASA-appointed SAG for Outer Solar System Missions examined additional outer planet mission concepts. In particular, scientific returns and design requirements of Jupiter atmo-spheric entry probes were analyzed. It was concluded that the heatshield technology likely to be available in the next several years would not protect a probe during a descent deep into

---

[51] "Pioneer 10," *National Space Science Data Center Master Catalog*, Spacecraft ID: 72-012A, 21 April 2000, *http://nssdc.gsfc.nasa.gov/cgi-bin/database/www-nmc?72-012A*; "Pioneer 11," *National Space Science Data Center Master Catalog*, Spacecraft ID: 73-019A, 29 January 1998, *http://nssdc.gsfc.nasa.gov/cgi-bin/database/www-nmc?73-019A*.

[52] Craig B. Waff, former JPL contract historian, telephone conversation with author, 22 May 2000; JPL, "Planetary Voyage," *Voyager—Celebrating 25 Years of Discovery*, *http://www.jpl.nasa.gov/voyager/science/planetary.html* (accessed 1 December 2003).

[53] "Voyager Project Information," *National Space Science Data Center*, 19 May 1999, *http://nssdc.gsfc.nasa.gov/planetary/voyager.html*; Waff, "The Struggle for the Outer Planets," pp. 49–51.

the Jovian atmosphere. In fact, the facilities to test an entry probe adequately under the heating conditions expected to be encountered at Jupiter would not be available until 1980.

The SAG was also concerned about radiation exposure and its effect on the spacecraft's operating systems and experiments. It recommended that not until the Pioneer 10 and 11 missions determined the severity of Jupiter's radiation environment should a Jupiter orbiter or probe mission be undertaken. With the above cautions in mind, the SAG recommended that other missions be undertaken in the 1970s, such as Jupiter flybys that then proceeded on to Saturn or out of the ecliptic plane. A 1979 atmospheric probe to Saturn was also recommended, because the conditions that the spacecraft would encounter there were presumed to be less severe than at Jupiter and within the bounds of what heatshield technology would be able to handle. The SAG also envisioned two Mariner Jupiter orbiter missions launching during 1981 and 1982.[54]

NASA authorized Ames, JPL, and their contractors to study the recommended mission concepts in more detail. In particular, JPL was directed to develop the concept of a Mariner Jupiter orbiter mission. Following up on a major new direction for a launching strategy, NASA directed JPL to make the first study of launching a planetary mission from the Space Shuttle.

An important idea that emerged from the study was of a "satellite tour," in which the trajectory of each satellite flyby would be designed so as to receive a gravity-assist that aimed the spacecraft toward its next satellite encounter. This would minimize fuel requirements and could extend the lifetime of the mission.[55]

The December 1973 results from Pioneer 10's Jupiter flyby indicated a less severe radiation environment than feared. Coupled with the progress being made in heatshield studies, it was thought that the atmospheric probes being designed for Saturn and Uranus might also be sufficiently durable for Jupiter. NASA considered adding an entry probe capability to a future Jupiter orbiter mission. The Agency was continuing to deal with tight budgets and, early in 1974, explored joint mission concepts with the European Space Research Organization (ESRO). Out of this study came the announcement in May 1974 that a 1980 NASA-ESRO Jupiter orbiter/probe mission using a Pioneer H spacecraft was now being actively considered (along with one of the Mariner Jupiter orbiter missions, which had been delayed to 1985).

Pioneer- and Mariner-series craft were both highly capable space vehicles that had been used for many years. One of the major differences between the two was in the method of attitude stabilization—a critical function if precise planetary measurements were to be taken. The Pioneer's stabilization was achieved through spinning the craft at (typically) 60 rpm. This provided dependable stabilization without requiring a propulsion system to maintain attitude. It also kept the spacecraft's weight and cost down. The Pioneer's spin gave its instruments a 360-degree view of its surroundings and a good platform from which to make field-and-particle measurements.

The Mariner, on the other hand, had a sophisticated system of gyros and jets that maintained stabilization in all three axes, a feature that could help maximize photographic resolution and the returns from remote sensing experiments. Three gyroscopes plus two sets of six nitrogen jets mounted on the ends of its solar panels controlled Mariner's attitude, which was measured relative to the Sun and the star Canopus. A special sensor was

---

[54] Craig B. Waff, "Jupiter Orbiter Probe: The Marketing of a NASA Planetary Spacecraft Mission" (paper presented at American Astronomical Society meeting, session on "National Observatories: Origins and Functions (The American Setting)," Washington, DC, 14 January 1990,) pp. 3–4.

[55] Waff, "Jupiter Orbiter Probe," p. 4.

mounted on the spacecraft to track the position of Canopus, while two primary and four secondary sensors recorded the Sun's position. Attitude information was also provided by an inertial reference unit and an accelerometer. The attitude stability that the Mariner craft offered was valuable, but its price was more weight and a higher economic cost than the Pioneer spacecraft. There were also more systems on the Mariner that could fail and end a mission.[56]

In some of its studies, Ames Research Center had recommended Pioneer craft for outer solar system work. JPL, on the other hand, was a strong advocate of three-axis-stabilized craft, and this position would ultimately affect space vehicle design for the Galileo mission.[57]

In summer 1974, the National Academy of Sciences Space Science Board's new Committee on Planetary and Lunar Exploration (COMPLEX) recommended a Mariner Jupiter-Uranus project, the 1980 Pioneer Jupiter orbiter and probe mission, and the 1985 Jupiter orbiter endeavor. Two months later, however, COMPLEX changed its position, stating that the current budget environment did not allow two large projects such as the Mariner Jupiter-Uranus and Pioneer Jupiter orbiter/probe missions to start simultaneously. A reassessment of outer planet exploration priorities was urged.

As a result, NASA organized working groups to study the various mission concepts and present recommendations to a Strategy for Outer Space Exploration (SOPE) advisory panel. By February 1975, after studying the working group reports, SOPE recommended a Mariner Jupiter-Uranus mission that would possibly include a Uranus atmospheric probe. SOPE also advised that a Pioneer Jupiter orbiter/probe mission precede the probeless Mariner Jupiter orbiter because atmospheric probes would initiate a whole program of outer solar system atmosphere exploration.

During the same month, National Academy of Sciences' COMPLEX also backed the Pioneer Jupiter orbiter/probe project, although it was divided over the Mariner Jupiter-Uranus mission. Acting quickly on the clear support for a Jovian orbiter and probe mission, NASA authorized Ames to initiate a "Phase B" development effort for the Pioneer project. By this time, ESRO had decided to drop out of the project. While the mission had the strong support of West Germany, other European countries were not so enthusiastic about participating.[58]

The major project that NASA was planning at this time was the Space Shuttle, which would have a human crew and would be partially reusable. It would have many uses and would eliminate NASA's need for most of its expendable launch vehicles such as the Titan/Centaur. Shuttle development costs were running higher than expected, and NASA activities were under budgetary pressures by the fiscally conscious Nixon and Ford administrations.

These were compelling factors for NASA to cease production of the expensive Titan/Centaur launch vehicles for missions beyond the already-approved Viking and Mariner Jupiter-Saturn (Voyager) projects in 1976 and 1977. But a Jupiter-Uranus mission using the Mariner spacecraft would need the Titan/Centaur. A Mariner spacecraft's mass was far more than the 321 pounds (146 kilograms) of the Pioneer. The Mariner space vehicles

---

[56] "Mariner 6," *National Space Science Data Center Master Catalog*, Spacecraft ID: 69-014A, 13 April 2000, *http://nssdc. gsfc.nasa.gov/cgi-bin/database/www-nmc?69-014A*; "Mariner 9," *National Space Science Data Center Master Catalog*, Spacecraft ID: 71-051A, 19 April 2000, *http://nssdc.gsfc.nasa.gov/cgi-bin/database/www-nmc?71-051A*.

[57] Torrance V. Johnson interview, tape-recorded telephone conversation, 31 July 2001; Waff, "Jupiter Orbiter Probe," p.7.

[58] Waff, "Jupiter Orbiter Probe," pp. 4–6.

used for the Voyager missions, for instance, each weighed 1,590 pounds (722 kilograms). The fact that a Mariner mission required a launch vehicle such as Titan/Centaur, as well as the lack of unified support from National Academy of Sciences' COMPLEX for a 1979 Mariner Jupiter-Uranus mission, may have contributed to the mission's exclusion from the FY 1977 budget request that NASA submitted in September 1975.[59]

JPL Takes Over the Management of the Jupiter Orbiter Probe Effort

NASA management considered various factors in deciding which of NASA's Centers would manage the Jupiter Orbiter Probe (JOP) project. According to John Casani, Galileo's first Project Manager, all of the Centers had certain roles that represented their core mission responsibilities. "Some centers were designated as 'research,'" said Casani, "and some as 'operational.'" JPL's role was operational—to conduct planetary missions—while the role of Ames Research Center, which also wanted the JOP project, was just what its name implied, a facility focused more on technology development than on planetary explora- tion. The distinction between the two Centers was not black and white, however. Ames had managed Pioneers 10 and 11 (which both flew by Jupiter), as well as the Pioneer Venus mission (involving the Pioneer 12 and 13 spacecraft). "But I don't think [Ames was] viewed as having the in-house capability to support a major planetary program activity," said Casani, referring to Galileo. "Or the charter for it."[60]

Restricted budgets were also factors that helped determine which Centers managed which NASA projects. NASA needed to consolidate its operations and cut costs. An Agencywide "roles and missions" study found that NASA could afford to maintain only one Center devoted to planetary mission development. The Center it recommended was JPL, which had originally fulfilled the assignment during NASA's early years in the latter 1950s. JPL was currently managing the Viking and Mariner Jupiter-Saturn (Voyager) proj- ects, but these were scheduled to launch by 1977. Due to the elimination of the Mariner Jupiter-Uranus project, JPL would soon find its very talented staff without a major in-house development project on which to focus. Ames was not in this situation. It was conducting various large aerodynamics projects in fields such as astrobiology, flight simulation, and vertical takeoff and landing aircraft.[61]

In the fall of 1975, NASA chose JPL to manage the JOP mission. Ames was not cut out of the mission, however. JPL and Ames negotiated and committed to paper an agreement in which Ames was assigned the task of developing the atmospheric-entry probe for JOP. This assignment took advantage of one of Ames's strengths—atmospheric studies.[62] In his "State of the Lab" address on 1 April 1977, JPL Director Dr. Bruce Murray stressed that one of the major institutional accomplishments of the past year was to develop this working partnership with Ames that had not existed one year earlier. The JPL-Ames affiliation included synergistic efforts not only on JOP, but also on the Infrared Astronomy Satellite (IRAS) project, which was

---

[59] Ibid., pp. 6–7.

[60] John Casani interview, tape-recorded telephone conversation, 29 May 2001; *First to Jupiter, Saturn, and Beyond* (Pioneer project home page), 18 August 2001, *http://spaceprojects.arc.nasa.gov/Space_Projects/pioneer/PNhome. html.*

[61] Glen Bugos, Ames Research Center historian, telephone conversation, 1 June 2000; Waff, "Jupiter Orbiter Probe," p. 7.

[62] *The Pioneer Missions,* 20 April 2001, *http://spaceprojects.arc.nasa.gov/Space_Projects/pioneer/PNhist.html;* Waff, "Jupiter Orbiter Probe," p. 7.

slated to start in FY 1978. As a result of the JPL-Ames joint ventures, Murray thought that "both institutions are stronger, and NASA is much better off as a consequence."[63]

Development of the Inertial Upper Stage

One of the necessary steps in launching JOP, as well as other spacecraft, from the Shuttle was the development of an adequate propulsion technology to send the vehicles on their way after the Shuttle lifted them to Earth orbit. In fall 1975, the Air Force chose to modify the solid-fueled Boeing Burner II rocket for this purpose. The two-stage propulsion system was originally given the name Interim Upper Stage (IUS), because it was to serve only until NASA found the funds to develop a Space Tug. Initiation of the Tug project slipped further and further into the future, however, and the propulsion system's name was changed to Inertial Upper Stage (with the same IUS acronym).[64]

A Centaur rocket had also been under consideration, but it was rejected because of the increased fire danger that a liquid-hydrogen-fueled rocket carried in the Shuttle's payload bay presented. This was a difficult decision; the Centaur was the more powerful rocket and would make planetary missions with larger payloads possible.

The solid-fueled IUS rocket was reluctantly approved when it was shown to be barely powerful enough to launch JOP, which was the most demanding planetary mission on NASA's books at that time, during the December 1981–January 1982 launch window. This was the best window in terms of launch energy requirements for several years. If the JOP spacecraft was not ready by that time, it would not be able to launch in its then-current configuration until 1987. Also, if mission requirements demanded a heavier space vehicle, or if IUS development was delayed or its performance did not meet expectations, it would be difficult or impossible to launch JOP during the 1981–82 time slot.[65]

Growing a JOP Project

NASA's "Jupiter Orbiter with Probe Marketing Plan," written by Dan Herman around the end of 1975, outlined the steps that would be taken to support the start of a JOP project in FY 1978. These steps included the following:

- *Forming a science working group (JOPSWG) to define mission objectives and outline probe and orbiter payload requirements.* JOPSWG was chaired by James Van Allen and began meeting in the first half of 1976.

- *Issuing a Request for Proposal (RFP) for Phase-B Probe development.* After the RFP was issued in May 1976, two contractors were selected to prepare proposals: Hughes and McDonnell Douglas. Hughes, partnered with General Electric, ultimately won the Probe contract in June 1978.

---

[63] Bruce Murray, First Annual "State of the Lab" Talk, 1 April 1977, Bruce C. Murray Collection, 1975–1982, folder 31, box 2, JPL 216, JPL Archives.

[64] Waff, "Jupiter Orbiter Probe," p. 8; The Boeing Company, "IUS Introduction," *Inertial Upper Stage*, 2003, http://www.boeing.com/defense-space/space/ius/ius_Introduction.htm.

[65] Waff, "Jupiter Orbiter Probe," p. 8.

- *Initiating Phase-B Orbiter development.* During the first half of 1976, JPL investigated the use of a three-axis-stabilized space vehicle for the JOP mission, and TRW conducted studies of spin-stabilized craft. JPL eventually chose a "dual-spin" design that combined features of both types of space vehicles. Hughes had employed the dual-spin concept in communication satellites, but it had never before been used for planetary missions. The dual-spin configuration accommodated the needs of both particles-and-fields scientists, who preferred spinning spacecraft for their experiments, and planetologists, who wanted a stable base from which to perform photography and remote sensing studies.[66]

It is important to note that performing the above tasks did not ensure an official JOP mission start in FY 1978. For that, congressional approval was needed. A marketing strategy that was put together to accomplish this involved getting approval first from NASA's top management, then OMB, and finally Congress. Regarding the first step, Dan Herman said that obtaining NASA's strong commitment to a JOP mission was necessary because convincing OMB of the importance of JOP would involve communicating the "keen interest in the program by science advocates for the mission, the Space Science Board, and the aerospace industrial community." Similarly, the pitch to Congress would have to stress the "vital interest in the mission by the scientific community."[67] In other words, it was critical to build a unified backing for JOP by the scientific community and aerospace industry, which had not been done a few years earlier for the Grand Tour mission.

One other condition that had to be met for an FY 1978 start, Herman emphasized, was to recognize the importance of not competing with the Hubble Space Telescope project for either FY 1978 budget dollars or total project funding. Herman recognized that Hubble would be the NASA Space Science Office's highest priority new start for FY 1978, with a current total budget estimate of about $500 million. Hubble also had considerable support in Congress. According to Dick Malow, staff director of the House HUD–Independent Agencies Appropriations Subcommittee at the time, "Hubble had been in the development phase for some time . . . . There had been a considerable amount of auxiliary work that had been done on it . . . and there was a lot of attraction to Hubble too." To put Congress in an "either-or" position regarding the two programs might end up being fatal for JOP. To avoid this situation, Herman estimated that the maximum JOP mission cost should not exceed $200 million.[68]

The Office of Space Science was not successful in keeping estimated JOP costs down to the recommended amount. A project budget of $270 million was developed. It did receive approval from NASA top management and also passed the OMB hurdle. After these events, it was included, along with the Space Telescope, in President Ford's federal budget submittal to Congress in January 1977.

Congress's first actions on JOP were favorable. The project was approved by both the House and Senate authorizing committees. But then House HUD–Independent Agencies

---

[66] Ibid., p. 9.

[67] Ibid., pp. 9–10.

[68] Waff, "Jupiter Orbiter Probe," p. 10; Dick Malow, former staff director of the House HUD–Independent Agencies Appropriations Subcommittee, interview, tape-recorded telephone conversation, 19 September 2001.

Appropriations Subcommittee chairman Eddie Boland (D-Massachusetts) entered the picture. It was still January 1977 when Boland repeatedly asked NASA Administrator James Fletcher to prioritize between JOP and the Hubble Space Telescope. Fletcher refused to do this, wanting both programs to receive funding.

It was quite unusual for NASA to request two new program starts in the same year, although the main factors that determined JOP's fate in the subcommittee were fiscal constraints. Dick Malow remembered that the "general feeling on the subcommittee was that we should proceed with one [program], given the budget situation, and reserve judgment on the other. So Hubble went ahead, and then there was a major debate over JOP." Malow further explained that "beginning in the mid '70s, particularly in FY78, we were entering into a long string of government-wide budgetary problems that only grew worse . . . . These problems were also affecting NASA, because NASA was in the throes of trying to bring to fruition the Space Shuttle, and that was running into overruns . . . . NASA was not getting the same percentage of the budget as it was in the late '60s or early '70s. So there was a squeeze everywhere."[69]

Malow and other subcommittee members expected the initial budget estimates for Hubble and Galileo to go up, which was typical behavior for NASA programs. "Carrying two missions, especially if they were going to grow exponentially, which of course they both did . . . given the budget climate there may be some problems. In any case, it was Hubble that we chose to fund."[70] The subcommittee may also have assumed that NASA would fight the strongest for the Hubble Space Telescope, which had been eliminated from the previous year's budget. JOP likely appeared to the committee to be the easier target, with only a relatively weak and possibly divided planetary science community to defend it. The result was that Boland's House subcommittee voted on 4 May 1977 to eliminate all funding for JOP.[71]

Boland had radically underestimated the resolve of the planetary science community. It organized a massive lobbying effort that, only five weeks later, succeeded in obtaining the Senate HUD–Independent Agencies Appropriations Subcommittee's approval for both the Space Telescope *and* JOP. This was done over the objection of Chairman William Proxmire.[72] About the planetary community's effort, Dick Malow said, "Rarely have I ever seen such a successful lobbying campaign . . . . It was masterfully done."[73]

The House and Senate were now in disagreement over the JOP issue. A conference committee from the two congressional houses tried to resolve the matter on 12 July 1977 but could not. At this point, the matter was returned to the House. As Malow explained, "There's a very technical procedure you can go through to bring an item back . . . where the conference does not come to an agreement. And then you vote it up or down in the House, and that's what happened."[74] Much to Boland's displeasure, the House of Representatives approved JOP by a large margin. In the words of JPL Director

[69] Malow interview, 19 September 2001.

[70] Ibid.

[71] Waff, "Jupiter Orbiter Probe," p. 10.

[72] Ibid., p. 11.

[73] Malow interview, 19 September 2001.

[74] Ibid.

Bruce Murray, "We won that one handily, 280 to 131. It was an extraordinary victory and quite unexpected." The last preproject barrier had been crossed, and JOP officially kicked off on the first day of the new fiscal year, 1 October 1977.[75]

Critical to JOP approval was a planetary science community united in its backing of the project. JOP had been very astutely designed to offer value to all of the planetary science community's subdisciplines: planetary meteorology, satellite geology, particles-and-fields studies, and so on. But offering so much required a commitment to build one of the most complex and sophisticated spacecraft ever conceived.[76]

Dr. Bruce Murray expressed both the strong support for JOP and the challenge that now lay before NASA when he said, in an address to JPL management:

> So our job now is to organize and get off to a good start with this mission. It is credible and warrants the kind of endorsement we have had—not only from NASA, but from the entire country—as manifested in the Congressional vote.
>
> I . . . believe that JOP demonstrates that there *really* is a clear constituency for planetary exploration in this country. It is not a constituency that will frequently support very large, Viking-class missions . . . . We were fortunate to discover this in time to affect the outcome of JOP . . . . However, our happiness over the good news is tempered by the fact that we do have to deliver on difficult commitments."[77]

[75] Bruce Murray, *Semi-Annual Report to the Laboratory*, 17 October 1977, Bruce C. Murray Collection, 1975–1982, folder 32, box 2, JPL 216, JPL Archives.

[76] Johnson interview, 31 July 2001; Waff, "Jupiter Orbiter Probe," p. 11.

[77] Murray, *Semi-Annual Report to the Laboratory*, 17 October 1977.

Galileo will be the primary source of information for the chapter on Jupiter and its moons in the atlas of the solar system read by our grandchildren.[1]

—John R. Casani

Chapter 3
THE STRUGGLE TO LAUNCH GALILEO: TECHNICAL DIFFICULTIES AND POLITICAL OPPOSITION

W HEN THE JUPITER ORBITER PROBE (JOP) PROJECT received congressional approval to begin operations on 1 October 1977, NASA planned to build, test, and launch the spacecraft in just over four years. JOP was to be transported up to Earth orbit in the Space Shuttle, then propelled onto a direct Earth-Jupiter trajectory by means of a solid-fueled Inertial Upper Stage (IUS) launch system.[2] Using the direct trajectory, the spacecraft was supposed to reach Jupiter by November 1984.[3]

Initial Project Planning

NASA designed the JOP mission to conduct a comprehensive exploration of Jupiter, along with its atmosphere, physical environment, and satellites, employing an orbiting space-craft and an atmospheric entry probe. The mission was to build upon Pioneer 10 and 11 and Voyager 1 and 2 results, but to focus more on long-duration, in situ investigations. NASA decided that the mission's primary objectives were to determine the following:

---

[1] "The National Value of Galileo," in John R. Casani to K. Kaesmeier and Dr. K. O. Pfeiffer, "GLL 8th Quarterly Report Agenda: February 26, 1981," John Casani Collection, Galileo Correspondence 2/81, folder 30, JPL 14, JPL Archives.

[2] W. J. O'Neil, "Project Galileo" (paper no. AIAA 90-3854, presented at the American Institute of Aeronautics and Astronautics (AIAA) Space Programs and Technologies Conference, Huntsville, AL, September 1990), p. 10.

[3] *NASA Program Approval Document, Research and Development,* Code Number 84-840-829, 19 November 1977, Galileo Documentation, folder 5138, NASA Historical Reference Collection, Washington, DC.

- The chemical composition and physical state of Jupiter's atmosphere.

- The chemical composition and physical state of Jupiter's Galilean satellites.

- The topology and behavior of the Jovian system's magnetic field and energetic particle fluxes.[4]

Early in his tenure as Galileo's first project manager, John R. Cassani solicited a more inspirational title for the endeavor than "Jupiter Orbiter Probe." Many people associated with the mission submitted suggestions, and the name "Galileo" received more votes than any other title. Casani commented that naming the mission after the 17th-century Italian astronomer was especially appropriate, for Galileo "was the first person to view the planet Jupiter through a telescope, and in so doing discovered the four largest moons of Jupiter, now known as the Galilean satellites." One of Casani's staff, Dave Smith, also pointed out that in the *Star Trek* series, Galileo was the name of the *Enterprise*'s shuttle craft that Spock piloted for planetary excursions.[5]

On 1 July 1976, before the mission had received congressional approval, NASA issued an "Announcement of Opportunity for Outer-Planets Orbiter/Probe (Jupiter)," in which proposals were solicited from 1) principal investigators and co-investigators for experiments to be performed by either the Orbiter or Probe, 2) individual investigators seeking membership on a NASA-formed team that would use subsystems such as the radio or imaging subsystems, and 3) individuals wanting to participate in the mission as interdisciplinary scientists or theorists.

Proposals were reviewed by an ad hoc advisory subcommittee of the Space Science Steering Committee, which was appointed by NASA's Associate Administrator for Space Science. The proposals considered to have the greatest scientific merit were further reviewed by the JPL project office (for Orbiter experiments) and Ames Research Center (for Probe experiments) using engineering, integration, management, cost, and safety criteria. Additional reviews were carried out by NASA Headquarters, the Agency's Office of Space Science Program Office, and the Space Science Steering Committee. Based on these reviews, NASA's Associate Administrator for Space Science appointed the investigators and selected the mission experiments. The Associate Administrator picked six science investigations for the Probe atmospheric entry vehicle (see table 3.1), with 29 scientists participating, including five from Germany and one from France. The Associate Administrator also selected 11 science investigations for the Orbiter (see table 3.2). These would employ 82 scientists that included nine from Germany, five from the Netherlands, and one each from France, Sweden, and Canada. In addition, the Associate Administrator identified 14 interdisciplinary scientists to conduct analyses that drew data from two or more investigations.[6]

---

[4] *NASA Program Approval Document.*

[5] J. R. Casani to distribution, "A Name for Project Galileo," 6 February 1978, JPL Interoffice Memo GLL-JRC-78-53, folder 5139, NASA Historical Reference Collection, Washington, DC.

[6] John R. Casani, "Testimony for: Space Science and Applications Subcommittee, Committee on Science and Technology, United States House of Representatives," 1 March 1978, JPL Archives; "Announcement of Opportunity for Outer-Planets Orbiter/Probe (Jupiter)," 1 July 1976, A. O. No. OSS-3, folder 18522, NASA Historical Reference Collection, Washington, DC.

Table 3.1. **Planned Probe investigations of Jupiter's atmosphere.**

STUDY OBJECTIVES	INSTRUMENT	PRINCIPAL INVESTIGATOR AND ORGANIZATION
Determine abundance ratio of helium to hydrogen.	Helium interferometer	Ulf von Zahn, University of Bonn
Obtain temperature, pressure, and density profiles; atmospheric mean molecular weight; wind velocities and wind shear; and turbulence intensity and scale.	Atmospheric structure instrument	Alvin Sieff, NASA Ames Research Center
Determine chemical composition and physical state; measure vertical variations.	Mass spectrometer	Hasso Niemann, NASA Goddard Space Flight Center
Measure vertical distribution of solar energy and planetary emissions, locate cloud layers, and use infrared to study cloud and aerosol opacities.	Net flux radiometer	Robert Boese, NASA Ames Research Center
Determine vertical extent, structure, and particle sizes of clouds.	Nephelometer	Boris Ragent, NASA Ames Research Center
Verify existence of lightning and measure basic physical characteristics; measure scale of cloud turbulence; study electrification; identify evidence of precipitation, sources of heat, and acoustic shock waves; and measure radio-frequency (RF) noise levels and magnetic fields.	Lightning instrument	Louis Lanzerotti, Bell Labs

Table 3.2. **Planned Orbiter investigations of Jupiter and its satellites.**

STUDY OBJECTIVES	INSTRUMENT	PRINCIPAL INVESTIGATOR(S) AND INSTITUTION(S)
Obtain high-resolution images of Jupiter, satellites, and targets of opportunity.	Imaging system (remote sensing instrument)	Michael Belton, Kitt Peak National Observatories, Arizona
Satellites: Map and identify chemical species; relate chemical to geological regions. Jupiter: Map atmosphere, record temporal changes in cloud morphology and vertical structure.	Near infrared mapping spectrometer (remote sensing instrument)	Robert Carlson, NASA JPL

Table 3.2. **Planned Orbiter investigations of Jupiter and its satellites.** (continued)

STUDY OBJECTIVES	INSTRUMENT	PRINCIPAL INVESTIGATOR(S) AND INSTITUTION(S)
Satellites: Study surfaces. Jupiter: Study cloud and haze properties, radiation budget, and atmospheric dynamics.	Photopolarimeter radiometer (remote sensing instrument)	Andrew Lacis, NASA Goddard Institute for Space Studies
Satellites: Study high neutral atmospheres, determine loss rates, and determine geometry of extended atmospheres. Jupiter: Study high neutral atmosphere, mixing ratios of ammonia and ultraviolet (UV)-active trace constituents, and auroral emissions.	Ultraviolet spectrometer (remote sensing instrument)	Charles Hord, University of Colorado
Characterize magnetic fields of Jupiter and satellites, map magnetosphere and analyze its dynamics, investigate magnetosphere-ionosphere coupling, and measure magnetic fluctuations.	Magnetometer (fields-and-particles instrument)	Margaret Kivelson, UCLA
Identify sources of Jovian plasmas and investigate their interactions with satellites and their role as sources for energetic charged particles, characterize equatorial current sheet, and evaluate impact of rotational forces and field-aligned currents on Jovian magnetosphere.	Plasma subsystem (fields-and-particles instrument)	Louis Frank, University of Iowa
Collect data on electron density and temperature, measure electron saturation current collected by spacecraft, and study conduction current of electromagnetic and electrostatic waves.	Electron emitter instrument (fields-and-particles instrument)	Rejean Grard, European Space Research and Technology Centre (ESTEC)
Study distribution and stability of trapped radiation and its interaction with satellites, solar wind, and particles.	Energetic particles detector (fields-and-particles instrument)	Donald Williams, National Oceanic and Atmospheric Administration (NOAA)
Study electromagnetic wave phenomena generated by magnetosphere, atmosphere, and satellites.	Plasma wave subsystem (fields-and-particles instrument)	Donald Gurnett, University of Iowa

Table 3.2. **Planned Orbiter investigations of Jupiter and its satellites.** (continued)

STUDY OBJECTIVES	INSTRUMENT	PRINCIPAL INVESTIGATOR(S) AND INSTITUTION(S)
Determine physical and dynamic properties of small dust particles, including their mass, flight direction, and charge.	Dust detector (fields-and-particles instrument)	Eberhard Grun, Max-Planck-Institute fur Kernphysik, Heidelberg, Germany
Using Galileo's radio telecommunications system, investigate gravitational fields and internal structures of Jupiter and satellites, as well as structures of their atmospheres. Also, search for Very Low Frequency (VLF) gravitational radiation.	Spacecraft radio system (fields-and-particles instrument)	John Anderson, NASA JPL

The First Design Changes and Delays

Galileo engineers identified the need for changes to the spacecraft design almost immediately after the mission began. The first project review in October 1977, as well as subsequent reviews, revealed that the combined Orbiter-Probe spacecraft required modifications that would make it significantly heavier than originally planned. For instance, project engineers decided that it was necessary to build a vented rather than pressurized Probe atmospheric entry vehicle in order to enhance reliability and reduce costs, even though this would add 100 kilograms to the craft's planned 1,500 kilograms. NASA staff also identified the need for structural improvements that would add 165 kilograms. These weight changes would have ramifications for both the Space Shuttle and the Inertial Upper Stage (IUS). It was the Shuttle's job to lift the Galileo spacecraft and IUS to low-Earth orbit; after that, the IUS would propel Galileo onto a Jupiter trajectory. Galileo's projected weight gain would require the IUS to take on additional fuel—but the Shuttle's maximum payload weight had severe limits as well. As a result, the IUS would not be able to carry sufficient fuel to launch the now-heavier Galileo space vehicle on a direct ballistic trajectory to Jupiter. A more fuel-conservative trajectory to Jupiter had to be found. JPL quickly identified a trajectory that would get Galileo to Jupiter through the use of a Mars gravity-assist, but it would add five months to the journey.[7]

The added weight issues were among the first of many problems that had to be addressed before the Galileo spacecraft was ready to fly. Table 3.3 provides a chronology of the steps taken to develop a launch-ready space vehicle.

Since the Galileo spacecraft depended on the Space Shuttle to carry it up to low-Earth orbit, delays in Shuttle development were threatening to postpone Galileo's launch. In August 1979, Thomas O'Toole, a reporter for the *Washington Post*, published an article

---

[7] U.S. General Accounting Office (GAO), "Fact Sheet for the Chairman, Subcommittee on Science, Technology, and Space, Committee on Commerce, Science and Transportation, U.S. Senate," in *Space Exploration: Cost, Schedule, and Performance of NASA's Galileo Mission to Jupiter*, GAO/NSIAD-88-138FS (GAO, May 1988), p. 23; Craig Waff, "Jovian Odyssey: A History of Project Galileo," unpublished outline, 9 December 1987 revision, folder 18522, NASA Historical Reference Collection, Washington, DC.

expressing serious concern over whether a Shuttle engine that could lift the combined weight of Galileo and the IUS into Earth orbit would be ready in time. O'Toole noted that the Shuttle engine-testing program was way behind schedule, and development had not even begun on the special advanced Shuttle engine needed for Galileo and its IUS. The advanced engine was termed the "109-percent" engine because during takeoff, it would burn at a higher temperature than normal, delivering 109 percent of its rated thrust. To protect the engine against the higher temperatures, a more elaborate cooling system needed to be developed.[8]

The Galileo mission was shooting for a 1982 launch window in which Earth, Mars, and Jupiter were in a favorable orientation that allowed the spacecraft to fly by Mars and use the planet for a gravity-assist to Jupiter. If the 109-percent Shuttle engine was not ready, launch might have to be delayed until 1983 or even later. This possibility introduced a serious problem: if the launch date was delayed and Galileo couldn't get a Mars gravity-assist or had to use additional fuel to reach Mars, then there would be insufficient fuel for the spacecraft to carry out all of its planned activities. Mission objectives would have to be abridged. One abbreviated mission scenario called for Galileo to reduce the number of its orbits of Jupiter and its moons from 11 to only 5. Another approach, which would allow a full Jupiter mission, was to substitute a more powerful engine for the solid-fueled IUS and possibly not use the Shuttle at all. For instance, NASA might employ an expendable, liquid-fueled Titan lower stage launch vehicle and a liquid-fueled Centaur upper stage. Such a plan, however, would add at least $125 million to the mission cost because it would require a rebuilt launchpad at Kennedy Space Center.[9]

A simpler option that did not require rebuilding a launchpad was to use the liquid-fueled Centaur as the upper stage but continue to use the Shuttle as the launch vehicle. Centaur had proven itself to be a reliable "mainstay upper-stage engine for most of NASA's large spacecraft and all of its planetary missions,"[10] although it had always been launched as part of a ground-based, expendable propulsion system that included initial stages such as the Titan. Centaur used liquid-hydrogen fuel, which, pound for pound, delivered significantly more thrust than the IUS's solid fuel. Centaur's weight had been kept down by using thin-walled stainless-steel fuel tanks with a "balloon" design, which relied on internal pressurization rather than heavy reinforcing elements to give them their strength.[11] The fueled Centaur upper stage was light enough for it and Galileo to be borne aloft by the Shuttle *without* need of its 109-percent engine. Use of the Centaur upper stage would increase the chances that the Shuttle could be used for a 1982 launch.[12]

NASA Associate Administrator Jesse Moore believed that the Centaur would become a "very integral, longtime part of the Space Shuttle program." A Shuttle/Centaur would combine the newest human space exploration program with "the world's most powerful upper stage rocket." Shuttle/Centaur appeared to be an effective way to send payloads into interplanetary trajectories.[13]

---

[8] Thomas O'Toole, "More Hurdles Rise in Galileo Project To Probe Jupiter," *Washington Post* (15 August 1979): A3.

[9] Ibid.

[10] Thomas O'Toole, "Problems Stall Plans To Launch from Shuttle," *Washington Post* (15 December 1980): A12.

[11] G. R. Richards and Joel W. Powell, "The Centaur Vehicle," *Journal of the British Interplanetary Society* 42 (1989): 99.

[12] O'Toole, "More Hurdles."

[13] Virginia P. Dawson and Mark D. Bowles, *Taming Liquid Hydrogen: The Centaur Upper Stage Rocket, 1958–2002* (Washington, DC: NASA SP-4230, 2004), p. 226.

There were downsides to the use of Centaur as the upper stage. It would have to be modified for transport in the Shuttle, and this modification would add to the cost of the mission. In addition, NASA engineers had concerns about the safety of such a venture. Centaur's liquid-hydrogen fuel presented an explosion danger to the Shuttle and the astronauts flying it.[14] The difficulty of safely handling such a low-density liquid, which boils at -253°C (-423°F) and can leak through minute cracks, made Galileo mission designers wary of mounting Centaur in the Shuttle cargo bay.[15] Safety features could be added, such as a system to vent hydrogen fuel into space in case of emergency. But NASA engineers estimated that reliable safety systems would cost $100 million or more. In addition, such systems would take considerable time to develop and test. NASA's Director of Planetary Programs, Angelo Guastaferro, projected that five years might be necessary to implement a venting system.[16]

Table 3.3. **Chronology of Galileo development: 1978 to 1986.**

DATE	EVENT
1978	JOP project is renamed Galileo.
January 1978	Project Science Group meetings commence. Group chooses Galileo science investigations.
1978	Need for heavier spacecraft is identified. NASA develops Mars gravity-assist trajectory to conserve fuel.
August 1979	Concerns are raised over 109-percent Shuttle engine readiness by Galileo launch date. Substitution of liquid-fueled Centaur upper stage for solid-fueled IUS is considered.
November 1979	NASA Administrator Robert A. Frosch opposes Centaur option for solving Galileo's weight problem. Congressman Boland pressures NASA to initiate Centaur development.
January 1980	NASA announces plans for separate Orbiter and Probe missions to solve weight problem and sets 1984 launch dates.
1980	Boeing cost overrun for developing IUS exceeds $100 million. NASA is forced to reconsider Centaur.
Late 1980	NASA cancels IUS development and initiates "wide-body" Centaur planning. Orbiter and Probe are recombined into one mission. Launch is postponed until April 1985.

---

[14] O'Toole, "More Hurdles."

[15] Richards and Powell, "The Centaur Vehicle," p. 99; U.S. Department of *Energy,* "Hydrogen Fuel," *Energy Efficiency and Renewable Energy Network* Web page, *http://www.eren.doe.gov/consumerinfo/refbriefs/a109.html,* February 2000.

[16] Thomas O'Toole, "NASA Weighs Deferring 1982 Mission to Jupiter," *Washington Post* (4 September 1979): A5.

Table 3.3. **Chronology of Galileo development: 1978 to 1986.** (continued)

DATE	EVENT
November 1980	President-elect Reagan's OMB director, David Stockman, indicates that NASA may be targeted for severe budget cuts.
February 1981	JPL learns of OMB "hit list" which includes Galileo cancellation. Casani drafts and circulates "Galileo Urgent to America" statement and begins campaign to keep Galileo alive.
February 1981	President Pro Tempore of Senate Strom Thurmond backs Galileo for its important military benefits.
November 1981	OMB proposes budget cuts which, if approved by Congress, could force NASA to eliminate most planetary exploration activities and shut down JPL.
December 1981	George Keyworth, head of White House Office of Science and Technology Policy, recommends halt to all new planetary missions for at least a decade. Space science community conducts campaign against the cuts.
December 1981	Department of Defense (DOD) and Aerospace Industries Association oppose proposed OMB cuts to NASA, fearing effect on U.S. space and aviation leadership and on American industry. President Reagan commits to support Space Shuttle development, which aids NASA's planetary exploration plans. OMB agrees to reinstate Galileo in FY 1983, but without Centaur.
January 1982	NASA makes plans to use Boeing's solid-fueled IUS, plus an additional "Injection Module" for increased propulsion. Launch is still planned for 1985, but reduced propulsion compared to that of the Centaur option will delay Jupiter arrival from 1987 to 1990. The ΔV-EGA spacecraft trajectory, which will use an Earth gravity-assist, will be implemented.
March–July 1982	Congress favors a return to Centaur upper stage propulsion system. Bill is passed in July blocking funds for any other type of upper stage development. Launch is eventually delayed until 1986, with Jupiter arrival in 1988.
July 1982	Probe is subjected to drop test simulating Jupiter atmospheric entry. Probe's parachute deploys, but later than planned.
March 1983	Electrical and mechanical integration of Galileo's components and scientific instruments begins at JPL's Spacecraft Assembly Facility.
April–July 1983	Probe's redesigned parachute system passes quarter-scale, half-scale, and full-scale tests.
September 1983	Integration of Probe and Orbiter begins at JPL.
September 1984	Full-scale structural model of Centaur tested at San Diego's General Dynamics Convair plant. System-level vibration and acoustic testing of Galileo spacecraft are completed.

Table 3.3. **Chronology of Galileo development: 1978 to 1986.** (continued)

DATE	EVENT
December 1984	John Casani proposes a close flyby option of the Amphitrite asteroid. NASA Administrator James Beggs endorses this option. Visiting Amphitrite could provide data on nature of the primordial nebula from which the Sun formed, but it would delay Galileo's Jupiter arrival from August to December 1988. Decision on exercising the flyby option would be made postlaunch.
April 1985	Space Shuttle *Atlantis*, scheduled to carry Galileo into orbit, is "rolled out" of Rockwell International's Palmdale, California, facility. Rollout ceremony denotes completion of the craft and readiness for flight.
August 1985	Centaur G-prime upper stage rolled out of General Dynamics Convair facility in San Diego.
October 1985	Environmental and electronic compatibility testing are completed. Problems arise with computer memory devices and with spin bearing assembly (SBA) connecting spun and despun section of Galileo.
November 1985	Galileo development and testing phase is completed; launch and flight operations phase begins.
December 1985	Galileo spacecraft is transported to Kennedy Space Center in Florida.
January 1986	First "tanking up" of the spacecraft with its hypergolic propellant takes place.
28 January 1986	Launch and loss of the Space Shuttle *Challenger*.

NASA Administrator Robert A. Frosch did not favor the Centaur option for solving Galileo's weight problem. In November 1979, he stated that integration of Centaur with the Shuttle would be unwise at that time. Instead, NASA was considering a mission strategy that would allow the IUS to be kept as the upper stage propulsion system. The mission could be split into two trips to Jupiter—one for the Orbiter and another for the Probe, thereby solving the weight problem but adding significant cost for the additional spacecraft that would have to be built.

Representative Edward P. Boland (D-Massachusetts), whose opposition to Galileo in 1977 had nearly prevented approval of the mission, was strongly against Frosch's option. Boland chaired the House HUD–Independent Agencies Appropriations Subcommittee of the House Appropriations Committee, and it was Boland's subcommittee that handled NASA's budget. Rather than allowing NASA to decide its own strategy, the Congressman influenced the House Appropriations Committee to order NASA to integrate the Centaur upper stage into the Galileo mission, should the spacecraft's weight problems cause a slip in the launch schedule, and to fly a combined Orbiter-Probe mission rather than two separate missions.[17]

Boland made it clear that the House Appropriations Committee, in pushing Centaur development, was not advocating IUS termination altogether. The IUS could still

---

[17] "House Panel Orders Shuttle-Centaur Integration If Galileo Slips," *Aerospace Daily* (30 November 1979): 142.

be used, for instance, for putting military payloads into Earth orbit. But Boland considered the IUS underpowered for planetary exploration and favored instead the Centaur, with its impressive past performance. Besides "being a known quantity," he said, "Centaur has 50% more payload capability for planetary missions than the IUS." Boland was very concerned about making the best choice of upper stage from a cost-benefit point of view, and he opposed retaining the IUS for a delayed Galileo launch because it would be "chasing good money after bad." Bringing IUS performance up to specification would involve costly improvements and necessitate "expensive and risky Shuttle weight reduction efforts." Because Centaur had greater payload capacity, Boland saw its use as insurance against the typical increases in spacecraft payload weight that often occurred as mission development proceeded. In short, he saw a Centaur-driven Galileo spacecraft as much more likely to become a "long-term national resource."[18]

Boland's position on the appropriate upper stage for Galileo was influenced by information collected by Dick Malow, the staff director for the Congressman's subcommittee. Malow made it his business to spend "a lot of time going around and talking to people" in the space science community about issues before Congress, and he regularly discussed his findings with Boland. Regarding the Galileo upper stage matter, Malow said, "It came to my attention that if we used the Shuttle/Centaur combination . . . you could launch to any of the outer planets, from launch date to arrival in *two years*, and that had a huge attraction . . . it gave you a lot of backup capability if you missed the launch opportunity. There's always a launch opportunity practically sitting there for you . . . that seemed to me to be a very logical way to go." Malow also mentioned that Boland was beginning to see the powerful Shuttle/Centaur combination as a potential asset for defense applications, and this influenced his backing of Centaur for use in the Galileo mission.[19]

When NASA received the committee's instructions to switch to Centaur in the event of another launch delay, the Agency questioned whether the orders had any legal standing. NASA could defy the orders by submitting an FY 1981 budget request that followed the original plan to use the IUS. Doing so could open up the issue for consideration by both houses of Congress. At a Senate hearing on Galileo several days later, Administrator Frosch said only that the Agency was considering its response to the Boland subcommittee's "order."[20] The order raised questions about NASA's autonomy in planning the details of its own missions and about whether Congress could or should micromanage the technical aspects of the space agency's projects.

At the end of 1979, NASA announced that delays in the Shuttle's development would indeed necessitate a delay in the Galileo launch. In JPL Director Murray's words, "The people who are in charge of the Shuttle have declared that they cannot confidently support the January '82 launch date of Galileo. Thus, we must reconfigure the mission for a later launch date, in one way or another, and achieve the primary objectives. The process of working out how that will be done is still underway."[21]

---

[18] Ibid., p. 142.

[19] Dick Malow, former staff director of the House HUD–Independent Agencies Appropriations Subcommittee, interview, tape-recorded telephone conversation, 19 September 2001.

[20] "House Panel," p. 143.

[21] O'Neil, "Project Galileo," p. 10; Bruce Murray, "Mid-Year Review," 5 October 1979, Bruce C. Murray Collection, 1975–1982, folder 43, box 2, JPL 216, JPL Archives.

NASA released details of its new mission plan in January 1980. The plan did not follow Congressman Boland's guidelines to use the Centaur upper stage alternative in order to work around the delayed development of the 109-percent Shuttle engine. NASA's plan also did not call for a Titan-Centaur launch vehicle that would not require the use of the Shuttle. Instead, the Agency decided to split the Galileo mission into two parts—one for the Jupiter Orbiter and one for the Atmospheric Probe—and set the target launch dates for early 1984.[22]

The planetary configuration during the originally planned 1982 mission would have permitted a very advantageous Mars gravity-assist. This would have reduced fuel requirements and allowed a single spacecraft, carrying both Probe and Orbiter and propelled by a solid-fueled IUS, to attain all of the project's objectives. Although a Mars gravity-assist was also possible on a mission launching in 1984, it would not have saved the spacecraft as much fuel, and this is why NASA planned to send the Probe and Orbiter into space as separate payloads on different Shuttle flights. NASA perceived this approach as safer than carrying the hydrogen-fueled Centaur aloft in the Shuttle's cargo bay. A split mission would, however, introduce serious budgetary and technical implications. In a Probe-Orbiter combined mission, the Probe would ride "piggyback" on the Orbiter. But a split mission would require that a separate rocket and carrier structure be built for the Probe, at a cost of over $50 million.[23]

NASA mission planners scheduled the Orbiter launch for February 1984 and the Probe launch for March 1984. A 1983 launch was possible, but it would have been an expensive alternative, requiring a sole-source contract for part of the job. Delaying the launch until 1984 was perceived by NASA as more cost-efficient because it allowed time for a competitive bidding process for construction of the Probe rocket and carrier.

Galileo's Atmospheric Probe was light enough for the Shuttle plus the solid-fueled IUS to launch it on a direct ballistic trajectory to Jupiter. Getting the much heavier Orbiter space vehicle to Jupiter, however, was more difficult. Although the Orbiter part of the split mission would use a Mars gravity-assist, the solid-fueled IUS still wouldn't have enough power to get the spacecraft to Jupiter. As a result, Galileo would need an auxiliary upper stage in addition to the IUS to provide sufficient thrust.[24]

New problems arose with the mission. Boeing Company, the contractor charged with developing IUS versions for both the Air Force and NASA, experienced severe cost overruns and projected that completing the IUS would cost over $100 million more than previous estimates. Administrator Frosch considered this unacceptable. Boeing proposed scaling down the IUS in order to reduce the overrun, but Frosch worried about the consequences of such an action. He decided that the possibility of "readying the IUS to send an orbiter and probe to Jupiter for its scheduled launch were remote."[25] The best course, Frosh concluded, was to support the propulsion system that NASA had rejected the year before and that Senator Boland had favored, the Centaur engine.[26]

[22] O'Neil, "Project Galileo," p. 10.

[23] Bruce Murray, "Third Annual 'State of the Lab' Report," 26 March 1980, Bruce C. Murray Collection, 1975–1982, folder 44, box 2, JPL 216, JPL Archives; O'Toole, "NASA Weighs Deferring 1982 Mission."

[24] O'Toole, "NASA Weighs Deferring 1982 Mission"; GAO, p. 23.

[25] Dawson and Bowles, *Taming Liquid Hydrogen*, p. 240.

[26] O'Toole, "Problems," p. A12.

"No other alternative upper stage," he stated, "is available on a reasonable schedule or with comparable costs."[27]

In spite of its dangerous liquid-hydrogen fuel, Centaur had an impressive track record of reliability, with 53 operational flights and only two failures during 19 years of service.[28] NASA planners had rejected Centaur the previous year because they had thought that adapting it for safe use in the Shuttle would be too expensive. Also, such a project would divert many engineers whose main mission had been to complete the Shuttle. But the IUS cost overruns altered this picture. Frosch reexamined the Centaur and concluded that it could be implemented more simply and cheaper than he had thought; plus, it might actually be the best alternative,[29] for it would offer "both to commercial customers and to national security interests a highly capable launch vehicle with growth potential."[30]

Many space scientists were concerned about the continued delays in Galileo's launch and did not welcome yet another postponement resulting from a new design change. They worried that the White House, OMB, and Congress would refrain from supporting any Shuttle-launched solar system exploration after Galileo, such as missions to Venus, Saturn, and Halley's Comet.[31] Nevertheless, in late 1980, NASA decided to cancel development of the IUS and instead build a "wide-body" Centaur upper stage, modified to be carried aloft inside the Shuttle rather than on top of an expendable launch vehicle. The launch was postponed from 1984 until April 1985 in order to give NASA time to develop the wide-body Centaur, as well as an Orbiter-Centaur interface.[32]

By changing from the solid-fueled, limited-thrust IUS to a liquid-fueled, higher energy Centaur, JPL was able to recombine Orbiter and Probe into a single payload that could be launched in a direct, rather than Mars gravity-assisted, trajectory to Jupiter that would enable the mission to be carried out in less time. The Centaur had other technical advantages as well. It would deliver a gentler thrust than the IUS. Solid-fueled rockets such as the IUS typically had a "harsh initial thrust," which could possibly damage delicate payloads. Liquid-fueled rockets such as Centaur developed thrust more slowly. In addition, liquid-fueled rockets were more controllable, in that they could be turned on and off as needed. Not so with a solid-fueled rocket: once ignited, it would burn until its fuel was used up.[33]

Planetary Exploration and Reagan's Austerity Program

During the early 1980s, Galileo project management struggled not only with technical problems associated with the spacecraft, but also with the Reagan administration's less than enthusiastic support for planetary exploration. Though Reagan voiced interest in revitalizing the space program, it never became a key national policy as had been the case

---

[27] Dawson and Bowles, *Taming Liquid Hydrogen*, p. 240.

[28] Craig B. Waff, "Jovian Odyssey: A History of NASA's Project Galileo," chap. 11, photocopy of unpublished draft, JPL, 17 March 1989, pp. 1, 43–44, folder 18522, NASA Historical Reference Collection, Washington, DC.

[29] O'Toole, "Problems."

[30] Dawson and Bowles, *Taming Liquid Hydrogen*, p. 176.

[31] O'Toole, "Problems."

[32] GAO, pp. 23–24; O'Neil, "Project Galileo," p. 10.

[33] Dawson and Bowles, *Taming Liquid Hydrogen*, p. 176.

in the 1960s. The new administration proposed dramatically reduced government spending as a means of combating the country's growing economic problems that had begun in the 1970s, and this policy greatly affected the scope of NASA's activities.[34] In November 1980, even before Ronald Reagan was inaugurated, his designated OMB Director, Dave Stockman, indicated that NASA might be targeted for severe budget cuts. In February 1981, JPL learned of a Stockman "hit list" for NASA that included possible cancellation of Galileo. OMB was seeking drastic funding reductions from the levels set in the Carter administration's FY 1982 budget, which OMB called "incompatible with a program of across-the-board (federal government) restraint." OMB proposed an immediate $96-million cut for NASA during the remainder of FY 1981 and a massive $629-million reduction in the FY 1982 budget.[35]

Galileo Project Manager John Casani and his team responded immediately upon learning about the OMB "hit list," drafting a "Galileo Urgent to America" statement containing seven strong reasons why the project had to continue:

- The science that Galileo will perform will be exceptional and has the strong support of the National Academy of Sciences and the general scientific community.

- Congress has given Galileo its strong support in each of its reviews for the past three years.

- Public interest and support are unusually strong.

- Galileo is the only U.S. planetary exploration project currently under development.

- The project is multinational, involving commitments to the West German government.

- The $230 million already committed to Galileo would be lost if the project were canceled.

- The U.S. industrial community has committed significant resources to its Galileo contracts. Cancellation of those contracts would cost the government $35 million.

The next day, a more formal JPL statement expanded on these points, underlining the outstanding success of the U.S. planetary exploration program since the "beginning of the space age" and the national pride it has engendered, as well as the innovative technologies it has fostered. JPL argued that cancellation of Galileo would effectively

---

[34] Lyn Ragsdale, "Politics Not Science: The U.S. Space Program in the Reagan and Bush Years," in *Spaceflight and the Myth of Presidential Leadership*, ed. Roger D. Launius and Howard E. McCurdy (Urbana and Chicago, IL: University of Illinois Press, 1997), pp. 135–136.

[35] Waff, "Jovian," pp. 1, 9–10.

terminate the U.S. planetary exploration program at a time when the Soviet Union, Europe, and Japan were all "vigorously pursuing" their planetary programs.[36]

JPL also issued statements demonstrating that Galileo was so far along in development that terminating it would cause damage far overshadowing the money saved. Al Wolfe, Galileo's Deputy Project Manager, stressed that the mission was on schedule, within budget, and in the final stages of engineering after three years of mission and operations planning and spacecraft system design. Major problems had been resolved, Wolfe reported, including the development of reliable radiation-hardened microprocessors and peripherals. Ninety percent of the long-lead-time electronic components had been delivered to JPL. Issues with sensitivity of the imaging system and data transfer between the spinning and nonspinning sections of the Orbiter had been worked through, leaving no current Orbiter problems threatening the launch schedule. Probe development was also on schedule, Wolfe reported. The Probe had passed its preliminary design review, as well as an important parachute test. These statements were all incorporated into a document presenting arguments against the termination of Galileo.[37]

JPL took another tack as well in its campaign to save Galileo, and it may have been the critical one. JPL staff focused on the one government department that had received large budget increases rather than cuts—the Department of Defense (DOD)—and portrayed Galileo as vital to military goals. It was the opinion of JPL Director Bruce Murray that the "silver bullet" that saved Galileo was a letter sent to David Stockman on 6 February 1981 by Strom Thurmond (R-South Carolina), the new President Pro Tempore of the Senate and a member of the Senate Armed Services Committee. In this letter, which was actually drafted at JPL, Thurmond argued that the military applications that could result from Galileo made it unwise to cancel the program. For instance, the Air Force needed satellites with a "survivable autonomous capability" to remain operational during time of war. This was especially so because ground control stations were not "hardened" to survive nuclear attacks. Defense satellites had not achieved the autonomous capability of which Thurmond wrote, but he hoped that it would be derived from the technology "that will be developed and demonstrated as part of the Galileo project." In fact, he noted, JPL had already undertaken the task of developing and applying such autonomous technology for the Air Force Space Division, in parallel with JPL's development of the Galileo spacecraft.[38]

The Air Force had turned to JPL for such help because of the Laboratory's expertise in developing highly autonomous spacecraft, according to David Evans, a manager in JPL's work for the Air Force. JPL's spacecraft needed to be autonomous because of the communication distances and round-trip communication times involved during planetary exploration. The craft needed to take care of themselves when they were not in close touch with ground stations. JPL was designing several autonomous features into the Galileo spacecraft that were of potential interest to the Air Force. The craft would be able to determine its attitude from any orientation, using only on-board

---

[36] John Casani memo, "Galileo Urgent to America (paraphrased excerpts)," undated (circa 1981), Galileo—Meltzer Sources, folder 18522, NASA Historical Reference Collection, Washington, DC; T. Johnson, "Cancellation of Galileo—Reclame Statement," 5 February 1981, in a set of John Casani Collection papers that begin with "GLL 8th Quarterly Report Agenda," Galileo Correspondence 2/81, folder 30, box JA370, JPL 14, JPL Archives; Waff, "Jovian," pp. 11–13.

[37] A. Wolfe, "Galileo Status Statement," draft, 5 February 1981, in a set of John Casani Collection papers that begin with "GLL 8th Quarterly Report Agenda," Galileo Correspondence 2/81, folder 30, box JA370, JPL 14, JPL Archives; Waff, "Jovian," pp. 13–15.

[38] Waff, "Jovian," pp. 16–18.

systems. Typical spacecraft were only able to do this within a narrow range of angles; otherwise, they had to rely on ground-based data processing. Other subsystems and instruments aboard Galileo, such as its dust and plasma analysis instruments, also operated far more independently than in typical spacecraft. In addition, JPL was designing Galileo to withstand the intense radiation surrounding Jupiter, and such radiation hardening was also of high interest to the defense establishment.[39]

Strom Thurmond made additional statements during February 1981 that indicated his strong support of NASA's space program, as well as of missions such as Galileo. He recognized that the U.S. space program was at a "critical turning point" and that our country was retreating from space during a time when "Russian and Soviet bloc cosmonauts come and go like weekend tourists . . . [and] the Russians, West Europeans and Japanese will visit Halley's Comet while we sit home and watch." As a result of this situation, he believed that the United States needed to "keep its flight manifest full with the military and scientific payloads that will help to make America first in the world again and keep America first."[40]

The Attacks Continue

The Galileo program stayed alive, but the Reagan administration's onslaught against NASA's planetary and other space science programs went on. Projected cost growth in FY 1983 for the Space Shuttle was $300 million to $500 million, and this deeply troubled the administration. NASA managers and scientists grew very concerned that the price tag for adapting Centaur to Shuttle-launched solar system missions, which could run as high as $500 million, might convince the White House to oppose its use for planetary exploration. Hans M. Mark, NASA's new Deputy Administrator in 1981, considered Centaur "very precarious politically," especially since Galileo was the only planetary mission for which it was definitely scheduled to be used.[41]

Mark tried to convince other Shuttle customers to orient their designs toward using Centaur, in order to "spread support base" for the adaptation of the upper stage. At stake was the Galileo mission itself. Delay or abandonment of Centaur would in turn delay and possibly kill Galileo. To prepare for this eventuality, NASA began developing fallback mission options that did not rely on Centaur. But such scenarios could not ensure the continuance of NASA's planetary program. Hans Mark saw a bleak future for Galileo and other solar system exploration missions under the Reagan administration, whose main priorities were revitalizing the country's economy and making sure that the U.S. remained the foremost military power in the world. The Reagan administration would not clearly commit to completing the Shuttle, which was vital for launching planetary missions, and this deeply worried Mark.[42]

---

[39] Waff, "Jovian," pp. 17–18; J. R. Casani to J. N. James, "Galileo and the ASP Connection," Interoffice Memo GLL-JRC-81-697, 10 July 1981, and two attachments to this memo: "A.F. Merchandising Plan for Galileo" and "Galileo Improvements Over Voyager in Autonomous Technology," Galileo Correspondence 6/81–7/81, folder 27, box 3 of 6, JPL 14, John Casani Collection, JPL Archives.

[40] All quotations and information in this paragraph come from the NASA Kennedy Space Center History Program, *Chronology of KSC and KSC Related Events for 1981*, KHR-6, 1 September 1983, p. 65, *http://www-lib.ksc.nasa.gov/lib/archives/chronologies/1981CHRONO1.PDF* (accessed 11 September 2005).

[41] Craig Covault, "Shuttle Costs Threatening Science Programs," *Aviation Week & Space Technology* (6 July 1981): 16.

[42] Ibid., p. 16.

Over the latter months of 1981, NASA forged a new planetary exploration policy in an attempt to salvage a "viable but more limited" ability to study the solar system within an ever grimmer budget environment. Under the policy envisioned by the key U.S. scientists and mission planners on NASA's Solar System Exploration Committee, exploration objectives would remain the same but would be spread over a larger number of more limited and less expensive missions. The exploration objectives would take possibly decades longer to be accomplished, and the total runout costs would go up because more spacecraft would have to be built. But peak-year funding would be reduced to within budget limitations.[43]

While envisioning more constrained future missions, all of the Committee members remained very concerned about preserving Galileo objectives in the face of Reagan budget cuts.[44] After OMB announced its proposed cuts to NASA's budget of $1 billion in both FY 1983 and FY 1984, however, it was not at all clear whether the Committee's vision for planetary exploration could be attained. In October 1981, NASA Administrator James M. Beggs scheduled a meeting with Counselor to the President Edwin Meese to argue that the cuts proposed by OMB would kill broad areas of U.S. aerospace capability. NASA would have to consider closing JPL and terminating not only Galileo and Centaur, but also all other U.S. planetary spaceflight.[45]

In spite of Beggs's meeting with Meese, OMB delivered proposed budget strictures to NASA on 24 November 1981 that would, if approved by Congress, force the Agency to virtually cease its planetary exploration activities as of FY 1983. Cancellation of Galileo would result. The Venus Orbiter Imaging Radar (VOIR) project (which became Magellan),[46] had been penciled into NASA's budget projections as a new start for 1984, but it would also have to be canceled. The only mission that would not be affected by the budget cuts was Voyager 2, already on its way to Uranus and Neptune. NASA had little time to appeal the OMB decision because its FY 1983 budget had to be ready for submittal to Congress in January 1982. OMB also leveled its sights on NASA's aeronautics program, recommending that its FY 1983 budget be cut by 50 percent.[47]

In December 1981, George Keyworth, head of the White House Office of Science and Technology Policy, echoed OMB's position by recommending a halt to "all new planetary space missions for at least the next decade." He believed that the White House would support this position. Keyworth had worked closely with OMB in formulating a pared-down NASA budget and favored a shift away from planetary exploration and toward Shuttle-launched experiments such as a space telescope. Although he thought that a great deal had been learned during NASA's 12 years of solar system exploration, he also thought that new missions would be nothing more than "higher resolution experiments."[48]

Keyworth's statement enraged many members of the space science community. The chairman of National Academy of Sciences' Subcommittee on Lunar and Planetary

---

[43] Craig Covault, "NASA Moves To Salvage Planetary Program," *Aviation Week & Space Technology* (2 November 1981): 16.

[44] Ibid., p. 16.

[45] Craig Covault, "NASA Assesses Impact of Budget Cut Proposal," *Aviation Week & Space Technology* (12 October 1981): 26.

[46] "Chapter 2: The Magellan Mission," in *The Magellan Venus Explorer's Guide*, JPL Publication 90-24, August 1990, *http://www.jpl.nasa.gov/magellan/guide2.html*.

[47] M. Mitchell Waldrop, "Planetary Science in Extremis," *Science* 214 (18 December 1981): 1322; Philip J. Hilts, "Science Board To Advise President Proposed," *Washington Post* (2 December 1981): A25.

[48] Hilts, "Science Board," p. A25; Waldrop, "Planetary Science," p. 1322.

Exploration, Eugene Levy of the University of Arizona, said that such a position "does not stand up to rational scrutiny" and insisted that "there are fundamentally important objects, the comets and asteroids, that we haven't even approached yet. They hold primitive, undisturbed material. Not only would [their exploration] enhance our understanding of the origin of the solar system, but of stars in general." Levy's point was that sending spacecraft to explore the solar system was very complementary to developing the space telescope experiments that Keyworth envisioned. It was "intellectually naïve" to try to separate one kind of research from the other.[49]

Galileo Project Manager John Casani expressed the "keen feeling of disappointment" that he would have "should Keyworth's views indeed be adopted by the White House." Casani went on to say, "It is difficult to accept that this country would abdicate by Presidential policy, leadership in a field of exploration where our accomplishments have been a source of pride and inspiration to people all over the world."[50] Casani responded to Keyworth's attack on planetary exploration by organizing a campaign of Galileo supporters, urging them to register their views on solar system exploration directly with the President or with Keyworth.[51]

James Van Allen also joined the fray. He was an influential University of Iowa physicist famous for his discovery of Earth's "Van Allen radiation belts," and he was also a Galileo interdisciplinary investigator. Van Allen initiated a letter-writing campaign to Keyworth to save Galileo. In a speech to the National Academy of Sciences' Space Science Board, he called Galileo's research the most exciting physics that the U.S. was conducting in the solar system and said that its loss would be devastating.[52]

Galileo's loss would hamper advances in space science, as well as the capability of U.S. scientists to remain the world leaders in solar system exploration. An editorial in *Aviation Week & Space Technology* expressed the effect that the loss of Galileo would have, in particular, on JPL:

> Without Galileo, there is little for the 1,200 program scientists of the Jet Propulsion Laboratory to work on. The disappearance of Galileo would disperse U.S. planetary capability. There is no way to put Jet Propulsion Laboratory on hold for two to three years while Reagan's budget director, David Stockman, leads the country out of the economic wilderness with candor and off-the-record interviews.[53]

The future looked grim indeed at the end of 1981 for Galileo and other U.S. solar system missions. But in December, NASA received some much-needed support from DOD, American industry, and, surprisingly, the White House. In a new Space Shuttle policy directive, President Reagan voiced a commitment to NASA for Shuttle development support, an action that helped all of NASA's planetary exploration plans. The Department of Defense strengthened its backing for the conversion of the Centaur

[49] Waldrop, "Planetary Science," p. 1322.

[50] John Casani to Melvin M. Payne, 4 December 1981, letter no. 230-JRC:db-778, John Casani Collection, Galileo Correspondence 11/81–12/81, folder 24, box 3 of 6, JPL 14, JPL Archives.

[51] John R. Casani to multiple addresses, 4 December 1981, John Casani Collection, Galileo Correspondence 11/81–12/81, folder 24, box 3 of 6, JPL 14, JPL Archives.

[52] "Washington Roundup," *Aviation Week & Space Technology* (7 December 1981): 17; Craig B. Waff, untitled chronology of Galileo events, 1979–82, JPL, undated, folder 18522, NASA Historical Reference Collection, Washington, DC.

[53] William H. Gregory, "Bean-Counting the Solar System," *Aviation Week & Space Technology* (14 December 1981): 15.

into a Shuttle-transported upper stage propulsion system. Such a conversion would further DOD goals, as well as efforts to maintain a U.S. solar system mission capability. In addition, top officials of DOD and the Aerospace Industries Association attacked OMB's proposed slashes to NASA's budget, fearing that, if implemented, they could be a major factor in removing the U.S. from its position of space and aviation leadership. Secretary of Defense Caspar Weinberger and Air Force Research and Development Chief Richard DeLauer made their protests directly to David Stockman, Director of OMB. The Aerospace Industries Association met with President Reagan himself, claiming that such a "bare bones" NASA budget would put the industry's important long-haul programs at risk.[54]

NASA also conducted meetings with critical Reagan administration staff. The Agency's top managers sat down with White House chief of staff James A. Baker and OMB Director David Stockman in order to underline how crucial to the space agency a planetary exploration program was. NASA staff not only talked about the important scientific data that would be lost if planetary exploration were eliminated, but also stressed the impact that the loss of 1,200 JPL jobs and considerable international prestige would have on the country.

Galileo Is Reinstated, but at a Cost

The NASA, aerospace industry, and DOD lobbying efforts against OMB's proposed funding cuts had an effect. Days before the end of 1981, OMB agreed to reinstate Galileo into the FY 1983 budget but recommended that Centaur development be killed. OMB was firm on this position, in spite of the support that Centaur had from DOD.[55] According to John Casani, "This change was driven solely by budget pressures and has resulted in a net reduction of about $150M in FY82 and 83 combined . . . ." He noted, however, that although the cost of developing and launching Galileo was reduced, total costs that would be charged to the mission over its entire lifetime would actually increase. Due to the elimination of Centaur, NASA had to foot the bill for design changes to the spacecraft and its propulsion system, as well as for two or more years of additional operating expenses, because the spacecraft would not be able to reach Jupiter and complete its mission as quickly. The additional operating expenses, however, would not have to be paid until years after the launch.[56]

NASA set to work immediately on restructuring Galileo for a mission without Centaur, making plans to use Boeing's solid-fueled IUS, augmented with a solid-fueled Injection Module (also called a "kick stage") for increased propulsion. Even with the Injection Module, the spacecraft would not receive as much thrust as with Centaur, and this required JPL to plot out a different, gravity-assisted trajectory to get Galileo to Jupiter. The launch date would remain in 1985, but Galileo would not reach Jupiter until 1989 or

---

54 "Washington Roundup," *Aviation Week & Space Technology* (14 December 1981): 15.

55 Craig Covault, "Galileo Reinstated in the Budget," *Aviation Week & Space Technology* (28 December 1981): 10; Waff, untitled chronology of Galileo events, 1979–82.

56 J. Casani attachment GLL-JRC-82-810 to letter from Al Diaz to John R. Casani, 19 February 1982, John Casani Collection, Galileo Correspondence 2/82–4/82, folder 21, box 2 of 6, JPL 14, JPL Archives.

1990—an increase of 24 to 30 months in trip time over what a Centaur-driven spacecraft could have achieved.[57]

Although cutting Centaur would save considerable funds through FY 1984, both NASA and the Air Force realized that upper stages more powerful than Boeing's IUS would eventually need to be built. By 1987 or 1988, heavy military payloads currently under development would need to be boosted up to geosynchronous orbit. High lift capability might also be required to send components of a proposed military space platform into orbit.[58] Opinions differed, even within agencies, as to the best way to prepare for the heavy lift requirements. Although loss of Centaur would reduce near-term capability of the Shuttle and upper stage, some NASA advanced planners thought this the best course to take. Retention of Centaur would, in their opinion, push the development of a more modern, possibly reusable upper stage further into the future. Others in NASA lobbied the Department of Defense to intensify its advocacy of Centaur, hoping that development of the liquid-fueled upper stage could eventually be salvaged.[59]

The Air Force was split on what to do. There was strong sentiment among its uniformed brass, with which the White House at this time concurred, to avoid putting development funds into the 20-year-old Centaur. Their belief was that this might "lead down a dead-end street on both cost and hardware utilization." A new high-energy upper stage should instead be built from scratch, implemented into Shuttle applications in the late 1980s, and used through the 1990s. Top-level civilians in the Air Force, however, were considering a different plan—an early transition to Centaur as the Shuttle upper stage, cutting off development of the IUS once Centaur was available. The civilian Air Force view was driven by estimates that IUS operating costs would be higher than expected—as much as $70 million more per mission if the IUS was manufactured in small quantities. The Air Force and NASA began meeting in January 1982 to further discuss this issue and lay out possible courses of action. They reportedly focused their discussions on how to insert Centaur funding into the FY 1983 budget.[60] Eventually, however, the Air Force and NASA decided against a joint development of Centaur due to funding constraints and a limited user base for the Centaur until the late 1980s. They did form a review team, though, to thoroughly "reexamine the entire Shuttle upper-stage issue."[61]

JPL scheduled Galileo's launch for August 1985, four months later than for the Centaur-driven craft, and adopted a "ΔV-EGA" (Earth gravity-assist) trajectory (see figure 3.1). Under this plan, the Shuttle would attain Earth orbit, after which Galileo would be injected into a two-year elliptical orbit around the Sun. In mid-1986, near the spacecraft's aphelion, or furthest point in the orbit from the Sun, the Orbiter's propulsion system would fire and impart a change in velocity, or "ΔV," of over 500 meters per second. This change would alter the orbit such that Galileo would cross inside Earth's orbit on the spacecraft's way to perihelion, the point of closest approach to the Sun. When Galileo

[57] John R. Casani to Galileo Project Science Group, "Galileo Baseline Change," 5 January 1982, John Casani Collection, Galileo Correspondence, 1/82, folder 23, box 2 of 6, JPL 14, JPL Archives; Covault, "Galileo Reinstated," p. 10; Al Diaz to John R. Casani, 12 February 1982, John Casani Collection, Galileo Correspondence 2/82–4/82, folder 21, box 2 of 6, JPL 14, JPL Archives.

[58] "Washington Roundup," *Aviation Week & Space Technology* (15 February 1982): 15.

[59] Covault, "Galileo Reinstated," pp. 10–11.

[60] Craig Covault, "New Shuttle Stage Aims at Manned Use," *Aviation Week & Space Technology* (28 June 1982): 137; "Air Force Mulls IUS-to-Centaur Switch," *Aerospace Daily* (13 January 1982); Waff, untitled chronology of Galileo events, 1979–82.

[61] "Washington Roundup," *Aviation Week & Space Technology* (15 February 1982): 15.

reencountered Earth in June 1987, the spacecraft would receive just the right EGA to accelerate it on the proper trajectory toward Jupiter.[62]

The June 1987 EGA would put Galileo on a trajectory to Jupiter that was nearly identical to one that would have been followed if a Centaur-driven craft had launched in June 1987. In other words, it would take an IUS-driven Galileo spacecraft almost two years (from launch in August 1985 to EGA in June 1987) to attain the trajectory that a Centaur-driven craft could have attained immediately because of its greater thrust.[63]

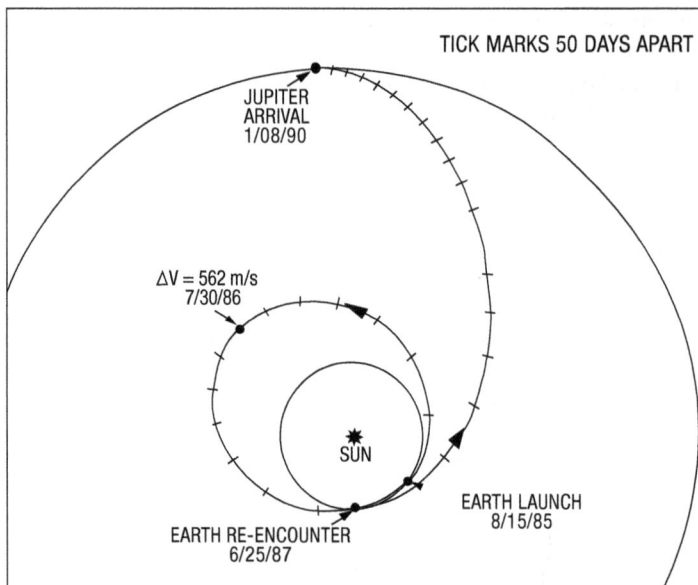

Figure 3.1. The ΔV-EGA trajectory employed an Earth gravity-assist to place Galileo on a Jupiter trajectory. (This figure was redrawn from a diagram found in JPL archives: Folder 23, box 2 of 6, JPL 14)

The IUS-driven Galileo would reach the Jovian system in January 1990. Because a significant fraction of the craft's fuel would have been used for the near-aphelion ΔV maneuver, Galileo would arrive at Jupiter with less propulsive capacity than it would have with Centaur as an upper stage. As a result, only six Jovian satellite encounters might be attainable, compared to the 11 encounters that had been envisioned previously.[64]

Continued Changes to Galileo's Propulsion System

In March 1982, *Aerospace Daily* reported that Congress was considering implementing yet another change in Galileo's propulsion system. Many in Congress wanted to restart the

---

[62] GAO, p. 24; Thomas O'Toole, "Budget Squeeze Stretching Out Journey to Jupiter 6," *Washington Post* (6 January 1982).

[63] J. R. Casani to distribution, "ΔV-EGA Plans," 20 January 1982, John Casani Collection, Galileo Correspondence 1/82, folder 22, box 2 of 6, JPL 14, JPL Archives.

[64] Casani, "Delta VEGA Plans"; John R. Casani to Galileo Project Science Group, "Galileo Baseline Change."

development of Centaur for use in the Shuttle, even though Centaur had, just months before, been rejected by OMB as an inappropriate propulsion system for Galileo.[65] These continual flip-flops in the Galileo mission plan upset NASA; each time the propulsion system was changed, the space agency's engineers had to backtrack and redesign a large part of the project. Several entities outside NASA, including Congress, the White House, and OMB, were trying to micromanage NASA's missions, down to the choice of propulsion systems used. NASA managers and engineers, who had the most intimate knowledge of mission needs, were often kept out of the decision-making. In a letter to NASA Administrator James Beggs, JPL Director Bruce Murray expressed his laboratory's frustration with the repeated changes in direction and with the threat these changes posed to the mission's continuance:

> . . . The project has had to redo spacecraft designs and imple-
> mentation plans, redirect contractors, redeploy people and generally
> reeducate and remotivate many people and organizations toward new
> plans and goals on three major occasions. Most seriously the project has
> been made vulnerable to and, in several instances, actually threatened
> by cancellation as a result of the program changes.
>
> The present plan [to launch in 1985 on a ΔV-EGA trajectory,
> using a solid-fueled IUS plus an Injection Module], although non-optimum,
> is believed to be realistic, doable, and credible. It is very important for the
> success and viability of the project to permit it to settle down on and to
> proceed to carry out a realistic and stable implementation plan.[66]

Murray went on to enumerate the threats that the propulsion system changes posed to the mission. At the top of his list was that Congress might react to the continual delays and budget modifications by canceling the mission (even though Congress itself had been responsible for some of the changes). He also worried about the morale of the project's staff, who perceived some of the OMB- or Congress-imposed modifications as totally unrealistic. He voiced concern about keeping his project team together and retaining subcontractor capabilities during the long delays. Losing members of his team or subcontractor capabilities would impair the "specific knowledge and competence required" for a successful launch and smooth in-flight operations. Finally, Murray feared that Centaur development would be so expensive that it would drain FY 1982 and FY 1983 funds needed for Galileo spacecraft development.[67]

Others in the space science community shared Murray's concerns regarding more propulsion system changes. Robert Allnut, NASA's Deputy General Counsel for Policy Review, said in an interview with *Aerospace Daily* that a switch to Centaur would delay Galileo's launch for at least a year.[68] Air Force Major General Jasper A. Welch, Jr., cautioned at an American Institute of Aeronautics and Astronautics (AIAA) meeting that significant, costly modifications would be required to reconfigure the Galileo mission from IUS to Centaur.[69]

[65] "Congress Considering Attempt To Save Centaur Upper Stage," *Aerospace Daily* (1 March 1982); Waff, untitled chronology of Galileo events, 1979–82.

[66] Bruce Murray to James M. Beggs, 26 February 1982, Galileo History files, folder 5139, NASA Historical Reference Collection, Washington, DC.

[67] Ibid.

[68] "Use of Centaur Upper Stage Would Delay Galileo," *Aerospace Daily* (26 May 1982); Waff, untitled chronology of Galileo events, 1979–82.

[69] Waff, untitled chronology of Galileo events, 1979–82.

In spite of the concerns, Congress favored a return to Centaur. On 11 May 1982, on the recommendation of Senator Harrison Schmitt, chairman of the Senate Space Subcommittee of the Commerce, Science and Transportation Committee, the U.S. Senate added $150 million to NASA's FY 1983 budget to fund Centaur development. Days later, Representative Bill Lowery from San Diego, home of major aerospace contractor General Dynamics Convair, urged Centaur development during a House floor discussion of the FY 1983 NASA budget. In July 1982, an Urgent Supplemental Appropriations Bill for FY 1982 (H.R. 6685) was proposed, which included a provision directing NASA to restart Centaur development for Galileo as well as the International Solar Polar Mission. The language of the bill indicated that no more funds were to be obligated for any other upper stages.

Both Secretary of the Air Force Orr and NASA Administrator Beggs urged Congress to eliminate the Centaur provision. The Air Force warned that switching to Centaur for the Galileo upper stage would involve significantly more expense than was generally recognized. Nevertheless, the House and Senate passed the bill containing the Centaur provision on 15 July 1982, and President Reagan signed it into law three days later. The bill allocated $80 million for the design, development, and procurement of the Centaur upper stage. Lewis Research Center in Ohio would manage NASA's multi-Center Shuttle Centaur program. The Agency sent letters to Boeing and the Air Force informing them that they had to stop work on the IUS for Galileo and the Solar Mission.[70]

Why did Congress reinstate Centaur, especially considering all the objections? John Casani believed that there were several reasons:

- Congress had high regard for NASA's planetary exploration programs and was concerned that future missions would be damaged if they were limited to using underpowered launch vehicles.

- While the Air Force would require a higher energy upper stage by the late 1980s, Congress was reluctant to embark on a major new propulsion system development program, with its attendant cost risks and uncertainties, when the very reliable and proven Centaur was available. Centaur had a long history of performance, as well as development investment. It was a mature technology that could be counted on. So funding was better spent adapting Centaur to the Shuttle than developing a new launch vehicle from scratch.

- Congress wanted to give NASA responsibility for upper stage development. This might not have happened if, instead of reinstating Centaur, Congress had directed the Air Force to develop a new high-energy upper stage to be ready in the late 1980s. Such a launch vehicle's development would be heavily controlled by Air Force requirements and needs.

---

[70] "Centaur for Shuttle Would Cost Too Much, Air Force Complains," *Aerospace Daily* (14 July 1982); John Casani, "From the Project Manager," *Galileo Messenger* (August 1982): 1, 4; Waff, untitled chronology of Galileo events, 1979–82; "NASA Directs Air Force, Boeing To Stop Work on IUS for Galileo and International Solar Polar Mission," *Aerospace Daily* (22 July 1982); Craig B. Waff, "Searching for Options, Part 4: Centaur—To Be or Not To Be?" (outline of chap. 12 of unpublished history of Galileo project), folder 18522, NASA Historical Reference Collection, Washington, DC; Dawson and Bowles, *Taming Liquid Hydrogen*, pp. 248–249.

- Finally, Europe's development of its own commercial space program worried Congress. It saw an "exodus of commercial customers to the European Ariane vehicle." The vehicle's parent company, Arianespace, had already captured nearly 30 percent of the market. Casani referred to estimates indicating that Centaur's recurring costs would be only half those of the IUS. But Centaur had twice the launch capability. Congress, Casani believed, thought that using Centaur was the best way to keep or recapture the largest commercial market share.[71]

Casani believed that, all things considered, the change back to Centaur was a good one. Centaur could get Galileo from Earth orbit to Jupiter faster than the IUS and with more propellant left in its tanks, which meant, in Casani's words, that there would be a "higher assurance of obtaining our science and mission objectives." The task now at hand was to adapt to the changed mission plan, adhere to Orbiter's and Probe's development and test schedules, and make Galileo a success.[72]

On the Road to a Launch

True to Robert Allnut's prediction, NASA delayed the new launch a year, to May 1986, to allow ample time to complete Centaur development, design and build the necessary Galileo-Centaur interfaces, and develop a new trajectory.[73] As the months went by without more changes, it appeared that a mission plan to which all parties could subscribe had finally been found. NASA and its contractors set to work designing the modified mission and developing the spacecraft for its 1986 launch.

For the next several years, the challenges to project staff became more technical and management-oriented than political. Some of the tasks before the project team included the following:

- Modify the Galileo spacecraft design to interface with Centaur.

- Reestablish documentation, management, and personnel interfaces with NASA's Lewis Research Center (now called Glenn Research Center), which would direct the development of Centaur for use in the Shuttle.

- Develop a direct trajectory to Jupiter for the 1986 launch because the extra power of Centaur would not require gravity-assists.

- Stop design modifications that had been required by the Earth gravity-assist in the ΔV-EGA trajectory.

- Revalidate the strategy for inserting Galileo's Orbiter into a Jupiter orbit.

---

[71] John Casani, "From the Project Manager," *Galileo Messenger* (August 1982): 4.

[72] Ibid., pp. 1, 4.

[73] GAO, p. 24.

- Reassess the pre- and postlaunch development plan.

- Close out all activity on Injection Module development because it would not be needed with Centaur.[74]

Other JPL tasks included modifying Galileo's subsystems to decrease their sensitivity to cosmic radiation, refining flight software, fabricating memory components, and improving the craft's spin bearing assembly, which would separate the rotating segment of Galileo from the nonrotating part.[75] In addition, John Casani planned for a second spacecraft, termed a "proof-test model," to be built—something that could serve as "a source of spares for the flight spacecraft."[76]

The original Galileo mission design had envisioned a "limited-spares" concept, but that design had been done when an early-1982 launch had seemed possible. A JPL environment had been envisioned in which the skills and facilities necessary to maintain and repair flight hardware would be readily available to support Galileo and other new projects. No other new starts had materialized, however, and Galileo's launch date had been delayed more than four years, with further delays possible if problems with Centaur development were encountered. As a result, Casani was concerned that JPL's ability to support a limited-spares concept might be seriously eroded by the time of the launch.[77]

Casani planned to address the spares issue through a combination of upgrading engineering models of Galileo and fabricating new parts in order to construct a proof-test model spacecraft. His intent was to manufacture the new hardware during 1983 and 1984, while the flight spacecraft was undergoing system-level integration and testing. Much of the fabrication would be done at JPL in order to avoid paying various subcontractors to maintain their abilities to manufacture Galileo parts, and also to optimize the use of the mission's engineering staff, who needed to be kept in place to provide support through Galileo's launch. Also, the fabrication work would help preserve the JPL staff's expertise in spacecraft maintenance and repair.

Casani envisioned acceptance testing of the flight spacecraft to be completed in late 1984, after which it would be put on an extended "burn-in"[78] schedule and the assembly of the proof-test model would be started using the same project team. Testing of the proof-test model would be completed by the end of 1985, and both spacecraft would be shipped to Kennedy Space Center in January 1986 to support the May 1986 launch.[79]

---

[74] A. E. Wolfe to J. R. Casani, "Project Impacts of 1986 Centaur Launch," 25 May 1982, John Casani Collection, Galileo Correspondence 5/82, folder 19, JPL 14, JPL Archives.

[75] GAO, pp. 24–25.

[76] John Casani interview, tape-recorded telephone conversation, 29 May 2001.

[77] J. Casani, "Impact of Switching to Centaur and Delay to '86" (draft), 25 May 1982, John Casani Collection, Galileo Correspondence 5/82, folder 19, JPL 14, JPL Archives.

[78] Years of empirical data have shown that electronic parts tend to fail early in their lifetimes if they contain defects. A "burn-in" period consists of continuous operation of the electronic parts for as many hours as possible in order to screen out the unreliable ones before actual use. Many of Galileo's parts were new designs that had been built using new manufacturing approaches, and so burn-in was a very essential part of spacecraft testing. "Q&A," *Galileo Messenger* (July 1985): 5.

[79] Casani, "Impact of Switching."

Atmospheric Entry Probe Testing

NASA engineers designed Galileo's 4-foot-wide Probe to separate from Orbiter 150 days out from Jupiter and eventually plunge into the planet's atmosphere. It would enter the atmosphere on a shallow trajectory of approximately 9 degrees below horizontal and gradually slow from a velocity of 107,000 mph (48 kilometers per second) to Mach 1. The Probe's thermal heatshield would enclose and protect the scientific instruments during this time. Once Mach 1 was reached, a parachute system would separate from the forward heatshield and control the rate of descent. Meanwhile, the scientific instruments would collect data that the Probe's transmitter would send to Orbiter, which would then relay the information to Earth.[80]

One of the critical tests in preparing the Probe was to confirm proper parachute operation. On 17 July 1982, NASA scientists placed the 460-pound Probe vehicle in a gondola attached to a 5-million-cubic-foot-capacity, helium-filled balloon and launched it from Roswell, New Mexico. A total of 830 pounds of ballast had been added to the Probe's forward thermal shield to get the test vehicle to the required Mach number and dynamic pressure that it would experience in the Jovian atmosphere. After 4 hours of flight time and at an altitude of 97,000 feet, the Probe was dropped from the balloon above New Mexico's White Sands Missile Range. The Probe's pilot chute deployed, followed by the removal of the aft heatshield and the deployment of the large main parachute. The Probe's Project Manager, Joel Sperans of Ames Research Center in Mountain View, California, reported that from on-board camera data, the entire descent sequence appeared to have been carried out successfully. Speeds of descent and dynamic pressures were virtually identical to those expected to be encountered in Jupiter's atmosphere.[81]

Although Sperans reported that a major milestone in Probe development had been attained, the test did not go flawlessly. Some of the Probe's ballast failed to separate. More importantly, the main parachute opened several seconds late. Its function was to slow the Probe and help separate its descent module, which contained the scientific package, from the deceleration module that performed a heat-shielding function. Instead, the main parachute did not fully open until the descent module had already separated.

Although project staff did not think these were serious problems, the parachute delay was painstakingly analyzed during the following months. Pictures of the test suggested that the wake created as the Probe sped rapidly through the air may have delayed the chute's opening. To fix this, project staff considered extending the length of line that attached the parachute canopy to the Probe. This would place the parachute farther from the Probe, where wake effects were less intense. The staff decided to delay the release of deceleration module components until Probe velocity decreased further, which would also reduce wake effects. Hughes Aircraft Company, the contractor developing the Probe, scheduled wind tunnel tests to examine the effects of these modifications.[82]

During the following months, Hughes integrated most of the descent module's instruments and components into the Probe flight vehicle and conducted electrical and other tests. Project staff also scheduled a series of environmental stress tests and more

[80] Bruce A. Smith, "Deployment Problem Forces Galileo Test Reschedule," *Aviation Week & Space Technology* (20 September 1982).

[81] Peter Waller, "Galileo Probe Drop Test Successful," NASA news release, Ames Research Center, 30 July 1982, pp. 1–2; Smith, "Deployment," p. 106.

[82] Waller, "Galileo Probe," pp. 2–3; Smith, "Deployment," pp. 106–107.

drop tests.[83] In April 1983, Lewis Research Center used its Transonic Dynamics Facility to test a quarter-scale model of the Probe's new parachute assembly in order to assess the impact of the parachute's position and distance from the Probe body on its behavior. Lewis investigated parachute distances ranging from 5.5 to 11 body diameters behind the Probe; results showed that chute behavior was normal at all distances except the original design position of 5.5 diameters. Lewis performed additional tests to optimize chute performance that included varying the chute's porosity, the Probe's ballast configuration, and the Probe's angle of attack when entering the Jovian atmosphere. At the completion of these tests, Lewis performed half-scale parachute tests.[84]

In late July 1983, NASA carried out a drop test of the redesigned full-scale Probe. The successful test verified that the parachute modifications avoided the slow-deployment problems of the year before.[85]

Component, Spacecraft, and Intersystem Testing

Exhaustively testing each component of the Probe and Orbiter and then integrating it into the spacecraft required almost as much time as Galileo's primary mission did. Each component, as well as the complete spacecraft, had to be carefully examined and subjected to a rigorous series of functional and environmental tests to make sure everything met design requirements. Staff also tested "intersystem functions," such as those between ground-based mission control and the spacecraft, or between the spacecraft and the Shuttle.[86]

Component Testing. Testing of the spacecraft's many subsystems began at the various contractor facilities that developed and fabricated them. Before their delivery to JPL, Galileo's components and scientific instruments had to be qualified and accepted for flight.

Integrating Flight-Qualified Components. After the components arrived at JPL, project staff integrated them mechanically and electrically into a functioning spacecraft. These integration operations began in March 1983 at JPL's Spacecraft Assembly Facility (SAF) in cleanrooms environmentally controlled to remain at 72°F and 50 percent humidity. Special procedures eliminated the possibility of electrostatic charging of personnel and equipment that could come in contact with and damage the spacecraft's sensitive electronics.[87]

Integration operations necessitated mating the Probe to the Orbiter. Hughes, which had built the Probe, shipped it to JPL in September 1983, and these two parts of the Galileo spacecraft were connected for the first time. Project staff needed to verify the operation of the interfaces between each Probe and Orbiter component and the rest of the spacecraft, as well as testing all the components using Orbiter power, telemetry, and commands. The staff electrically coupled the Probe to the Orbiter, then set up a data link between them to verify overall data flow from the Probe to the Orbiter's radio relay

---

[83] Smith, "Deployment," pp. 106–107.

[84] "Galileo Probe Parachute Tests," NASA Daily Activities Report, 28 April 1983.

[85] "Galileo Probe and Orbiter To Be Mated in September," *Aerospace Daily* (24 August 1983): 298.

[86] "SAF Activities," *Galileo Messenger* (June 1983): 1.

[87] "SAF Activities," pp. 1–2; "Status/Highlight Summary," in a set of John Casani Collection papers that begin with "GLL 8th Quarterly Report Agenda," Galileo Correspondence 2/81, folder 30, box JA370, JPL 14, JPL Archives.

hardware, command and data system, modulation/demodulation system, and radio transmitter. The transmitter then sent test data to Earth-based tracking systems, to Orbiter and Probe ground data systems, and to scientists for analysis.[88]

Environmental and System Testing. Once all flight components had been integrated and tested, the Probe was sent back to Hughes for environmental testing. Orbiter system testing also began, including the following:

- Interference analysis to determine whether the operation of any component had adverse effects on other components.

- Mission profile tests to simulate key mission phases such as launch, trajectory correction maneuvers, cruise operations, Probe release, and insertion into Jupiter orbit. During these simulations, operational capabilities were analyzed and areas for improvement identified.[89]

Probe-Orbiter System Testing. After Hughes completed Probe's environmental tests, it returned the Probe to JPL and project staff carried out operational tests of the combined Probe-Orbiter system. These were followed by system environmental tests, which included vibration, acoustic, pyrotechnic, electromagnetic, vacuum, and solar radiation tests. Vibration analyses sought to verify that the spacecraft would operate properly after being shaken and jostled during liftoff, ascent, and injection into its interplanetary trajectory. Project staff bolted the spacecraft to a 30,000-pound shaker table in JPL's Environmental Test Laboratory, then attached 29 accelerometers and strain gauges so that vibration levels could be monitored and controlled. To simulate flight conditions, vibrations ranged in frequency from 10 to 200 Hertz, producing accelerations of up to 1.5 g's. Staff added an additional 100 sensors for monitoring during the test. Fifty-nine of these were part of an automatic shutdown system to protect the spacecraft in case vibrations exceeded specified limits.

Acoustic testing simulated the high noise levels to which Galileo would be subjected during liftoff in the Shuttle payload bay. Project staff placed Galileo in JPL's 10,000-cubic-foot acoustic chamber, a concrete room containing two 4-foot-square "feed horns" to generate the sound. Numerous microphones were placed around the spacecraft to monitor noise levels during the testing. Preparation for the testing took several days, although the test itself lasted only 1 minute, with noise levels of 142 decibels. By comparison, the noise level of a typical conversation is about 50 decibels, while heavy traffic generates about 75 decibels. Each 6-decibel increase indicates a doubling of sound pressure.[90]

After the actual launch, explosive devices were set off to deploy booms, release Centaur, separate the Orbiter's spun and despun sections, and jettison instrument covers. Pyrotechnic testing assessed impacts on the spacecraft of concussions similar to those that were generated by these small explosions. Electromagnetic compatibility

---

[88] "Galileo Probe and Orbiter To Be Mated in September," *Aerospace Daily* (24 August 1983); "SAF Activities," p. 2.

[89] "SAF Activities," p. 2.

[90] "Environmental Tests," *Galileo Messenger* (July 1985): 2.

tests examined impacts of the radio-frequency environment that surrounded the craft during and after launch.[91]

To assess the impact of vacuum on Galileo, project staff employed JPL's large space simulator. This 25-foot-tall chamber was evacuated rapidly to simulate the pressure reduction Galileo would experience during the Shuttle's ascent from the launchpad. Pressure in the simulated space environment was brought down as low as 1×10^{-5} torr.[92] Although testing at this level of vacuum was able to build a degree of confidence in Galileo's ability to withstand actual vacuum during the mission, the test chamber could not attain the vacuum of deep space, which can be as low as 1×10^{-14} torr.[93]

Project engineers conducted the tests in hot and cold modes to represent both the inner solar system part of the mission, when Galileo would be closest to the Sun, and the times when the craft would be out at Jupiter and hidden behind the planet, receiving no warming solar radiation at all. Liquid nitrogen cooled the inner wall and floor of the room to temperatures as low as -196°C (-321°F) to simulate the frigid parts of the mission, and a focused array of 20- to 30-kilowatt arc lamps approximated the times when Galileo would receive large doses of solar radiation.

The testing needed to validate Galileo's ability to function under the extreme conditions it would encounter. Spacecraft thermal controls had to manage heat from the Sun as well as from the craft's internal radioisotope thermal generators (RTGs), various heater units, and complex electronic devices. Galileo was designed to maintain thermal equilibrium by minimizing variable solar input as much as possible, distributing heat to those areas of the craft needing to be maintained at higher temperatures, and controlling the loss of heat to space. The craft employed louvers similar to Venetian blinds. It also used multilayer thermal blankets, which gave some protection against micrometeorites and electrostatic discharge, as well as helping to retain heat.[94]

Galileo's electric heaters could be turned on and off, but its RTGs always generated thermal energy (because their plutonium fuel was constantly decaying and emitting energy). Orbiter contained a total of about 70 heating devices, while Probe contained 34. For safety's sake, the environmental testing did not use the actual RTGs, with their radioactive fuel, but instead employed electrically heated, simulated RTGs.

Most of Galileo's subsystems were designed to operate between 5 and 50°C but were tested at temperatures that ranged from -20 to 75°C (-4 to 167°F). Some components, however, had narrower restrictions, and it was important to determine whether these could be maintained during the environmental tests. The temperature of the craft's propellant fuel, for instance, had to be strictly maintained because overpressures could cause its storage tank to rupture. In addition, fuel lines could never be allowed to freeze. The hydraulic system that deployed the Orbiter's booms was also very sensitive to temperature. In order to maintain acceptable viscosities in the system's hydraulic fluid, temperatures had to be kept between 20 and 30°C (68 and 86°F).[95]

---

[91] Ibid., pp. 2–3.

[92] One torr of pressure is equal to the pressure of one millimeter of mercury, or 1/760 of an atmosphere (the mean pressure on the surface of Earth).

[93] "Environmental Tests," pp. 3–4.

[94] Ibid., pp. 4–5.

[95] Ibid., p. 5.

After testing the Probe and Orbiter together, staff performed similar tests on Orbiter alone to assess its expected performance after Probe was released near Jupiter. During the final months of testing, spacecraft power was maintained 24 hours per day in order to log the maximum number of operating hours on Galileo's electronics assemblies.[96]

Adapting Centaur to the Shuttle

Design studies integrating Centaur with the Shuttle date back to the start of the Shuttle program, although when Boeing's solid-fueled IUS was the favored upper stage, prospects for a Centaur-driven Galileo spacecraft seemed very remote. When General Dynamics finally received the go-ahead in 1982 for full-scale Centaur adaptation to the Shuttle, it had to fulfill both NASA's requirements and those of the Air Force. The result was two versions of the upper stage: the G type for the Air Force and the G-prime for NASA. Both were shorter and wider than the version used as an upper stage for Atlas, which was over 9 meters long and 3 meters in diameter (about 30 feet by 10 feet), but they also differed markedly from each other. The Air Force needed space in the Shuttle cargo bay for its larger satellites, so the Centaur G was made only 6 meters long but 4.3 meters in diameter (roughly 20 feet by 14 feet). Its liquid-oxygen tanks and two engines were left unchanged, but its liquid-hydrogen tank was shortened and widened.[97]

In the design of NASA's G-prime Centaur, the need for more thrust took priority over extra space in the Shuttle cargo bay. The G-prime was less than a foot shorter than the 9-meter Atlas upper stage and was as wide as the G version, or in other words, 40 percent wider than the Atlas upper stage. The G-prime's liquid-hydrogen and liquid-oxygen tanks were longer than those of the G type, although its engines and electronic systems remained the same. The performance characteristics of a Shuttle/G-prime Centaur combination were very impressive when compared to other launch system combinations. The Atlas-Centaur combination could lift a payload of only 1,900 kilograms (4,200 pounds) to geostationary orbit. The Shuttle plus the Air Force's G-type Centaur could move a satellite weighing 4,800 kilograms (10,600 pounds) to geostationary orbit, while the Shuttle plus NASA's G-prime Centaur could transfer a 6,350-kilogram (14,000-pound) satellite into geostationary orbit.[98]

Inside the Shuttle's payload bay, Centaur was attached to a support structure and tilt-table. The support structure also brought electrical connections to Centaur that were necessary for supplying power, transferring data, and controlling fueling and emergency fuel-jettisoning operations. Deployment of Galileo and Centaur would involve opening the Shuttle's doors and rotating the tilt-table to swing the forward end of the spacecraft out of the payload bay, then using a shaped charge[99] to sever the spacecraft's connection to the Shuttle. After

---

[96] Ibid., pp. 1, 4–5.

[97] Curtis Peebles, "The New Centaur," Spaceflight 26 (September/October 1984): 358; G. R. Richards and Joel W. Powell, "The Centaur Vehicle," *Journal of the British Interplanetary Society* 42 (1989): 115.

[98] Peebles, "The New Centaur," p. 358.

[99] A shaped charge is an explosive charge shaped to focus the energy released during detonation along a single line, making the charge's delivery of energy very accurate and controllable. Lawrence Livermore National Laboratory, "Shaped Charges Pierce the Toughest Targets," *Science & Technology*, June 1998, http://www.llnl.gov/str/Baum.html.

this, 12 steel springs would push Galileo-Centaur away from the Shuttle. Ignition would be delayed for approximately 45 minutes, until the Shuttle flew to a safe distance away.[100]

The long list of tests to be performed on Galileo before launch day included a very thorough examination of the Centaur rocket, which had never before been deployed from the Space Shuttle. The hydrogen-fueled Centaur had been used for years as an upper stage of robotic rockets, but during Galileo's launch, it would be carried aloft inside the cargo bay of a craft with a human crew. Because Centaur's liquid-hydrogen and liquid-oxygen fuel was potentially explosive, elaborate safety features had to be developed before it was deemed safe for use in the Shuttle. These features included the capability to dump Centaur's fuel overboard quickly in the event of an emergency. NASA staff had designed and installed a dump system that could jettison all of Centaur's propellant in 250 seconds in the event of a launch abort or failure to deploy in orbit. The staff had also built in the capability of waiting to fuel Centaur until the Shuttle was on its launchpad.[101]

The Amphitrite Asteroid Option

In December 1984, John Casani proposed that an additional task be added to the Galileo mission: to fly by the asteroid Amphitrite (the last part of its name rhymes with "flighty"). Because the spacecraft would traverse the solar system's asteroid belt on its way to Jupiter, it would be in an excellent position to make close observations of the asteroids. Casani and other Galileo staff thought that such a study would add significantly to our understanding of the solar system and its origins and that it fell within the stated mission of NASA's solar system exploration program. Primitive bodies such as asteroids and comets appeared to be "the best preserved remnants of the early solar system and most representative of the overall composition of protoplanetary/protosolar nebula material." This was important, because evidence of early solar system processes is not easily extracted from larger planetary bodies.[102]

The narrow region between Mars and Jupiter known as the asteroid belt was also the boundary between the rocky inner planets of the solar system and the more volatile, gas-rich outer planets. Over 3,000 asteroids have been identified in this belt. Amphitrite, whose name refers to a queen of the sea and one of Poseidon's wives in Homer's *Odyssey*, is a rocky little body 200 kilometers (120 miles) in diameter. It was the 29th asteroid discovered (and is often referred to as 29 Amphitrite). Albert Marth identified it on the night of 1 March 1854 from William Bishop Observatory in London. Later work determined that the asteroid revolved in a nearly circular orbit around the Sun at an average distance of 230 million miles and rotated on its axis once every 5.4 hours. Beyond these basic facts, little was known about Amphitrite. It was chosen because it was in the right place at the right time.[103]

The idea of a Galileo flyby of Amphitrite arose at JPL in 1983 after engineers planning the spacecraft's course through the asteroid belt finished a series of computer simulations looking for possible navigational hazards. The asteroids in the general area of

---

[100] Richards and Powell, "The Centaur Vehicle," p. 115.

[101] William Harwood, "NASA Current News," from United Press International (UPI), 31 December 1985; Peebles, "The New Centaur," p. 358.

[102] J. R. Casani to W. E. Giberson, "Addendum to Project Plan 625-1, Rec. C, Section III," 3 December 1984, John Casani Collection, Galileo Correspondence, 11/84–12/84, folder 1, box 1 of 6, JPL 14, JPL Archives.

[103] Casani, "Addendum"; Paul Stinson, "Why Amphitrite," *Space World* (December 1985): 24.

Galileo's path turned out to be far enough away to not present a collision danger. Other asteroids in this group besides Amphitrite, such as 1219 Britta or 1972 Yi Xing (the number indicates the order of discovery), were only a few kilometers in diameter. For safety reasons, Galileo's planners did not want to send the spacecraft closer than 10,000 kilometers (about 6,000 miles) from any asteroid, because no one knew what the surrounding environment would be like. Space scientists considered it possible that an asteroid could be enveloped by a cloud of fine dust particles from numerous collisions with other asteroids. If Galileo were to fly through such a cloud, its optics could be severely damaged and the spacecraft itself might be destroyed.

At the "safe encounter" distance, not much could be learned from Galileo's cameras' observation of a small asteroid only several kilometers in diameter. Amphitrite, on the other hand, was among the largest 1 percent of all asteroids and huge compared to Britta or Yi Xing. A flyby from even 20,000 kilometers, twice the safe encounter distance, could yield significant data.[104]

Because of the attractive opportunity, John Casani requested that NASA management approve a postlaunch flyby option as a secondary mission objective. Under this plan, a trajectory modification could be implemented such that Galileo would pass at a distance of 10,000 to 20,000 kilometers from Amphitrite and arrive at Jupiter about three months later than planned, in early December 1988. But if the spacecraft was not "healthy" when flying by Amphitrite—that is, if all systems did not fully check out—then no encounter operations would be performed and no pictures or measurements of the asteroid would be taken.[105]

A successful encounter with Amphitrite was viewed by NASA as quite significant because no U.S. mission to the asteroids was planned until at least a decade later. In addition, experts in the taxonomy of asteroids thought that Amphitrite would be especially valuable to study because it was an S-type asteroid. Such asteroids were believed to be either remnants of a larger planetary body (perhaps part of its stony iron core), or fragments of chondrites—bodies that coalesced directly from the primordial nebula out of which the Sun and planets formed. During the two (Earth) days of Galileo's closest approach to Amphitrite, the asteroid would turn 10 times on its axis, allowing Galileo to repeatedly map its entire surface and discern features as small as 600 feet across. Photographs with such resolution of the Martian surface from Mariner 9 resulted in a radically changed understanding of the planet. In addition, with Galileo's infrared, visual, and ultraviolet sensing systems, a detailed geological survey of Amphitrite could be carried out. Atmospheric instruments would also search for any signs of a tenuous atmosphere, and Galileo's dust detector would look for evidence of debris orbiting the asteroid. Changes in Galileo's velocity due to Amphitrite's gravitational field would allow the body's mass and average density to be calculated for the first time.[106]

On 6 December 1984, NASA Administrator James M. Beggs endorsed the Amphitrite option and supported a change in Galileo's trajectory and in its Jupiter arrival date from 27 August 1988 to 10 December 1988, with the understanding that Amphitrite

---

[104] Paul Stinson, "Why Amphitrite," and Alan Stern, "Amphitrite, Maybe," both from *Space World* (December 1985): 23–24.

[105] J. R. Casani to W. E. Giberson, "Galileo Asteroid Flyby," 11 September 1984, and R. J. Parks to Geoffrey A. Briggs, 20 September 1984, both in John Casani Collection, Galileo Correspondence, 11/84–12/84, folder 1, box 1 of 6, JPL 14, JPL Archives; Casani, "Addendum"; Stern, "Amphitrite, Maybe," p. 22.

[106] Stern, "Amphitrite, Maybe," pp. 22–23; JPL Public Information Office, press release no. 1062, 17 January 1985, *http://www.jpl.nasa.gov/releases/84-86/release_1985_1062.html.*

was to be a secondary objective and must in no way compromise the primary Jupiter objectives. Because no prelaunch funds were authorized for the Amphitrite option, and due to the political sensitivity of asking for funding to cover yet another Galileo mission change, Beggs stressed that no Amphitrite mission planning was to be undertaken until after launch. And no papers or articles were to be written on any aspect of the Amphitrite option until after the actual encounter.[107]

Down-to-the-Wire Problems

By October 1985, project staff had completed all electronic compatibility and characterization testing of the spacecraft, as well as environmental testing. But Galileo was still missing its spin bearing assembly, the mechanical and electrical interface between the spun and despun sections of the craft. The bearing assembly used a system of metal slip-rings and brushes to transmit electrical signals from one section of the spacecraft to the other. Project staff had discovered contamination on the slip-rings and sent them and the brushes to be remanufactured.[108]

Analysis of the bad parts and their manufacturing process revealed that a whole chain of circumstances had led to the problem. These circumstances involved the coatings on the parts, the effect on the coatings of heat from soldering operations, the chemical used to clean the parts, and the impact of vacuum on the system. Heat from soldering had made the coating material more porous, and when the parts were dipped into a chloro-fluorocarbon (CFC) solution for cleaning, some of the CFC was absorbed into the pores in the coating. During environmental testing, the low-vacuum conditions pulled the CFC out of the pores, and it mixed with the fine, powdery debris generated as the brushes and slip-rings slowly wore down. This debris normally did not cause problems, but when mixed with the CFC, it clumped up, eventually accumulating under the surface of a brush and opening up the contact between it and the slip-ring. The result was an intermittent power transmission problem.[109]

A second problem involved Galileo's command and data system. Project staff noticed that the performances of certain memory devices deteriorated in an electromagnetic radiation environment. The origin of this problem, too, was traced back to the manufacturing process. But after JPL acquired and installed several thousand remanufactured memory devices, they were found to have yet another problem called "read-disturb." Reading the contents of a memory cell in a device could disturb the contents of adjacent cells. The project team determined that the processing changes made to repair the memory cells' sensitivity to electromagnetic radiation also caused them to operate faster, and it was these parts that were susceptible to the read-disturb problem. Project staff had to construct a special screening test, and one-third of the memory devices had to be removed and changed. This was a very labor-intensive operation because of the parts' small size and close spacing.[110]

[107] J. R. Casani to distribution, "Amphitrite Option," 17 December 1984, John Casani Collection, Galileo Correspondence, 11/84–12/84, folder 1, box 1 of 6, JPL 14, JPL Archives.

[108] Bruce A. Smith, "Galileo Undergoes Final Tests Prior to Shipment to Kennedy," *Aviation Week & Space Technology* (21 October 1985): 135.

[109] Ibid.

[110] Ibid.

In spite of the problems, Casani believed that Galileo had a good record for parts quality. Components were carefully examined and burned in at elevated temperatures. All spacecraft equipment and spare parts received at least 2,000 hours of operation prior to launch. NASA designed Galileo to operate for at least five years, including two and a half years of travel to Jupiter and sufficient time to carry out its experiments and observations in the Jovian system. This "design life" was dictated by the propellant capacity of the spacecraft, as well as expected degradation due to the radiation environment in which it would operate. Project officials believed that damage from radiation would gradually deteriorate electronic components and eventually bring about the end of the mission.[111]

The First 3,000-Mile Leg of a Half-Billion-Mile Journey

"Everything seemed on the home stretch, when at 3:00 a.m. on December 19, 1985, Galileo began the first 3,000-mile leg of its half-billion-mile journey," announced the narrator of WGBH's *NOVA* television program on Galileo, "The Rocky Road to Jupiter." JPL staff had been readying Galileo for a high-speed truck convoy from Pasadena, California, across America to Kennedy Space Center, Florida. When the convoy rolled out of JPL in the middle of the night, its drivers did not yet know the exact route they would be taking. They only knew that they would drive all night and all day, stopping only for food and fuel. Galileo management was concerned that the plutonium units that would power spacecraft functions would draw antinuclear protesters or, even worse, a terrorist attack.[112]

After being repeatedly criticized by Congress, the American press, and the antinuclear movement, Galileo finally appeared ready to take off. If it survived the truck journey to Florida and another battery of grueling tests, it would leave Earth in May 1986, unless some new, unforeseen problem arose. But if Galileo missed the May launch date, it would have to wait 13 months for another "window."

The physicist James Van Allen thought of Galileo's history as the "Perils of Pauline." He remembered four instances when the whole project had been formally canceled and many times when it was "just sort of hanging by its fingernails." Time and again, as in the *Perils of Pauline* silent film serial,[113] a hero would come by in the nick of time and rescue Galileo from oblivion.

James Fletcher, NASA Administrator, raised the question whether so much time had gone by without a launch that there might now be a better mission to carry out in Galileo's place. But when NASA seriously considered this question, the answer that came back was, "No, Galileo is still the best thing we can do at this time."

Galileo would be the first planetary mission to launch from the Shuttle, and "routine emergencies" were expected in the weeks before launch. To deal with these incidents, JPL engineers would stay at Kennedy Space Center until Galileo was on its way to Jupiter. In NASA's major past missions, such as Voyager, the Agency typically launched two similar spacecraft. But Galileo's cost precluded this strategy, making it

---

[111] Ibid.

[112] Transcript, "The Rocky Road to Jupiter," *NOVA* no. 1410 (Public Broadcasting System television program), 7 April 1987, WGBH Transcripts, Boston, MA.

[113] In the *Perils of Pauline* silent film episodic serial (circa 1914), the heroine would weekly evade attempts on her life perpetrated by pirates, Indians, gypsies, rats, sharks, and her dastardly guardian. Her most well-known plight was being tied to railroad tracks in front of a rapidly approaching train. From *The Greatest Films* Web site, *http://www. filmsite.org/peri.html* (accessed 29 June 2003).

essential that the launch take place without problems. Seventeen principal investigators from six nations would be flying experiments on board the spacecraft. Over one hundred scientists would be working closely with them. Some of the staff had been planning for this mission since the 1960s.

As the trucks rolled through the night on their way to Florida, some staff worried about the possibility of a road accident. Tom Shain, an electrical engineer on the project, commented, "Think of the repercussions—we get sideswiped, driving along here, something like that. And there's an awful lot of people worked an awful lot of long hours on this thing, y'know, for something like that to happen . . . . Maybe that's one of the reasons I'm having a hard time sleeping."[114]

John Casani was worried about crises a little further down the line. For instance, could the spacecraft be brought back home for repairs if trouble arose soon after launch? He concluded, "We have looked at the problem of recovering the spacecraft from orbit. It is virtually an impossibility. There are no provisions on the spacecraft for mechanical attachment or recapture . . . we're going to Jupiter or we're not going anywhere."[115]

For three and a half days, the truck convoy rolled. Spacecraft technicians became truckers and talked on the Citizens Band (CB), learning its special language. Escorted by police, state troopers, and other, less obvious guards, the caravan crossed America without incident or mishap, reaching Kennedy Space Center on 23 December. Once there, the staff breathed a large sigh of relief and took some vacation time over Christmas.[116]

Ahead of the team lay the integration of Galileo with Centaur and a host of final tests. One of the activities of late January 1986 that could be performed only at Kennedy was the first "tanking up" of the spacecraft. To power its course-correction jets, Galileo used hypergolic propellant, two different substances that ignited upon contact with each other. Loading such fuel into the spacecraft's propulsion module could be very hazardous, and so project staff evacuated the area around Galileo for a distance of 200 yards. The crews took one full day to load each of the four tanks, working short, carefully monitored 80-minute shifts. Each crew member donned a special suit for protection that was equipped with self-contained breathing equipment. In those suits, the crew resembled astronauts.[117]

On 24 January, NASA threw a large party to introduce JPL staff to their new cohorts, the Kennedy personnel who would integrate Galileo into the Shuttle *Atlantis*. The night of the party, the weather was growing noticeably chilly. A cold front was moving in. The cold front did not dampen the spirits at the NASA party, but it would have a dire effect on the Space Shuttle that was at that time sitting on a nearby launchpad. This was not the Shuttle *Atlantis*, which was scheduled to bear Galileo aloft. It was the *Challenger*, and it was due to launch on its next mission in four days.

---

[114] "Rocky Road," pp. 2–5, 11.

[115] Ibid., pp. 15, 16.

[116] Ibid., pp. 16–17.

[117] Ibid., pp. 17–18.

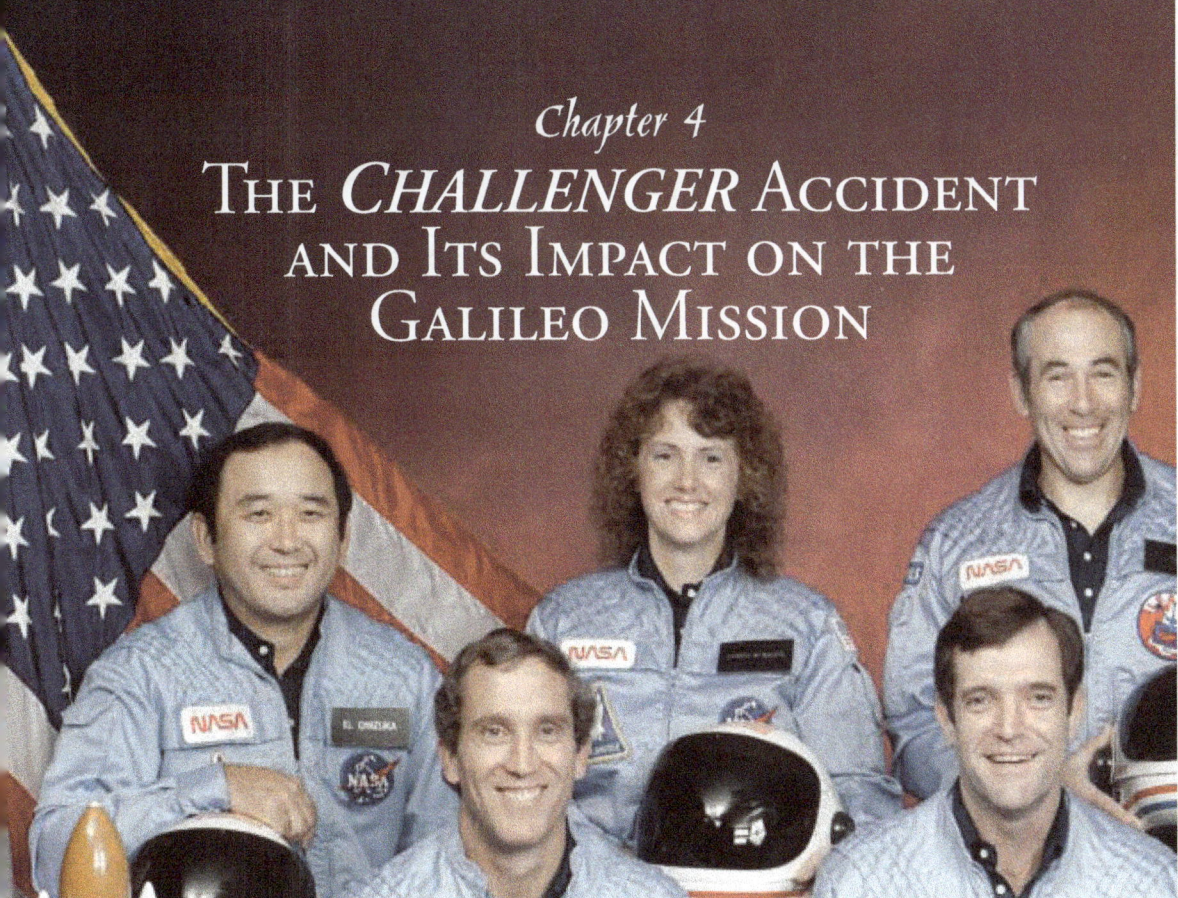

Chapter 4
The *CHALLENGER* Accident
and Its Impact on the
Galileo Mission

PREPARATIONS FOR THE LAUNCH OF Shuttle mission 51-L were not unusual, although they had been made more difficult by changes in the launch schedule. The sequence of complex, interrelated steps preceding the mission were being followed closely, as always. Flight 51-L of the *Challenger* had originally been scheduled for July 1985, but it was postponed to late November to accommodate payload changes. The launch was subsequently delayed further and finally rescheduled for late January 1986. NASA had successfully launched the Shuttle 24 times before and expected this launch to be successful as well.[1]

Events Leading up to the *Challenger* Disaster

In spite of NASA's confidence in the Shuttle, signs of malfunction had periodically arisen over the years. Morton Thiokol built the Shuttle's solid-fueled booster engines, and years before flight 51-L, its personnel had seen evidence of problems with the O-rings that sealed the booster engines' field joints. These were joints between sections of the engine that were assembled at the launch site rather than at the factory. As far back as the second flight of the Shuttle, high-pressure gas penetrating the joints had caused erosion

---

[1] William P. Rogers, chairman, "Chapter II: Events Leading Up To the Challenger Mission," in *Report of the Presidential Commission on the Space Shuttle Challenger Accident*, vol. 1 (Washington, DC: Government Printing Office, June 1986).

to the O-rings. The thinking at the time, however, was that even with a small amount of erosion, the joints could reliably contain combustion gases.[2]

The primary and secondary O-rings in the field joints constituted the main barriers against a loss of pressure once the booster motor fired. These rings could not by themselves withstand the high temperatures generated during a burn, however. A substance called "vacuum putty," containing asbestos and zinc chromate, was also packed into the joints. Occasionally, trapped air expanded so much during the burn that it blew a hole through the putty. This allowed hot gases to blow past the inner, primary O-ring and attack the field joint's outer, secondary O-ring, the last line of defense against a pressure leak.[3]

After a January 1985 Shuttle flight, one year before flight 51-L, Morton Thiokol engineers inspected the solid rocket motors (SRMs) and found blow-by that was much worse than had been thought. The January 1985 launch had been the coldest in Shuttle history up to that point, with 53°F air temperatures at Kennedy Space Center. Though this was chilly for Florida, the air temperature during the Challenger Shuttle flight 51-L launch a year later was even colder—36°F.[4]

27 and 28 January 1986

Flight 51-L was ready to fly on 27 January 1986, but NASA project managers and contractor support personnel (including Morton Thiokol staff) worried about the high crosswinds at the launch site. Shortly after noon, NASA managers scrubbed the launch for that day and polled project personnel, asking if there were any "constraints" against launching the next morning, when temperatures were predicted to be very low.[5]

Morton Thiokol management expressed concern about field joint performance at low temperatures. The company's vice president of engineering, Robert Lund, recommended waiting until O-ring temperatures reached at least 53°F. NASA challenged this recommendation. George Hardy, Deputy Director of NASA's Marshall Space Flight Center, stated that he was "appalled" by Morton Thiokol's proposal to delay the launch. According to Al McDonald, who was Morton Thiokol's SRM project director at the time, NASA had been under considerable pressure to reduce the frequency of its launch delays. The American press had been extremely critical of the Agency's apparent inability to launch Shuttles according to schedule.[6]

In response to NASA's challenge, Morton Thiokol upper management revisited the launch issue. Four of the company's vice presidents weighed various factors, then decided to go against the company's previous recommendation and approve a launch the next morning. They said that the SRM's primary O-ring could erode "by a factor of

[2] Allan J. McDonald, Morton Thiokol's solid rocket motor project director at the time of the *Challenger* launch, interview, Thiokol Propulsion, Corinne, UT, 23 August 2000.

[3] Ibid.

[4] Ibid.; Mark A. Haisler and Robert Throop, *The Challenger Accident: An Analysis of the Mechanical and Administrative Causes of the Accident and the Redesign Process That Followed*, Mechanical Engineering Department, The University of Texas at Austin, spring 1997, *http://www.me.utexas.edu/~uer/challenger/challtoc.html*.

[5] Rogers, p. 104.

[6] Ibid., pp. 105–108.

three times the worst previous case" before it would fail. And even if it did fail, they thought that the secondary O-ring would provide a backup seal.[7]

When Al McDonald was informed of his management's decision, he argued for scrubbing the launch and asked how NASA could rationalize launching below the qualification temperature. He stressed that there were three factors whose impacts were unknown, and they should preclude *Challenger* from taking off the next morning:

- What would happen to the O-rings at subfreezing temperatures?

- The ships tasked to recover Morton Thiokol's booster rockets after the launch would have to contend with 30-foot seas and high winds. Could they safely tow the boosters to shore under these conditions?

- There was ice everywhere around the launchpad. What effect would it have on Space Shuttle equipment?[8]

If anything happened to *Challenger*, McDonald told his bosses, he would not want to explain it to a board of inquiry. In response, Morton Thiokol management told McDonald that his comments would be passed on in an "advisory capacity" but that the decision to launch was not his concern.[9]

Between 7:00 and 9:00 the next morning, B. K. Davis, a member of the project's "ice crew," measured temperatures on the rocket boosters and found some of them to be even lower than predicted. The right-hand booster's aft region measured only 8°F. But there was no Launch Commit Criterion on booster surface temperatures, and so the ice crew did not even report its measurements.[10]

Photos taken the morning of the launch showed thousands of foot-long icicles hanging in curtains from the structures around *Challenger*, as well as water troughs frozen solid, even though antifreeze had been added to the water to prevent icing. The lower section of *Challenger*'s left booster rocket had sheet-ice deposits on it. At 9 a.m. on 28 January, NASA management met with project managers and contractors to discuss the icy conditions, but, according to the Rogers Commission report, there was "no apparent discussion of temperature effects on (the) O-ring seal." Never had the Shuttle taken off under such conditions. And yet, NASA's and Morton Thiokol's decision-makers were apparently so confident that the Shuttle had adequate safety margins built into its design that they still opposed another launch delay.[11]

At 11:38 a.m., 28 January 1986, *Challenger*'s three main engines and two booster engines were ignited. Al McDonald, who was watching from Kennedy Space Center's control room, thought that if a booster rocket O-ring blew, it would do so on the launchpad during the first seconds after ignition, as pressure was building up in the engines. Indeed, less than a second after booster rocket ignition, a puff

[7] Ibid., pp. 107–108.

[8] Ibid., p. 109; McDonald interview, 2000.

[9] Rogers, p. 109.

[10] Ibid.

[11] Ibid., pp. 109–113.

of gray smoke was seen above one of the right-hand booster's field joints. In the next couple of seconds, eight more puffs of increasingly blacker smoke were observed near the right-hand booster rocket. The black, dense smoke was a bad sign. It indicated that the grease, joint insulation, and rubber O-rings in the field joint were burning. But then the smoke stopped, *Challenger* lifted from the pad, and McDonald thought that *Challenger* might be in the clear. During the reconstruction of launch events that occurred later, investigators thought that aluminum oxide from the booster's fuel had formed a temporary ceramic seal over the leak.[12]

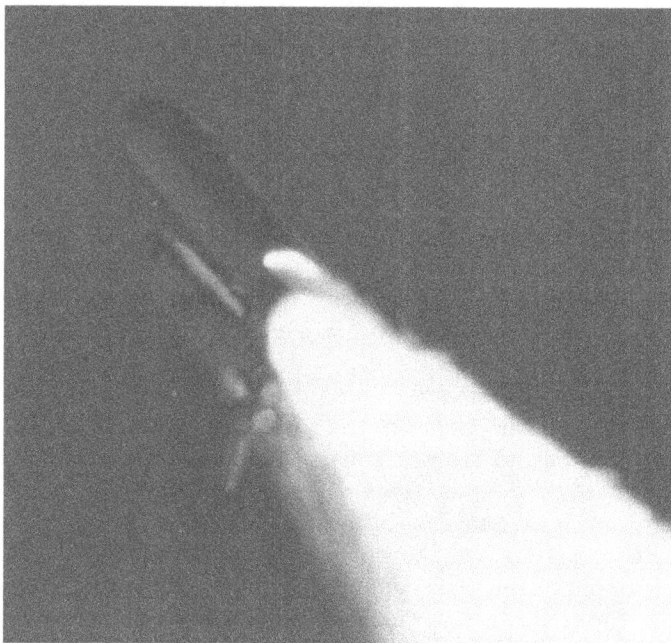

Figure 4.1. **Booster rocket breach. At 58.778 seconds into powered flight, a large flame plume was seen just above the right-hand booster engine exhaust nozzle. (Johnson Space Center image number S87-25373, 28 January 1986)**

A minute later, the right-hand booster motor sprouted a flame out of its side (see figure 4.1). The temporary ceramic seal had perhaps held only until pressure inside the engine rose and blew it out. Or possibly it was the high-altitude shear winds through which *Challenger* flew 37 seconds after launch that flexed the booster motor enough to break the ceramic seal. The pressure inside the right-hand booster started to diverge from that of the left-hand booster. This was potentially dangerous because radically different thrusts could drive the Shuttle onto a crash trajectory. That didn't happen, however. The booster engines' nozzles, which could swivel, were able to compensate for the different thrusts and keep *Challenger* on course.[13]

---

[12] "Sequence of Major Events of the Challenger Accident," *http://science.ksc.nasa.gov/shuttle/missions/51-l/docs/events.txt* (accessed 28 August 2000); Rogers, chap. 3; McDonald interview, 2000.

[13] "Sequence of Major Events"; McDonald interview, 2000.

The flame increased in size. Air racing past the ascending *Challenger* deflected the flame plume rearward. It hit a protruding structure that attached the booster motor to *Challenger*'s large external tank, which contained liquid hydrogen and liquid oxygen. The protrusion directed the flame onto the surface of the external fuel tank. At 65 seconds after launch, the shape and color of the flame abruptly changed. It was mixing with hydrogen leaking from the Shuttle's external fuel tank. A "bright sustained glow" developed on the underside of the *Challenger* spacecraft, between it and the external tank.[14]

Challenger began to skew off course. Hydrogen fuel tank pressures dropped, first slowly and then precipitously. Seventy-two seconds after launch, the lower strut connecting booster engine to external tank pulled away from the weakened tank, perhaps burned through by the flame. The booster was now free to rotate around an upper strut. Seventy-three seconds after launch, white vapor emerged from the external tank bottom. The tank then failed catastrophically, its entire aft dome dropping away and releasing a massive amount of hydrogen. As this was happening, the loosened booster motor smashed into the fuel tank's liquid-oxygen compartment. Hydrogen and oxygen mixed, burned explosively, and engulfed *Challenger* in a fireball (see figure 4.2).[15]

If the booster motor field joint alone had failed, but not the Shuttle's external fuel tank, the disaster might have been averted. Al McDonald was watching the launch, and he could see that the swiveling booster motor nozzles were compensating for the sideways thrust from fuel spurting out through the field joint. But the field joint failure had the bad fortune to occur inboard, near the external fuel tank. The spurt of flame from the booster was directed onto the external tank wall, eventually burning through it and liberating the hydrogen within. Still, if hydrogen had not mixed with oxygen, the disaster might not have occurred. There was no explosive burning immediately after the hydrogen leak was detected. The explosive burn started around the time that the booster motor crashed through the top of the fuel tank's liquid-oxygen compartment, and it was this that sealed *Challenger*'s fate.[16]

As *Challenger* broke apart, shock and disbelief swept through Kennedy Space Center's control room. Some people started to cry; others could barely talk. Orders were shouted out: "No one leave the room!" "Disconnect all telephones." "Don't talk to the press." The staff members were kept in the control room for hours, until events could be sorted out. Vice President George Bush flew down to Kennedy to talk with staff members and head the investigation into the cause of the disaster.[17]

Impact of the Disaster on Galileo

After the destruction of flight 51-L, NASA suspended all Shuttle flights until it could determine the root causes of the tragedy. Within a week of the crash, the White House formed a Presidential Commission, led by William Rogers, that had four months to submit a report to the President. Galileo would not be allowed to take off on its planned May 1986 launch

---

[14] Rogers, chap. 3.

[15] "Sequence of Major Events"; Rogers, chap. 3.

[16] McDonald interview, 2000; Rogers, chap. 3.

[17] McDonald interview, 2000; Carlos Byars, "Challenger Explodes, Shuttle Falls into Ocean, Crew Apparently Killed," *Houston Chronicle* (28 January 1986): 1.

date. It would have to wait at least 13 months until the next suitable launch window, if not longer. The delay would eliminate the possibility of flying by the Amphitrite asteroid. This was especially disappointing to the project team because the chances of finding another asteroid as large as Amphitrite (125 miles in diameter) when Galileo finally did launch were very small.[18]

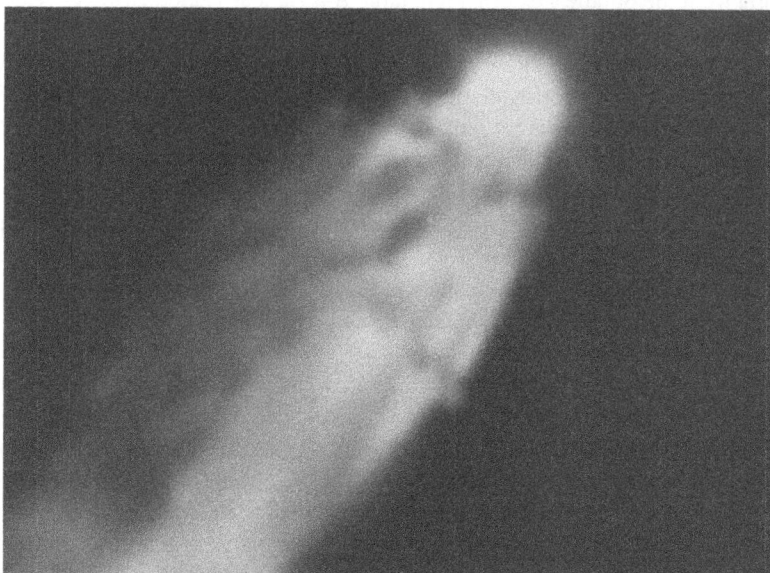

Figure 4.2. **Liquid-oxygen tank rupture. The luminous glow at the top is attributed to the rupture of the liquid-oxygen tank. Liquid oxygen and hydrogen mixed and engulfed *Challenger* in a fiery flow of escaping liquid propellant. (Johnson Space Center number S87-25390, 28 January 1986)**

No firm launch date could be set until after the Rogers Commission made its report to President Reagan. The Galileo and other project teams wanted answers to vital question such as these:

- When would Shuttles be allowed to fly again?

- How many missions would be conducted each year?

- Which missions would be placed at the top of NASA's priority list?

Galileo's managers worried that safety modifications that would probably be made to both the Shuttle and the Centaur upper stage might significantly increase their weight. This would reduce their payload lift capability and could compromise their ability to send Galileo on a direct trajectory to Jupiter. In this regard, Neil Ausman, Galileo's mission operations and engineering manager, said, "If we lose too much performance, we have no mission."[19]

---

[18] Boyce Rensberger, "Space Shuttle Explodes, Killing Crew," *Washington Post* (29 January 1986); Jay Malin, "NASA Postpones Next Three Shuttle Missions," *Washington Times* (11 February 1986).

[19] "On Hold: Probes to Jupiter and the Sun," *Science* (28 April 1986): 20–21.

As of April 1986, Galileo remained in a processing facility at Kennedy Space Center, where it was being coupled to the Centaur upper stage. NASA's plan was to complete the coupling operation, then put the craft into "active storage" at Kennedy until the new launch date.[20]

Concerns About Plutonium

As NASA and the Rogers Commission struggled to sort out the details of the *Challenger* tragedy, a new concern arose about missions such as Galileo that would carry plutonium fuel to power on-board functions. A periodical called the *Nation* expressed the opinion that "far more people would have died" if the explosion that destroyed *Challenger* had occurred during Galileo's launch, for in that case its plutonium fuel might have rained down throughout the area.

The *Nation* made much of a draft NASA Environmental Impact Statement that said that a Galileo launch would involve a very small risk of a plutonium-238 release because of the possibility of a Space Shuttle malfunction. In closing, the *Nation* article expressed the opinion that the indefinite suspension of the Galileo project should be made permanent.[21]

The 1981 review draft of the *Environmental Impact Statement for Project Galileo* did address the risk of a plutonium release. Statistical calculations performed during the preparation of the draft Environmental Impact Statement had indicated that a Shuttle mishap resulting in a plutonium release was very unlikely. The total probability of failure of a radioisotope thermal generator, or RTG (the plutonium-bearing source of on-board power), resulting from an unsuccessful Shuttle launch was estimated at one in ten thousand. But even in the event of a plutonium release, it was not considered likely that "a person would accumulate enough plutonium to exceed current radiation exposure standards." The risks of human fatality or injury from "accidental reentry of Galileo payload components" were estimated at one in a million and one in ten thousand, respectively.[22]

In spite of the low estimated probability of health damage from a plutonium release, NASA requested that JPL study, for a second time, the armoring of Galileo's RTG. In a previous study, JPL had conducted shock tests on the RTG containment capsule, subjecting it to a pressure of 2,000 pounds per square inch (psi) without a failure in the capsule. These were pressures three times higher than those to which the RTG might be subjected during an explosion on the launchpad, according to a risk analysis. Because of such analyses, NASA had considered the RTG to be safe. Armoring, which would have added several thousand pounds to Galileo's launch weight, had been rejected. But this decision had been made before the *Challenger* explosion.[23]

NASA now had reason to revisit its decision. Another of its statistical studies had estimated that the chance of losing the Shuttle and its crew was quite low, between

---

[20] Ibid.

[21] "The Lethal Shuttle," *Nation* (22 February 1986): 16.

[22] R. W. Campbell, *Environmental Impact Statement for Project Galileo*, Review Draft, JPL, 24 July 1981.

[23] Michael A. Dornheim, "Galileo Probe Managers Review Jupiter Mission Launch Options," *Aviation Week & Space Technology* (3 March 1986): 20.

one in one thousand and one in ten thousand. Since just such a tragedy had occurred on only the 25th Shuttle flight, NASA had little confidence in its previous statistical study and asked JPL to investigate the armoring issue again.[24]

Representative Edward Markey (D-Massachusetts) also requested a study, asking that the Department of Energy (DOE), which had built Galileo's plutonium-bearing RTGs, estimate the impacts of the most probable Shuttle launch accident involving plutonium releases. Department of Energy analysts estimated that such an occurrence might result in over 200 cancer deaths and contaminate almost 370 square miles of land. The cancer estimate was based, however, on the somewhat unlikely case that people would remain in the contaminated launch area for a year with no decontamination effort and no protective actions conducted. Representative Markey, whose House Energy and Commerce energy conservation subcommittee was investigating the risks posed by Shuttles carrying radioactive materials, advised that no such missions be launched until public health and safety could be assured.[25]

New Launch Scenarios

By April 1986, NASA planners had come up with likely scenarios for resuming Shuttle flights. One possible schedule called for a 12-month "standdown," with the first post-*Challenger* flight taking off in February 1987, followed by only three more flights that year. Another schedule delayed the first post-*Challenger* launch until July 1987, followed by three more flights in 1987. The critical factor, Shuttle experts believed, was the length of time it would take to implement reliable design changes in the Shuttle's solid-fueled rocket boosters. In the 18-month delay scenario, the spacecraft that would launch in July 1987 would carry a Shuttle tracking and communications satellite to replace the one lost with *Challenger*, as well as an experiment to develop heat-rejection techniques for use in NASA's planned space station. The Shuttles that would launch in September and October 1987 would carry military surveillance experiments and classified cargo. Galileo would launch on 1 December 1987 aboard the Space Shuttle *Atlantis*.[26]

This schedule was by no means final. Galileo was in competition with other missions for space on the Shuttle. National security was the first priority in assigning launch dates, and DOD had many payloads that it wanted to send into space. The next priority in line after defense, according to Chester Lee, NASA's director of customer services, was to complete the satellite relay system needed for the space telescope. NASA was also deeply committed, after it recovered from the *Challenger* disaster, to pushing toward its long-term goal of building and running a space station, which would require many Shuttle missions.[27]

---

[24] Ibid.

[25] "Launch Hazards of Atomic Satellites Studied," *Washington Post* (19 May 1986): A11.

[26] "Tentative Flight Plan Is Drafted for Shuttle," *New York Times* (14 April 1986).

[27] Douglas B. Feaver, "The Spreading Ripples of Challenger Disaster—Scientific Ventures Losing Momentum," *Washington Post* (16 June 1986).

Implementing the Rogers Commission Report

The Presidential Commission on the Space Shuttle Challenger Accident, led by William Rogers, issued its report on 6 June 1986. Much of the blame for the disaster was placed on NASA's seriously flawed decision-making process. A better system would have "flagged the rising doubts about the Solid Rocket Booster joint seal" and the strong opinions expressed by most of the Morton Thiokol engineers, and, in all likelihood, the launch would have been delayed under such a system. The Rogers Commission also sharply criticized the "propensity of management at Marshall [Space Flight Center] to contain potentially serious problems and to attempt to resolve them internally rather than communicate them forward." This too contributed to the disaster. Although it was not revealed what Marshall Space Flight Center Shuttle managers said to Morton Thiokol managers on the night of 27 January, the Rogers Commission concluded that Morton Thiokol managers reversed their position in favor of delaying the launch and overrode the strong recommendations of their solid rocket motor project director because they wanted to keep a very major client happy.[28]

On 13 June 1986, President Reagan directed NASA to implement, as soon as possible, the Rogers Commission's nine recommendations for returning the Space Shuttle to safe flight status. NASA was to report in 30 days or less how it would implement those recommendations; it would also and give milestones to measure progress. At the top of the list of recommendations was to reengineer the SRM design. Other recommendations targeted the Shuttle's management structure and procedures for communication among all levels of management. Also targeted were methods of hazard analysis and safety assurance.[29]

NASA's view was that it would take at least until early 1988 to meet the Rogers Commission recommendations and prepare the Shuttle to fly again. Galileo would probably have to wait longer, depending on where it ended up on the flight priority list. It would have to compete with military payloads and high-priority civilian missions such as the Hubble Space Telescope and the Ulysses project to orbit and study the Sun.[30]

Centaur Is Canceled

On 19 June 1986, NASA Administrator James Fletcher delivered a huge blow to American space science and to Galileo in particular by canceling the $1-billion Shuttle/Centaur program. This decision ended the development of the hydrogen-powered satellite booster that was supposed to send Galileo and other spacecraft from low-Earth orbit into a planetary trajectory. The venerable Centaur, with its track record of reliability, was deemed unsafe for use in the Shuttle due to concerns regarding its volatile, highly flammable fuel. The final decision to cancel Centaur use on board the Shuttle was made on the basis that even implementing the modifications identified by the ongoing

---

[28] Rogers, chap. 5.

[29] *NASA's Actions to Implement the Rogers Commission Recommendations After the Challenger Accident,* http://www.hq.nasa.gov/office/pao/History/actions.html (accessed 28 August 2000). Excerpted from executive summary of *Actions to Implement the Recommendations of the Presidential Commission on the Space Shuttle Challenger Accident* (Washington, DC: NASA, 14 July 1986).

[30] *NASA's Actions to Implement the Rogers Commission Recommendations;* Kathy Sawyer, "Quickness, Safety Again Conflict as NASA Builds New Shuttle," *Washington Post* (29 November 1986): A3.

reviews, Centaur would still not meet the safety criteria being applied to other cargo on the Space Shuttle.[31]

It is ironic that Representative Edward Boland, who, just a few short years before, had argued so vociferously for using Centaur in NASA's planetary exploration program, now was one of the agents of its demise. An independent study conducted by the staff of Boland's House Appropriations subcommittee led congressional leaders to pressure NASA to terminate the Shuttle/Centaur program.[32]

A major concern was the ability to jettison Centaur's 22 tons of highly volatile fuel in the event of an emergency. If the Shuttle and its astronauts had to abort the mission and return to Earth for a forced landing with Centaur still in the cargo bay, could Centaur's liquid hydrogen and oxygen be dumped quickly and safely? In the months before *Challenger*, NASA engineers had reported trouble with the spacecraft's "dump valve," and this raised serious concerns about carrying Centaur inside the Shuttle.[33]

There were other, less obvious reasons for canceling the Centaur program. As Dawson and Bowles point out in their book on the Centaur, NASA management realized that returning the Space Shuttle to flight would require a huge expenditure of money and personnel. The Agency did not think there would be sufficient resources both to accomplish this *and* to solve the remaining technical issues of the Centaur. Another reason for the cancellation was the lower tolerance for risk that many NASA personnel and the public had after the *Challenger* tragedy. Astronauts especially opposed flying with the hydrogen-filled Centaur inside the Shuttle's cargo hold, considering it too hazardous.

Before *Challenger*, the risks associated with a Shuttle/Centaur were known but considered acceptable in the face of the benefits to be derived from successful missions. These benefits included an enhanced U.S. capability to explore and exploit space for both civilian and national security purposes. The Shuttle/Centaur design had initially been accepted based on extensive technical arguments, safety features, and social benefits. But after the devastating loss of life resulting from the *Challenger* explosion, strong emotional factors came into play as well. The result was "a new era of conservatism and a diminished acceptance of risk."[34]

JPL's Alternative Launch System Concepts

Jet Propulsion Laboratory staff had anticipated that Centaur might be scrapped for use in the Shuttle and had already initiated a series of alternative launch plans for Galileo, as well as for Ulysses, which would orbit and study the Sun, and Magellan, which would map Venus. The alternative plans included the following elements:

Enlarge NASA's Inertial Upper Stage. NASA's Inertial Upper Stage (IUS), a solid-fueled rocket that the Agency had developed for use with the Shuttle in launching

---

[31] *NASA's Actions to Implement the Rogers Commission Recommendations.*

[32] John Noble Wilford, "NASA Drops Plan to Launch Rocket from the Shuttle," *New York Times* (20 June 1986): A1.

[33] "Centaur Program Canceled by NASA Over Safety Issues," *Washington Times* (20 June 1986): 3A; Wilford, "NASA Drops Plan."

[34] Virginia P. Dawson and Mark D. Bowles, *Taming Liquid Hydrogen: The Centaur Upper Stage Rocket, 1958–2002* (Washington, DC: NASA SP-2004-4230, 2004), pp. 215–219.

communications satellites, could be made powerful enough to launch planetary craft such as Galileo from the Shuttle.[35]

Rely on an Expendable Lower Stage. Planetary spacecraft could be launched using an expendable lower stage such as Titan 34D-7 in place of the Shuttle, with either a Centaur upper stage or the enlarged IUS.[36] The Titan 34D-7 was a more massive version of the expendable launch vehicle that had successfully sent Viking and Voyager into space. Unfortunately, the 34D-7 was still under development in 1986 and was not expected to be ready until at least January 1991. The Titan 34D-7 alternative was also an expensive one. JPL estimated that building three of the vehicles to launch the Galileo, Ulysses, and Magellan missions, plus adapting the spacecraft to the new launch system, would cost up to $1.2 billion.[37]

Modify Orbital Science Corporation's Upper Stage. Orbital Science Corporation was a company that had developed a solid-fueled rocket engine for commercial applications. JPL considered developing a more powerful version of it and implementing it as the upper stage in planetary missions. It could be used with either a Titan or Shuttle launch vehicle.[38]

Add Propulsion Stages to NASA's Existing IUS. Rather than design and develop a more massive IUS vehicle, NASA could also achieve the additional power needed to send large spacecraft such as Galileo into planetary trajectories by adding two propulsion stages to the existing IUS. The combination vehicle could be launched from either a Titan or a Shuttle lower stage.[39]

Waiting for a New Launch Date

While waiting for a new launch date, the Galileo project team continued with spacecraft testing and integration procedures at Kennedy Space Center. The spacecraft's software was verified, and an end-to-end data-flow analysis was conducted. RTGs were mated to Galileo. The Orbiter received a thorough electrical system exam. Several of Galileo's spare subsystems had been transferred to the Magellan spacecraft, and new ones were fabricated. They included Galileo's spare attitude and articulation control, as well as its power subsystem and command and data subsystem. Galileo's retropropulsion module oxidizer tanks had been filled for the May 1986 launch and needed to be drained and decontaminated. Safety analyses also continued. Shields were fabricated for additional protection of the RTGs, in the event that this should be required. The Probe was sent to Hughes, where it had been built, for an upgrade of its scientific instruments.[40]

---

[35] J. R. Casani, "From the Project Manager," *Galileo Messenger* (July 1986); Wilford, "NASA Drops Plan"; "Centaur Program Canceled."

[36] Centaur was only banned for use in the Shuttle, because of the danger to its crew. It was not banned for use with robotic launch vehicles such as Titan or with robotic spacecraft such as Galileo.

[37] Wilford, "NASA Drops Plan"; "Centaur Program Canceled."

[38] Wilford, "NASA Drops Plan."

[39] Ibid.

[40] Casani, "From the Project Manager," *Galileo Messenger* (July 1986).

The Galileo team hoped that during the fall of 1986, a new launch date would be selected.

Selecting a New Trajectory

Without Centaur, a direct flight to Jupiter was impossible. The available solid-fueled upper stage options simply weren't powerful enough to propel Galileo on a direct Earth-Jupiter trajectory. Gravity-assists were required for the spacecraft to attain a trajectory that would get it to Jupiter.[41]

Because of the need for gravity-assists, JPL analyzed new trajectory options in conjunction with its investigation of launch vehicle alternatives. Using planetary gravity fields to adjust Galileo's velocity reduced on-board fuel requirements. But gravity-assists also resulted in longer travel times because the trajectory to Jupiter would be much less direct and would cover many more kilometers. Parts would age, and on-board power systems would be depleted more than initially estimated. Also, some trajectories would bring Galileo closer to the Sun, resulting in additional thermal stresses. The aim was to balance the benefits of various gravity-assist options against their deficits and select the trajectory–launch vehicle combination that would result in the highest possible scientific return.[42]

One of the promising trajectory strategies that NASA considered was termed ΔV-EGA (and was discussed in chapter 3). It made use of an Earth gravity-assist (EGA) to change velocity and had been successfully employed in sending the Voyager spacecraft to the outer planets. If Galileo used ΔV-EGA and launched in December 1989, one of the "windows" for the trajectory, it would take almost five years to reach Jupiter. If Centaur could have been used in conjunction with the Shuttle, Galileo could have reached Jupiter in only two years.[43]

Two discriminating factors in selecting the best trajectory–launch vehicle combination for Galileo were the "injection energy" needed and the "interplanetary ΔV" required. Injection energy is the amount of kinetic energy a spacecraft will eventually have when it escapes Earth's gravity field onto its interplanetary trajectory.[44] A spacecraft's injection energy needs to be supplied by its launch vehicle. The choice of interplanetary trajectory for Galileo was thus constrained by the injection energy that its launch vehicle could provide.[45]

Galileo's "interplanetary ΔV requirements" were the changes in velocity it would need for trajectory-shaping maneuvers between Earth and Jupiter. The changes could not be so large that overall propellant requirements for the mission exceeded the capacity of Galileo's on-board propulsion system, which contained 925 kilograms of propellant. This fuel would have to be used not only for ΔV maneuvers during the long flight to Jupiter, but also for the spacecraft's attitude control, trajectory corrections, insertion into a Jupiter

---

[41] J. R. Casani, *Galileo Messenger* (December 1986).

[42] Ibid.

[43] Casani, "From the Project Manager," *Galileo Messenger* (July 1986).

[44] A more technical definition: a spacecraft's injection energy is measured in terms of the square of its asymptotic speed with respect to Earth on its departure hyperbola. This is excerpted from Louis D'Amario, Larry Bright, and Aron Wolf, "Galileo Trajectory Design," *Space Science Reviews* 60 (1992): 25.

[45] Louis A. D'Amario, Larry E. Bright, and Aron A. Wolf, "Galileo Trajectory Design," *Space Science Reviews* 60, nos. 1–4 (1992): 25.

orbit, and tour of the planet's satellites. The propellant needed for these other maneuvers was significant and severely restricted the amount of fuel available for interplanetary ΔV requirements. This constraint eliminated all but one of the trajectories that JPL considered for Galileo. Given the spacecraft's propellant capacity, only the Venus-Earth-Earth (VEEGA) trajectory would work.[46]

In this trajectory, Galileo would swing toward the Sun after leaving Earth's gravitational field. A launch in late 1989 (which Galileo staff were using as a target date even though NASA had not made its final decision) would be followed by a flyby of Venus about three and a half months later. Gravitational forces experienced during this and other flybys would alter Galileo's velocity without the need for burning more fuel (see figure 4.3). Assisted by Venus' gravity, Galileo would head back toward Earth, flying by it approximately 10 months later.[47]

Galileo's first Earth flyby would send the craft on a two-year elliptical orbit around the Sun. Midway through this orbit, Galileo's engine would fire and use much of the spacecraft's ΔV fuel allotment for trajectory shaping. At the end of the two-year orbit, Galileo would once again intercept Earth, performing its second and last flyby. This time, interaction with Earth's gravity would swing Galileo in a new direction, toward Jupiter. The total flight time from launch to arrival at Jupiter would be about six years.[48]

The new trajectory introduced new problems. Galileo had originally been designed to operate at distances between 1 and 5 Astronomical Units (AU)[49] from the Sun. Within this range, Galileo's temperature-control systems would be able to keep the spacecraft's instruments and experiments from getting too hot or too cold. But on the VEEGA trajectory, Galileo would travel as close as about 0.7 AU from the Sun, encountering twice the solar radiation intensity as that found near Earth.[50] VEEGA posed a severe thermal threat to the integrity of the spacecraft. To protect Galileo, blanket material and thermal shades and shields would have to be added.[51] Changes to Galileo's power management system would also be needed. Due to the extended trip time to Jupiter (roughly six years, versus the two years the trip would have taken if the Centaur upper stage had been used), the plutonium in the craft's RTGs would decay more than had originally been planned, thus reducing the generators' power output.

The May 1986 launch plan had included a chance to fly close to the asteroid Amphitrite. The later launch date would not make that possible, but the VEEGA trajectory would carry Galileo through the asteroid belt twice, and the mission might be able to target a new asteroid.

---

[46] Ibid., p. 25.

[47] Casani, "From the Project Manager," *Galileo Messenger* (December 1986).

[48] Ibid.

[49] One AU is the distance from Earth to the Sun, 150 million kilometers (93 million miles).

[50] Solar radiation intensity is proportional to the inverse square of the distance to the Sun.

[51] Casani, "From the Project Manager," *Galileo Messenger* (December 1986).

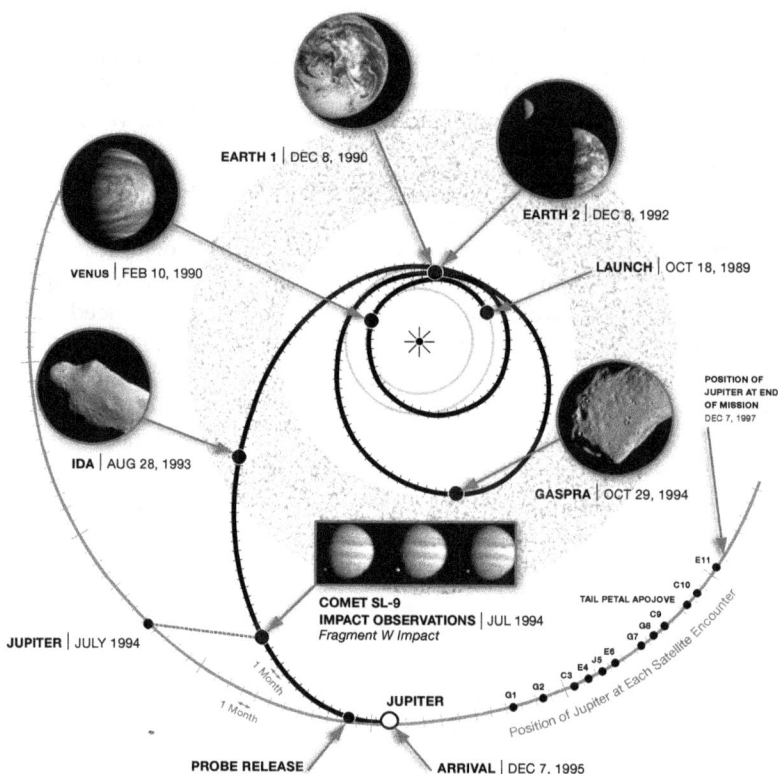

Figure 4.3. **The VEEGA trajectory. Galileo saved fuel by getting gravity-assists from Venus and Earth. (Image number P-48182)**

VEEGA was a very fuel-conservative trajectory. JPL staff calculated that after Galileo flew to Jupiter and completed all mission objectives, the craft would still have 80 kilograms (176 pounds) of propellant aboard. This surplus fuel would prove to be enormously important during the actual mission.[52]

Making the Shuttle Safe To Fly: The Need for New Management

One major step in getting the Shuttle safely back into space was to examine the management structures that had led to the *Challenger* disaster. The presidential investigating commission headed by William Rogers found that NASA managers at Marshall Space Flight Center had "repeatedly ignored warnings about the potential for the booster rupture that destroyed Challenger." Marshall had isolated itself from the rest of NASA's management structure, according to the panel. This attitude was nothing new. Marshall's leaders, beginning after World War II with German rocket pioneer Wernher von Braun, had built a proud, close-to-the-vest tradition of doing business. For decades, Marshall's attitude had

---

[52] Ibid.

been for Washington to give it a job and then go away. "Don't worry," they would say, "we'll do an excellent job." Such an approach was no longer acceptable.[53]

In the wake of *Challenger*, NASA Headquarters insisted that Marshall cease its insular tradition of project management. To ensure that this happened, NASA appointed James R. Thompson, the former head of the Shuttle's main engine development program, as Marshall's new Director. Thompson had spent 20 years at Marshall, shepherding new engine designs through years of frustrating explosions on test stands, delays, and cost overruns, eventually producing the world's most advanced and powerful liquid-fueled rocket—the Space Shuttle, which first took flight in 1981.[54]

Thompson was made Marshall's Director in part because he knew how to turn failures into successes, but also because he was "untainted" by the *Challenger* tragedy. He had left Marshall in 1983, taking a position at the Princeton Plasma Physics Laboratory. Thompson had also aided NASA in its internal investigation of the crash, serving as vice chairman of the task force and supervising day-to-day collection and analysis of accident data. As the new Director of Marshall, Thompson's plans were to open up lines of communication among various management levels and return to the rigorous system of checks and balances that he felt had been missing over the last few years. Thompson saw the need for aggressive, top-down probing of NASA technical centers, programs, and individuals. He also vowed to "beef up ground testing of high-risk hardware, and pay for it out of other programs if necessary." Thompson appeared to be perfect for the directorship of Marshall Space Flight Center. He had not been involved in the mistakes leading to the *Challenger* disaster, and he was familiar enough with Marshall operations to implement a new Shuttle management system.[55]

Another action aimed at breaking the "parochialism" and autonomous decision-making of Marshall, as well as in other technical centers, was to shift overall authority for the Shuttle program from Houston to Washington. Establishing strong direction from NASA Headquarters made the Shuttle program's new management approach more like the highly respected approach of the Apollo Moon landing program. NASA appointed Arnold Aldrich director of the Shuttle program and gave him "full responsibility and authority" to run it. Now the Shuttle program was in the hands of someone whose only job was to lead the program. In appointing Aldrich, NASA gave him many of the responsibilities of NASA Associate Administrator for Space Flight Richard Truly. The Associate Administrator had many other issues on his plate besides the Shuttle and supported this management change, stating that the Shuttle program "deserves a full-time leader."[56]

The Government's Role in the Disaster

Initial probes of the *Challenger* disaster placed the blame largely on NASA and Morton Thiokol management. But lingering questions remained as to the role other parts of the U.S. government played in pressuring NASA to end its long string of delayed launches and get *Challenger* up into space. On 7 October 1986, the House Science and Technology

---

[53] Kathy Sawyer, "Old Hand Takes Up a Mandate to Break Space-Center Tradition," *Washington Post*, The Federal Page (30 September 1986): A13.

[54] Ibid.

[55] Ibid.

[56] Kathy Sawyer, "Shuttle Management Shifted from Houston to Washington," *Washington Post* (6 November 1986): A3.

Committee issued a report that placed a share of the blame on both the White House and Congress for launch schedule pressures that led to the tragedy. The government had pushed NASA too hard to meet schedules, while at the same time starving the Agency of sufficient resources. The White House, with the support of Congress, had directed NASA to "achieve a launch schedule of 24 flights per year in order to make its costs competitive internationally."[57]

The House committee report expressed concern that NASA managers still had not figured out how a fatal hardware flaw had been missed in the Agency's elaborate testing procedures. The House committee questioned whether NASA's level of technical expertise was adequate for the decisions it had to make. How, the House committee asked, would NASA be able to protect against similar breakdowns in its checks-and-balances system in the future?[58]

Although the House committee generally concurred with the Rogers Commission findings, it reached a different conclusion as to the root cause of the disaster. The Rogers Commission thought that the cause was poor communication and inadequate quality-assurance procedures, while the House committee felt that the crash had been due to "poor technical decision-making over a period of several years by top NASA and contractor personnel."[59]

Getting the Shuttle Back into Space

By the fall of 1986, NASA had set its sights on launching the first post-*Challenger* Shuttle in February 1988. Al McDonald of Morton Thiokol and other engineers thought that this goal was highly "success-oriented," meaning that the only way it could be met was if no major equipment tests ended in failure and no major design changes were needed. Only an uninterrupted string of successes in the reengineering of the Shuttle launch system would allow a safe February 1988 launch. By March 1987, NASA and Morton Thiokol would need to have completed 90 percent of the solid rocket booster redesign, with ground tests continuing until the end of 1987.[60]

Some NASA watchdog groups were uneasy about the importance the Agency placed on meeting its new schedule versus observing safety considerations. The House Science and Technology Committee reported that pressure for an "unrealistic number of flights" continued to exist within NASA, a condition that jeopardized a "safety first" attitude.[61]

Members of the President's Rogers Commission were angered to learn that NASA was proceeding with a redesign strategy chosen largely because it could save time in getting the Shuttle back into space. NASA's strategy involved using 72 steel booster rocket casings that had been ordered back in 1985. Only one company, the Ladish Company in Cudahy, Wisconsin, was equipped to forge these large pieces of hardware, and Ladish needed one to two years to manufacture and deliver them. A new casing design would involve retooling costs and long delays.[62]

---

[57] Kathy Sawyer, "New Challenger Report Faults Hill, White House," *Washington Post* (8 October 1986): A12.

[58] Ibid.

[59] Ibid.

[60] Kathy Sawyer, "Quickness, Safety Again Conflict as NASA Builds New Shuttle," *Washington Post* (29 November 1986): A3.

[61] Kathy Sawyer, "Redesign of Shuttle Combines Pressures for Care and Urgency," *Washington Post* (10 November 1986): A1; Sawyer, "Quickness, Safety."

[62] Sawyer, "Redesign of Shuttle."

NASA defended its decision to use existing designs of key hardware by saying that it was better to stick with the known in this case because radical changes would be complex; would require extensive analysis; and, most important, might introduce new, unforeseen risks along with their advantages. Morton Thiokol managers strongly supported using the current field joint design, saying that even though it failed, it did many things well, and with certain modifications, it would be as good as it needed to be. But the Rogers Commission thought that NASA should have more seriously considered new ways of thinking. A radical redesign of the booster rockets might have produced a safer, more reliable spacecraft in the long run.[63]

Disturbing opinions on the reliability of NASA's new designs were voiced by some within the Agency itself. Marshall engineer Frank Ledbetter thought that if a redesigned joint near the booster rocket exhaust nozzle was not further altered, it carried a "potential for disaster." John Thomas, director of the redesign effort at Marshall, summarized some of his engineers' concerns as, "If you look at the numbers over 50 to 100 flights, the possibility [of failure] may be higher than one would want to accept."[64]

The Rogers Commission was seriously bothered by what NASA considered to be sufficient safety factors and backup systems to minimize the chances of technology failures, and it worried that NASA would continue to sacrifice safety to schedule and budgetary pressures. The Commission recommended that NASA's Administrator request the National Research Council to form an independent solid rocket motor design oversight committee to implement the Commission's design recommendations and oversee the design effort. After this oversight committee was formed, one of its criticisms was of Marshall's review and approval structure for new designs. The same office that approved new designs also set the testing and analysis requirements for them and had the power to waive those requirements if the hardware failed to meet them. The oversight committee strongly urged NASA to separate these conflicting responsibilities.[65]

Once the Shuttle did launch, DOD estimated that it would have a backlog of about 20 military payloads waiting to go up. The first major science payload, the Hubble Space Telescope, was scheduled for the fifth post-*Challenger* flight, which was then planned for November 1988. If that date was met, this mission would be the first science-oriented flight in about three years.[66]

Al McDonald, one of the Morton Thiokol engineers who had tried to stop the *Challenger* launch, now led the task force charged with fixing the flawed booster rocket design. He was one of the many NASA and contractor staff walking the line between implementing better, safer designs and trying to meet mission schedule pressures. Two questions that McDonald had to ask constantly were 1) what aspects of the booster rocket design still need to be reexamined and possibly modified and 2) when have we done enough analysis and redesign?[67]

The Morton Thiokol task force, along with the NASA managers overseeing it, were taking a high-stakes gamble on the field joint redesign. Rather than start from scratch and

[63] Ibid.

[64] Kathy Sawyer, "Launch Delays Darken Future of Space Science," *Washington Post* (26 January 1987): A1; Sawyer, "Quickness, Safety."

[65] Rogers, "Recommendations of the Presidential Commission"; Sawyer, "Quickness, Safety."

[66] Sawyer, "Quickness, Safety."

[67] Sawyer, "Redesign."

design a new type of field joint, they were betting that a complex modification of the old design that had failed on *Challenger* would now work. If their modified field joint ran into problems during ground testing, the task force would have to begin again, grounding the Shuttle for an additional one to two years and calling NASA's engineering judgment and credibility into question once again.[68]

Morton Thiokol was working on its redesign under the critical eyes of watchdog panels, business competitors, and reporters. Top NASA officials said that they would not "be stampeded by schedule pressures,"[69] which had been cited as a factor behind *Challenger*'s destruction. Nevertheless, pressures to get the Shuttle back into space weighed heavily in every design decision. It was ironic that the Shuttle was in a situation that NASA had always labored to avoid—grounded because of a single-point failure. NASA had a well-known policy of developing a backup technology for every mission-critical system. But it had built no backup for its flawed booster rocket field joints, and thus the Shuttle could not fly until a reliable joint was developed. In hindsight, NASA could also have developed backup launch vehicles such as expendable Titan stages, to be used in the event that the Shuttle ran into problems. Since the Agency had not done this, U.S. solar system exploration was now in the position of near-total dependence on the Shuttle. In the words of Torrence Johnson, Galileo project scientist, "We ain't got no rockets! It's humiliating."[70]

NASA critics urged that one or more alternative field joint designs be developed, to be used in the event that the chosen joint design exhibited major problems during testing. Such a strategy would enable the Agency to better meet its mission commitments in the future.[71]

Business rivals of Morton Thiokol were in fact drawing up alternative joint designs for NASA's consideration. But the enormous expense involved in bringing an alternative joint design from conception to production, an estimated $1 billion, would probably prevent NASA from funding such an effort—unless, of course, Thiokol's redesigned joint did not work. To determine the joint's reliability with a high degree of confidence, the National Research Council panel urged NASA to implement a greatly expanded ground-testing program. NASA also needed to ensure that safety requirements would not be compromised in order to accommodate unexpected test results.[72]

NASA and Morton Thiokol added several tests to their field joint program and were considering adding more, but a schedule deadline was fast approaching that would force them to decide how much testing was enough. In order to meet the February 1988 target date for the first post-*Challenger* launch, flight hardware would have to be ordered before the ground-test program could be completed. Was the Agency's faith in the new design sufficient to commit funds and order the hardware, or was it better to delay the launch until a higher degree of confidence was achieved?[73]

Ground testing was performed in the hills of Morton Thiokol's rocket-manufacturing site, in a new facility called the Joint Environment Simulator (see figure 4.4). The simulator

[68] Ibid.

[69] Ibid.

[70] Sawyer, "Launch Delays Darken Future."

[71] Sawyer, "Redesign of Shuttle."

[72] Ibid.

[73] Ibid.

duplicated a range of launch conditions, such as intense temperatures and pressures, using life-sized segments of the booster rocket. Powerful jacks on the simulator wall applied side stresses to the field joints while sleeves reaching out from the walls blew hot or cold air on the joints to simulate a range of weather conditions. During the test firing, 150 channels of data were telemetered to nearby engineers.[74]

Figure 4.4. **Joint Environment Simulator test. Booster motor sections needed to be rigorously tested under conditions approximating actual use. (Image number 8772121)**

Near the simulator was Morton Thiokol's full-scale test stand, where complete rocket motors, rather than just segments, were strapped down horizontally, fired, and analyzed. NASA had concluded that the horizontal test position of the motors better simulated the stresses experienced during launch. For safety, the motors were nosed up against 5,000 cubic yards of concrete and steel to make sure that they did not go anywhere when fired.

In previous trials, the 2-minute "hot-fires" of booster motors had blackened slopes adjacent to the test stand for a distance of 500 yards. NASA had decided to build a second, more advanced horizontal test stand, fitted with upgraded monitoring and simulation equipment and costing $19 million. This second stand would be able to more closely duplicate actual launch stresses when the booster rockets were connected to the rest of the Shuttle stack (which included the large external hydrogen-oxygen fuel tank and the Shuttle spacecraft).[75]

By January 1987, it became apparent to the National Research Council oversight committee that NASA would not have time to determine and incorporate all the substantial

---

[74] Ibid.

[75] Ibid.

design changes that might be needed in safety-critical items in time to meet NASA's February 1988 target for the first post-*Challenger* flight. Although the panel praised NASA's reviews and critiques of hundreds of items and procedures that could lead to another disaster, it felt that the Agency needed to base its reviews not only on the judgment of its experienced engineers, but also on quantitative risk analysis techniques. To accomplish this, the panel recommended that NASA rank mission-critical items on their likelihood of failure and evaluate Shuttle redesign activities in the light of this ranking. NASA needed to use such a risk-related analysis in order to determine the adequacy of the changes it was making to Shuttle hardware, software, and operational procedures.[76]

Days after making the above report, the National Research Council panel found additional reason to doubt that the Shuttle could be safely launched by February 1988. NASA had been on such a tight schedule to make the launch date that it could do so only if the results of each major equipment test were "expected and understood." There was little room in the schedule for modifying designs due to unexpected test outcomes. But unexpected outcomes were exactly what occurred in several tests. In one, the "aft skirt," a cone-shaped covering at the base of the booster rocket, cracked under stress sooner than predicted by the original 1970s stress analysis. This result called into question how reliable the stress analyses were for the rest of the Shuttle. In particular, the panel worried about stresses on the struts attaching the booster rockets to the Shuttle's external fuel tank. During the *Challenger* mission, a snapped strut had been involved in the series of events leading to the disaster.[77]

In another ground test, the booster's new O-rings, which had been designed to better withstand cold weather, deteriorated after extended exposure to rust-inhibiting grease. As a result, engineers decided to use the old O-rings and to install heaters in the field joints to keep the O-rings warm and resilient. But could this new system be finished and adequately tested by the February 1988 launch date? The panel noted that the heaters might adversely affect nearby adhesive bonds, which were known to weaken at elevated temperatures, and suggested that engineers instead try using different greases with the new O-rings.[78]

The panel also questioned the integrity of the booster motor's rear joint, where the rocket case met the exhaust nozzle. It bothered the panel that NASA was not developing a backup for the rear joint, which was critical for Shuttle safety. Few tests had been planned for the joint, and they would occur late in the test program. The panel believed that new designs would eventually have to be developed, not only for the rear joint, but for the entire booster. "The nation's manned space flight program," the panel said, "cannot afford to rely on older technology indefinitely into the future."[79] The panel thought that NASA should actively begin the design of future generations of rocket motors.

The February 1988 launch date grew increasingly doubtful, but it was not out of the question until new manufacturing mistakes were discovered. Morton Thiokol had applied insulation materials to the inside of the booster in a sloppy manner. As a result, there was a two-month delay in the first full-scale firing test of the rocket. NASA management was furious. J. R. Thompson, the new Director of Marshall Space Flight

---

[76] Kathy Sawyer, "Panel Questions Shuttle Launch Plans," *Washington Post* (14 January 1987): A8.

[77] Kathy Sawyer, "February '88 Shuttle Launch in Doubt," *Washington Post* (16 January 1987): A4.

[78] Ibid.

[79] Ibid.

Center, publicly berated Morton Thiokol. Although NASA did not immediately announce a change in launch date, the Agency switched its budget schedule to support an August 1988 return to space for the Shuttle, a six-month delay. The booster redesign team, in the view of some, had been working too hard and too fast. They were overburdened. A launch delay was definitely necessary.[80]

By the time the Space Shuttle was ready to fly again, approximately 200 safety-oriented changes had been implemented into its design. These included changes not only to its solid rocket motor, but also to other parts of the spacecraft. Modifications to NASA's system of managing the Shuttle program had also been made. Many of the major changes to the Shuttle program and spacecraft are listed in table 4.1.

Several of the changes addressed the number-one "necessary corrective action" on the Rogers Commission list, which stated that "the faulty Solid Rocket Motor joint and seal must be changed" and that the joint must be insensitive to environmental effects, flight loads, assembly procedures, and other conditions. Toward this end, the Solid Rocket Motor field joint was totally redesigned and included an additional O-ring as a thermal barrier, a heater for cold weather conditions, and a pressure-actuated J-seal considered far more fail-safe than the putty used in the old design. Case-to-nozzle joint modifications similar to those in the field joint were also implemented.[81]

Table 4.1. **Impacts of *Challenger* on the Space Shuttle Program and the Galileo spacecraft.**

MANAGEMENT CHANGES	COMMENTS
Managers that were held responsible for *Challenger* disaster were removed from jobs.	Both NASA and Morton Thiokol Managers were affected.
Shuttle program management structure was improved.	Overall authority was moved from Houston to Washington, DC, and NASA Headquarters.
	Aggressive oversight of NASA technical centers, programs, and personnel was instituted.
	"Parochialism" of local spaceflight centers was attacked.
TESTING PROCEDURES	**COMMENTS**
Equipment-testing procedures were improved.	Additional analysis of high-risk hardware was instituted.
Hazard-analysis procedures were improved.	NASA implemented procedures to prevent safety's being sacrificed to budget and schedule pressures.
HARDWARE MODIFICATIONS	**COMMENTS**
Changes were made to the Space Shuttle orbiter.	Maneuvering system valves were fitted with "sniffers" to detect contamination that could affect system operation.

---

[80] Kathy Sawyer, "NASA Assuming Further Delay," Washington Post (11 April 1987): A4.

[81] "Recommendations of the Presidential Commission," in *Report of the Presidential Commission* on the Space Shuttle Challenger Accident, vol. 1 (Washington, DC: Government Printing Office, June 1986), *http://www.panix. com/~kingdon/space/rogers11.html.*

Table 4.1. **Impacts of *Challenger* on the Space Shuttle Program and the Galileo spacecraft.** (continued)

HARDWARE MODIFICATIONS	COMMENTS
	Fuel cell electric heaters were replaced with safer temperature-control system.
	Redundant pipes were added to potable water system to prevent freeze-ups.
	Redundant water-purity sensors were added.
	Main landing gear was strengthened, and braking system's reliability was upgraded.
	Thermal protection system was improved.
	Wings were strengthened to increase safety margin.
	Computer system was upgraded to incorporate more capabilities with smaller size, lower weight, and less power required.
	Auxiliary power units with longer lifetimes were installed.
	Inertial navigation system was upgraded.
	In-flight crew escape system was installed.
	Middeck emergency egress slide was installed.
	Fuel valves between external tank and orbiter were upgraded to prevent inadvertent closure, which could cause catastrophic engine failure.
	Main engine durability and combustion chamber and turbine life were improved.
	Nose wheel steering system was made safer.
Solid rocket motor was redesigned.	Field joint was reengineered for greater reliability. New design ensured positive compression at all times on O-rings. Heater was added for cold-weather operation. Additional, inner O-ring was installed as a thermal barrier in case joint insulation was breached. Field joint internal insulation was now sealed with a pressure-actuated flap called a J-seal, rather than with putty, as in the old design.
	Case-to-nozzle joint modifications similar to those in field joint were implemented.
	Internal joints in nozzle were redesigned to incorporate redundant and verifiable O-rings at each joint.
	Insulation was added to engine casing's factory joints.
	Ignition system case was strengthened.
	Ground-support equipment was redesigned to minimize distortion of SRM during handling.

Table 4.1. **Impacts of *Challenger* on the Space Shuttle Program and the Galileo spacecraft.** (continued)

HARDWARE MODIFICATIONS	COMMENTS
	SRM certification now included testing over full range of operating environments and conditions. It included joint and nozzle joint environment simulation, as well as transient pressure tests in which SRM was subjected to ignition and liftoff loads in addition to maximum dynamic pressure structural loads.

A Date Is Set for Galileo

By April 1987, NASA had given the Galileo team an approximate launch date for its space-craft—October or November of 1989. Galileo had been given priority over the Ulysses mission to the Sun, which would not be launched until October 1990. NASA based its decision on optimizing data return from the two missions. Launching Ulysses first would have resulted in too long a wait before Galileo reached Jupiter and began transmitting prime data from the Jovian system.[82]

New Spacecraft Requirements

JPL would send a very different spacecraft to Kennedy Space Center in April 1989 from what had been sent three-plus years before. Mission programmers were modifying the complex software to accommodate VEEGA's new requirements. Because the VEEGA trajectory would carry Galileo so much closer to the Sun, the craft would require substantial thermal protection changes, which JPL would have to test exhaustively. The Galileo team was currently fabricating new Sun shields and blanketing. When Galileo traveled inside Earth's orbit, the craft's axis would need to be pointed toward the Sun in order to orient the new Sun shields so as to shade the craft's high-gain antenna (HGA) and other delicate equipment. As a result, radio communications with Earth would not be maintained by the HGA, which would be pointed in the wrong direction for that. Two low-gain antennas (LGAs) would be used instead. The LGAs were far less sensitive than the HGA but did not require protection from the Sun. The original design called for only one LGA, but an additional one had been added so that at any given time, at least one of them would usually be able to communicate with Earth. The LGAs could not transmit data back to Earth at the rate that the HGA could have, and as a result, much of the information collected during the Venus flyby would have to be stored temporarily on Galileo's tape recorder. It would be transmitted months later during the first Earth flyby.[83]

　　　　Galileo would now use a solid-fueled Inertial Upper Stage instead of Centaur to boost it from low-Earth orbit into its VEEGA trajectory. Because of this change, the hardware that had been built to connect Galileo to Centaur needed to be redesigned.[84]

　　　　The VEEGA trajectory presented new scientific opportunities for Galileo. During the six-year flight to Jupiter, the spacecraft would pass close to Venus, Earth, and the Moon.

---

[82] J. R. Casani, "From the Project Manager," *Galileo Messenger* (April 1987).

[83] Casani, "From the Project Manager," *Galileo Messenger* (April 1987): 1; "Galileo: Up to Date," *Galileo Messenger* (November 1989): 3.

[84] Casani, "From the Project Manager," *Galileo Messenger* (April 1987): 1.

Galileo would also fly twice through the asteroid belt, which would give it the opportunity to be the first spacecraft to make a close observation of an asteroid. The project team studied encounter opportunities with three asteroids: Ausonia, Gaspra, and Ida, ranging in diameter from 6 to 92 kilometers (see table 4.2).[85]

Table 4.2. **Potential asteroid opportunities.**

ASTEROID NAME	ASTEROID DIAMETER IN KILOMETERS (MILES)	POSSIBLE ENCOUNTER DATE	RELATIVE VELOCITY IN KILOMETERS PER SECOND (MILES PER SECOND)
Ausonia	92 (57)	9 April 1992	8.2 (5.1)
Gaspra	6 (3.7)	29 October 1991	8.0 (5.0)
Ida	13 (16.1)	27 August 1993	12.5 (7.8)

Source: J. R. Casani, "From the Project Manager," *Galileo Messenger* (April 1987).

VEEGA also presented challenges in the area of equipment reliability. Shelf lives and aging of equipment took on increased importance in JPL's testing program. Galileo's previous mission schedule had had an end date of October 1990. But with the current launch set for late in 1989 and a longer flight time to Jupiter, Galileo would, scientists hoped, be collecting data into 1998. The JPL team had to find the parts and components that needed to be augmented or replaced in order to keep Galileo functioning through that time.[86]

JPL staff examined components for a range of possible aging effects, including the following:

• Degrading flexibility of cable coverings.

• Adhesion loss of conducting tape.

• Metal corrosion.

• Chemical changes and aging of paints and coatings.

• O-ring fracturing.

The project team worried that exposure to Earth's atmosphere during the years the spacecraft was waiting to launch would corrode some of its magnesium alloys (once in space, the alloys would not corrode). Quality-assurance personnel needed to inspect all such surfaces before shipping Galileo back to Kennedy. Personnel also performed periodic visual inspections of painted surfaces in order to search for signs of peeling, flaking, or cracking. Specialized coatings such as uralane and humiseal could shrink over time, damaging the parts they were meant to protect. O-rings were subject to fracturing after long periods of disuse. To

---

[85] Ibid.

[86] Casani, "From the Project Manager," *Galileo Messenger* (April 1987).

[87] "Aging: Its Effect on the Spacecraft," *Galileo Messenger* (April 1987).

coatings such as uralane and humiseal could shrink over time, damaging the parts they were meant to protect. O-rings were subject to fracturing after long periods of disuse. To guard against this danger, project engineers made sure that all parts containing O-rings were "exercised" twice a year. Springs kept compressed for extended periods, such as those that would help deploy the spacecraft's magnetometer boom, had to be regularly checked for signs of stress. Even the testing of parts could itself introduce damage. Personnel might deposit dirt on the components or overstress flexible parts.[87]

Attrition of Critical Project Staff

Galileo mission delays not only aged spacecraft components and parts; they also aged the project team. For scientists and engineers who joined Galileo at the time it received congressional approval, or even before, half their research careers would be over by the time the spacecraft reached the Jovian system. For those who had joined the project in their 40s or 50s, time was running out. Many scientists were nearing retirement and hoped that they would still be working when their Galileo experiments started to return data.[88]

Tom Donohue, a University of Michigan professor who had an experiment on Galileo, was very apprehensive about the years of delay between solar system exploration missions. The last major U.S. scientific space probe had been Pioneer to Venus, which had been launched in 1978. Galileo was not going to launch until over a decade later, when Donohue was in his late 60s. Running his Galileo experiment was to have been a major effort during his last years in the field. But now Galileo would not even reach Jupiter until he was in his mid-70s.[89]

The Galileo project needed new blood, as did other solar system missions, but young talent was hard to find. With the long gaps between missions, during which limited solar system study could be performed, many young scientists were not staying in the space science field, or even entering it. A young space scientist with a newly earned Ph.D. typically had five years to conduct sufficient research and publish enough respected papers to earn a permanent research or academic position. This postdoctoral period was a feverish time for most new graduates, and if meaty, important projects could not be found in space science, the graduates would be under great pressure to find them in other fields. Fifty science missions had been planned for the Shuttle during the years from 1986 through 1992. Now, with the Shuttle's reduced schedule, only 17 were planned, and due to *Challenger*, they were all facing an average three-year delay. This was yet another legacy of *Challenger*—the loss of top talent from the space science field, and the threat to the field's future.[90]

The solar and space physics department of Baltimore's Johns Hopkins University was one of many places hit hard by the aftermath of *Challenger*. The department had been involved in four space science missions scheduled to fly in 1986. But after *Challenger*, Stamatios Krimigis, the department chairman, received notice from NASA of a 50-percent cut in funding between 1987 and 1989. "We will have to let go our young research associates," Krimigis said at the time, "and they will probably become computer

---

[88] Sawyer, "Launch Delays Darken Future."

[89] Douglas B. Feaver, "The Spreading Ripples of Challenger Disaster—Scientific Ventures Losing Momentum," *Washington Post* (16 June 1986).

[90] Sawyer, "Launch Delays Darken Future."

scientists or join other fields where they are in great demand. You can't do space science with no wages."[91]

Nor would there be the flow of small science payloads that had in years past been carried aloft by balloons and modest sounding rockets. These missions, as well as larger ones such as Galileo, had been vital in training graduate students and establishing them on a career path. But the Shuttle, which had been expected to carry the small payloads more efficiently, effectively had phased out many other launch vehicles. Diverse launch options had been consolidated into just one, and when that went down, it threw not only the Galileo mission, but also the whole space science field, into crisis.[92]

Antinuclear Groups Try To Block the Launch

As Galileo's scheduled October 1989 launch date approached, public concern about sending radioactive materials into space flared up into open opposition. Residents of the area surrounding Cape Kennedy, Florida, allied with environmental as well as disarmament groups from around the nation, focused their attention on stopping the launch. "Humans and technology can fail," said Bruce Gagnon, coordinator of the Florida Coalition for Peace and Justice. "We don't want Brevard County, the Space Coast or anyplace else made into a national sacrifice area. We don't want to be guinea pigs." Gagnon and thousands of others were worried about Galileo's exploding and raining "radioactive death" over eastern Florida, the "Space Coast." Gagnon didn't want people endangered to further NASA's planetary exploration program.[93]

NASA could not definitively say that this would not happen, any more than it could have predicted without a doubt whether *Challenger* was going to explode. However, NASA, other agencies, and the executive branch of the federal government could and did bring extensive analytical capabilities and oversight to bear in order to gauge the expected safety of the mission. Galileo's launch approval process was defined by Presidential Directive PD/NSC-25, which required that the President's Office of Science and Technology Policy approve the launch of any nuclear power source. The complete approval process involved approximately 100 personnel from federal agencies, private industry, universities, and the Executive Office of the President. A key part of the process was a series of independent safety evaluations by an Interagency Nuclear Safety Review Panel (INSRP) made up of representatives from the Department of Defense, the Department of Energy, and NASA, as well as observers from the Nuclear Regulatory Commission, the Environmental Protection Agency, and the National Oceanic and Atmospheric Administration.[94]

Three safety analysis reports (SARs) were developed during various stages of Galileo project development. The final safety analysis report (FSAR), the one most relevant

[91] Ibid.

[92] Ibid.

[93] Victoria Pidgeon Friedensen, 'The Protest,' chap. 4 in "Protest Space: A Study of Technology Choice, Perception of Risk, and Space Exploration" (master of science thesis, Virginia Polytechnic Institute and State University, 11 October 1999); Michael Lafferty and Chris Reidy, "NASA Tries to Catch Up in the PR War Over Space Probe," *Orlando Sentinel* (22 July 1989).

[94] NASA, *Mission Operation Report*, Report Number E-829-34-89-89-01, NASA Office of Space Science and Applications (1989), pp. 38–41.

to questions of RTG integrity and the radiological risks that it presented, was completed in December 1988 and reviewed by INSRP. After reviewing the FSAR, both Department of Defense and Department of Energy agency heads had to decide "whether to concur with or recommend against NASA's request for launch."[95] If both agency heads concurred, then NASA had to request approval for the launch from the President's Office of Science and Technology Policy, whose director could either give approval or refer the matter to the President. No launch could take place without approval from either the Office of Science and Technology Policy's director or the President.[96]

NASA's safety analyses did not do much to stop the protests. Perhaps this was because nuclear energy had become not only dangerous but also morally evil in many people's minds. To activists all over the country, it was connected to the evils associated with the military and big corporations, such as waging unjust wars and greedily reaping excessive profits. Nuclear material was controlled by entities such as the Department of Energy, Department of Defense, and electric power megacompanies that many people didn't trust. Bruce Gagnon, for instance, thought that the Department of Energy was forcing NASA to use plutonium aboard Galileo. He believed that the nuclear industry in turn controlled DOE and would stop at nothing to expand the market for radioactive fuels. When an accident involving nuclear fuel occurred, such as Three Mile Island or Chernobyl, or even an accident such as *Challenger*, in which nuclear materials had not been used but *might* have been used if it had been a later mission, it just confirmed many people's worst fear: nuclear energy could not be made safe. Its use potentially threatened everyone.[97]

Larry Sinkin was an influential opponent of nuclear power who was based in Washington, DC. He was the director of the Christic Institute, an interfaith center for just laws and public policy. The Christic Institute had been involved in the Karen Silkwood case and in legal action against the Three Mile Island power plant in Pennsylvania. Like Bruce Gagnon, Sinkin mistrusted the Department of Energy, stating that its long history of lies, deceptions, and cover-ups gave it no credibility. "This is the agency telling us that plutonium is safe. We simply don't believe them," said Sinkin, who was perhaps referring to DOE's controversial role in the management and cleanup of its radioactively contaminated sites such as Rocky Flats, Colorado; Savannah River, South Carolina; and Fernald, Ohio. Sinkin believed that most people were unaware of just how dangerous plutonium was. If the public was alerted to the threat that the material posed, popular opinion might demand that Galileo be kept on the ground.[98]

NASA stated that it used plutonium as a power source in its spacecraft for reasons of necessity, not because DOE insisted. Batteries could not supply enough power for Galileo's needs without being prohibitively heavy, and the spacecraft would travel too far from the Sun to use solar power. Because of Jupiter's distance from the Sun, 2,000 square feet of solar panels weighing over 1,000 pounds would be required, which was not feasible.[99]

Not all activist groups opposed Galileo. The Committee to Bridge the Gap was an organization that frequently lobbied against space nuclear power projects, yet it gave

[95] Ibid., p. 40.

[96] Ibid., p. 40.

[97] Lafferty, "NASA Tries to Catch Up"; personal communications with nuclear power protesters at Diablo Canyon nuclear reactor in California, 1980.

[98] Ibid.

[99] Ibid.; William Harwood, "NASA Current News," from UPI, 28 June 1989.

Galileo its blessing. The Committee was aware of the risks that the mission presented. In a paper analyzing Galileo's safety, the Committee's executive director, Steven Aftergood, identified several situations that could result in plutonium release. But Aftergood also believed that the "unique benefits" that would be derived from exploring Jupiter outweighed the nuclear safety risks. The Committee's view was not dissimilar to that of NASA's management, who recognized that space travel had certain dangers associated with it, both from the use of nuclear materials and from other factors. NASA's decisions to launch spacecraft in spite of known dangers said much about the Agency's values. Sometimes attaining an ideal outweighed certain risks.[100]

The risk presented by the plutonium aboard Galileo was *not* that it would turn into a nuclear bomb and explode. The isotope used in Galileo's RTGs was plutonium-238, a material unsuitable for nuclear weapons. In contrast to nuclear weapons or nuclear reactors, no fission process is involved in the operation of RTGs, nor was a fission process possible, given the type of plutonium used by Galileo and the design of its RTGs. But plutonium-238 poses other dangers due to its radioactivity.[101]

Unlike radioactive materials with intense gamma radiation, plutonium's alpha emissions are dangerous only at close range. Alpha rays do not generally pose much risk unless pieces of plutonium (even tiny pieces) enter the body. In that case, the alpha emissions can do intense damage. In the event of a Galileo accident, small fragments of the RTGs' plutonium might be released into Earth's environment. The largest human health risks would come from inhaling particles that could get into the lungs or ingesting plutonium-contaminated food.[102]

Although the RTG design included triple encapsulation to contain the plutonium fuel in case of an accident, NASA conceded that the safety records of previous missions using RTGs were not perfect. In fact, 3 of the 22 missions launched with RTGs seriously malfunctioned. In 1968, a Nimbus weather satellite did not attain orbit due to a launch system malfunction, and its two RTGs fell into the Santa Barbara Channel. But the RTGs were recovered intact, without loss of fuel, from the ocean floor five months later. Their plutonium fuel was reprocessed and used in a later mission. In 1970, the Apollo 13 spacecraft failed to land on the Moon, where it was going to leave its RTG. After the abort of the mission, the Lunar Module reentered Earth's atmosphere, falling into the South Pacific Ocean in the area of the Tonga Trench, where it still remains. Extensive air and water samples taken in the vicinity of the reentry, however, found no evidence of fuel release.[103]

Radioactive fuel did disperse in an incident that occurred before NASA implemented its RTG triple-containment system. In 1964, the Transit 5BN-3 navigational satellite failed and released radioactive fuel over a wide area of the upper atmosphere. The release of radioactive material was extensively studied by scientists from the Atomic Energy Commission in air- and soil-sampling efforts. As a result of the assessment, U.S. RTG

---

[100] Theresa Foley, "NASA Prepares for Protests over Nuclear System Launch in October," *Space Technology* (26 June 1989): 23, 25.

[101] "Facts About RTG Misconceptions," NuclearSpace.com Web site, *http://www.nuclearspace.com/facts_about_rtg. htm* (accessed 6 June 2005).

[102] Foley, "NASA Prepares," pp. 87–88.

[103] Ibid.; "Facts About RTG Misconceptions"; "RTG Fact Sheet: Past Releases of Radioactive Materials from US Nuclear Power Sources," *NASA SpaceLink*, 17 July 1991, *http://spacelink.nasa.gov/NASA.Projects/Human.Exploration.and. Development.of.Space/Human.Space.Flight/Shuttle/Shuttle.Missions/Flight.031.STS-34/Galileos.Power.Supply/ RTG.Fact.Sheet.*

design philosophy changed to full fuel containment. The objective was that in the event of an abort during launch or the on-orbit phase of a mission, the RTGs would retain all their fuel.[104]

NASA had learned much from these past incidents and had confidence in its current triple-containment RTG design, which was used in the Galileo spacecraft. The plutonium was divided up into little spheres about the size of golf balls, and each of these was welded into iridium spheres, which were intended to seal off the plutonium even at high temperatures. The spheres were packed into extremely strong cylinders designed to withstand the impact of a crash, and these in turn were placed inside 2-by-4-inch bricks made of graphite, the same sort of material used for ballistic reentry nose cones. In fact, the bricks themselves were designed to serve as little reentry bodies. Department of Energy tests on the RTGs included subjecting them to explosion and fire environments of varying intensities, as well as high-velocity impacts—situations intended to recreate the conditions that might occur during a Shuttle accident. The RTGs withstood all their trials very well, including pressures that exceeded 2,200 pounds per square inch. In defending its RTG design, NASA cited the Nimbus and Apollo incidents, holding that the RTGs "performed precisely as designed and maintained full containment with absolutely no release of fuel." NASA believed that these experiences demonstrated the RTGs' fail-safe reliability.[105, 106]

A final environmental impact statement (FEIS) issued by NASA in cooperation with the Department of Energy on the Galileo mission concluded that the RTGs posed only small health or environmental risks. The risk analysis performed during the preparation of the FEIS considered three accident scenarios:

- *Most Probable Case:* The highest probability accident in a mission phase leading to a release of plutonium.

- *Maximum Credible Case:* The accident in a mission phase that leads to a release of plutonium with the most severe impact on human health.

- *Expectation Case:* The probability-weighted sum of all accidents in a mission phase.

The FEIS stated that the Most Probable Case resulting in a plutonium release was an IUS failure leading to spacecraft breakup, reentry of the RTG modules, and impact of the modules on hard rock. The FEIS set the probability of this occurrence at 1 in 2,500.[107]

The Maximum Credible Case involved an accidental reentry during one of Galileo's Earth flybys (scheduled for 1990 and 1992); it was given a probability of occurrence of

[104] Foley, "NASA Prepares"; "RTG Fact Sheet: Past Releases of Radioactive Materials from US Nuclear Power Sources"; Regina Hagen, "Past Missions—a Chronology," chapter 3 in *Nuclear Powered Space Missions—Past and Future*, on the Web site of the Global Network Against Weapons and Nuclear Power in Space, 8 November 1998, http://www.space4peace.org/ianus/npsm3.htm.

[105] John F. Murphy, NASA Assistant Administrator for Legislative Affairs, to the Honorable Alfonso D'Amato, United States Senate, 30 April 1986, Galileo Documentation File, folder 5138, NASA Historical Reference Collection, Washington, DC; "Radioisotope Thermal Generators (RTGs)," *Galileo Messenger* (April 1984); "RTGs—A Plutonium Crap-Shoot?" *Space World* (June 1987): 24; Foley, "NASA Prepares," p. 87.

[106] William Sheehan, NASA Associate Administrator for Communications, to Officials-in-Charge of Headquarters Offices; Directors, NASA Field Installations; and Director, Jet Propulsion Laboratory, "Galileo and Ulysses Mission Information," 17 November 1988, Galileo Documentation File, folder 5138, NASA Historical Reference Collection.

[107] *Final Environmental Impact Statement for the Galileo Mission* (Tier 2), NASA No. N90-22147, May 1989, pp. iv–v.

one in nine million. In this very unlikely scenario, the RTG modules were subjected to intense reentry heating and a collision with a hard rock surface, resulting in a plutonium release that would expose over 70,000 people. Nevertheless, the health effects predicted for even this extreme case were small. Over a 70-year period, the exposed population would receive an average dose of less than 20 percent of the natural background radiation level. Without this radiation exposure from the RTGs, roughly 14,000 of the 70,000 would be expected to die of cancer. *With* the radiation exposure from the RTGs, only about nine more people would be expected to die of cancer. Thus, even for the Maximum Credible Case, the human health impact would be small.

The Expectation Case analysis, which represented a probabilistic combination of all accident scenarios, reached a similar conclusion. It estimated the risk of fatality to an individual exposed to radiation from any Galileo accident to be about four in one billion, which was hundreds to thousands of times lower than from "ordinary" causes such as motor vehicle accidents, falls, fire, disease, or natural disasters (hurricanes, tornadoes, or lightning).[108]

The Committee to Bridge the Gap, as discussed above, supported the Galileo mission. But even this activist group thought that the FEIS's numbers were surprisingly low, and it was not convinced that they really represented worst-case scenarios. "Since we have not launched, say, 100,000 missions, we cannot speak with high confidence about how many failures out of 100,000 launches can be expected," the Committee said, and pointed out that NASA's estimates were based on engineering judgments that contained a degree of subjectivity.[109]

As the Galileo launch date grew nearer, antinuclear activists stepped up their efforts. In September 1989, a group called the Maryland Safe Energy Coalition demonstrated outside NASA Headquarters in Washington, DC. The Florida Coalition for Peace and Justice began a 200-mile "peace walk" from Cape Canaveral, Florida, to Kings Bay, Georgia, to protest both the Galileo and the Trident missiles, which were flight-tested in Kings Bay. The Florida group also planned to enter Kennedy Space Center and sit on Galileo's launchpad in order to block the launch, said Bruce Gagnon, the group's leader.[110]

Because Galileo was going to carry nuclear material into orbit, its launch required an executive office decision. President George H. W. Bush was not expected to oppose the flight, according to Associated Press reports, and preparations for the launch continued as planned. On 12 September 1989, the crew that would fly the Shuttle *Atlantis* and its Galileo payload into orbit arrived at Kennedy Space Center. Two days later, the Shuttle's "Countdown Demonstration Test" began. The astronauts suited up and boarded *Atlantis* for the test.[111]

On 18 September 1989, NASA reported that the White House had approved the Galileo launch. In response, a coalition of antinuclear activists filed suit to stop the take-off. The coalition argued that a catastrophic launch failure would endanger the health of people in central Florida and that using the Shuttle to launch Galileo had a higher probability of an accident than using a robotic rocket. "NASA is in effect playing ecological roulette with the people of Florida," said Anthony Kimbrell, one of the lawyers asking

---

[108] Ibid., pp. v, 4-16 to 4-28.

[109] "NASA Says a Nuclear Generator on Satellite Will Pose Little Risk," *New York Times* (7 January 1989): 9.

[110] Associated Press, 12 September 1989.

[111] Ibid.; J. Kukowski, *NASA Headline News*, Message EJIJ-2835-7747, 12 September 1989.

for a court restraining order on the launch. Attorneys for the coalition maintained that as many as 700,000 cancer deaths might result from an accident and that the chances of this happening were greater than NASA believed.[112]

Steve Frank, a lawyer arguing for the Justice Department, countered the coalition's position by maintaining that all of its "tales of horror" were based on the fear that Galileo's plutonium fuel would get pulverized to the extent that people up and down the Florida coast would inhale little particles of it. Frank insisted that with NASA's triple layers of protection, the chance of pulverizing the plutonium pellets was virtually zero.[113]

Radically different numbers were thrown back and forth by both sides, but what the court focused on was not whether NASA's or the coalition's estimates of risk were more accurate, but whether NASA had broken the law. The coalition maintained that NASA had failed to follow National Environmental Protection Act requirements by filing an insufficient environmental impact statement that did not adequately address the launch dangers. It was this assertion that Judge Oliver Gasch of the U.S. District Court examined.[114]

In the midst of the court hearings, a petition was introduced that gave the hearings an international dimension. West Germany's Green political party joined the American coalition of citizens' groups in trying to get a restraining order on the launch. The Greens claimed that Galileo would violate a 1967 outer space treaty that both the United States and Germany had ratified, and they wanted the launch delayed so that the issue could be put before the International Court of Justice. A provision of the 1967 treaty stated that one country's space exploration activities were not to harm any other country. The plutonium carried by Galileo might, in the Greens' view, result in radioactive contamination of Germany. The Greens petitioned the U.S. District Court to issue a 10-day temporary restraining order on the launch. This would allow the Greens time to determine whether the world court would hear the case and whether to approach the German government to intervene formally.[115]

While the coalition and the Greens brought their suit to court, the Shuttle astronauts, having finished their countdown demonstration test, flew to Washington for a meeting with Vice President Quayle, who was the chairman of the National Space Council. Quayle threw his support behind Galileo by announcing that he would travel to Kennedy Space Center, where he planned to watch the launch from the control room.[116]

On 9 October 1989, as the final countdown began for a 12 October takeoff, protesters enacted a mock death scene at Kennedy Space Center and waited for the court's decision on their suit to stop the mission. They vowed that if Judge Gasch did not stop the mission, then they would carry out the earlier threat made by the Florida Coalition for Peace and Justice to sit on the launchpad so that Galileo couldn't fire its rockets.[117] On 10 October, Judge Gasch ruled that it was not the function of his court to decide whether the government's decision to launch Galileo was a good one or whether the launch would endanger this or other countries. The judge said that his *only* job was to examine the coalition's charge that NASA had violated the National Environmental Policy Act (NEPA)

---

[112] *NASA Headline News*, 18 September 1989; Jay Malin, "Galileo Survives Suit, but Mishap Delays Liftoff," *Washington Times* (11 October 1989): A3.

[113] Malin, "Galileo Survives Suit," p. A3.

[114] Ibid.

[115] "West Germany's Greens Want Shuttle Issue Before World Court," *Washington Times* (13 October 1989): A7.

[116] Suelette Dreyfus and Julian Assange, *Underground* (Australia: Mandarin, a part of Reed Books, 1997), p. 15; "Shuttle Countdown Begins Amid Protests," *Washington Times* (10 October 1989): A2.

[117] "Shuttle Countdown Begins," p. A2; Associated Press, 12 September 1989.

by filing an insufficient impact statement. The case record explained that the court needed to "ensure that the environmental impact statement contains a sufficient discussion . . . to allow the agency to take a hard look at the issues and make a reasoned decision on the matter." The case record went on to explain, "The court will not substitute its own judgment regarding the merits of the proposed action for that of the government agencies. NEPA itself does not mandate particular results, but simply prescribes the necessary process."[118] The court concluded that NASA had complied with the provisions of the act and rejected the coalition's request for a restraining order against the launch. Judge Gasch added that issuing such an order would have a "costly adverse" economic effect on the public interest if it caused Galileo to miss its present launch window, for it would cost NASA, and ultimately the public, $164 million just to maintain the Galileo program until the next launch opportunity 18 months later.[119]

Judge Gasch's decision had removed the legal hurdles blocking Galileo's planned liftoff on Thursday, 12 October, but an hour after the ruling was issued, a faulty computer in the Shuttle delayed the mission once again. Launch director Robert Sieck announced that the computer would have to be replaced, pushing liftoff forward by about a week. Meanwhile, lawyers for the three citizens' groups that had brought the Galileo matter to court—the Florida Coalition for Peace and Justice, the Christic Institute, and the Foundation on Economic Trends—announced that they would appeal Gasch's decision.[120]

If the protesters could succeed in delaying liftoff for just a few weeks, it would have a very significant impact on the mission. Galileo had to launch by 21 November 1989 or wait until 1991 for the planets to align properly again for the mission. The possibility of another delay alarmed many on the Galileo team. Neil Ausman, who would direct Galileo in flight, was "very, very frustrated" by the delays thus far, saying "I'm looking at an investment of 20 years" by the time the spacecraft was expected to reach Jupiter in 1995. Some of the scientists who'd initiated the mission had already retired, died, or moved on to other projects out of frustration, and Galileo could lose more good people if it was delayed yet again.[121]

Anticipating demonstrations, Kennedy Space Center security officials beefed up the protection of the site with high-tech equipment and armed patrols. Gary Wistrand, deputy director of Kennedy's security office, said that his force, which numbered 200 to 225 guards, was armed with M-16 assault rifles and 9-millimeter semiautomatic pistols and was ready to arrest anyone illegally entering the 140,000-acre space center. While Wistrand expected the trespassing attempts to be made on foot, his force was also on the lookout for other modes of entry, including by boat or plane. He thought that the terrain helped keep the site secure against a land invasion. Much of the space center boundary was swampy and infested with alligators, wild hogs, and rattlesnakes. In the past, would-be trespassers had kept to the high, dry places and had been easy to locate.[122]

While security forces were preparing to stop protesters, technicians began the laborious task of replacing the faulty Shuttle computer that had given erroneous fuel pump

[118] *Florida Coalition for Peace & Justice v. Bush*, Civil Action No. 89-2682-OG, United States District Court for the District of Columbia, 1989 U.S. Dist. Lexis 12003, decided and filed on 10 October 1989.

[119] Ibid.; Malin, "Galileo Survives Suit"; "Galileo," *The Acorn—Newsletter of the American Nuclear Society* 96-1 (February 1996): Oak Ridge/Knoxville Section, *http://www.engr.utk.edu/org/ans/newsletter/1996/acorn96_1.html*.

[120] Kathy Sawyer, "Faulty Engine Computer Delays Launch of Shuttle," *Washington Post* (11 October 1989): A3.

[121] Ibid.

[122] "Shuttle Guards Spare No Effort to Foil Foes," *Washington Times* (12 October 1989): A6.

pressure readings and delayed the mission. The computer was a critical piece of equipment because it controlled all functions of one of the Shuttle's three main engines during liftoff.[123] By Sunday, 15 October, a spare Shuttle engine computer had been installed and tested. Workers began closing up the engine compartment, removing work platforms and reinstalling heatshields. Meanwhile, NASA closed large sections of the space center to the public and mobilized its security forces, including boat and helicopter patrols. NASA scheduled the resumption of the launch countdown to begin the next day at noon, with the goal of launching Galileo on Tuesday afternoon, 17 October. The weather on Tuesday looked favorable, although there was a 10-percent chance that winds from the southeast would exceed the 23-mph launch limit.[124]

On Monday, 16 October, the last-minute appeal of the U.S. District Court's ruling on Galileo was rejected by the Washington, DC, Circuit Court of Appeals. Chief Justice Patricia Wald wrote in her concurrence that, contrary to the protesters' main legal claim, she was not able to find that NASA had improperly compiled its environmental impact statement on the Shuttle launch risks. Also on 16 October, eight protesters were arrested at the space center for trespassing. The arrests were made at a gate near Kennedy's headquarters. If the protesters had succeeded in approaching the Shuttle, they might have delayed the takeoff, because NASA guidelines prohibited a launch when anyone was within the "blast danger zone," a 4,000-foot-radius circle around the launchpad.[125]

Finally, Galileo appeared ready to be launched. It had survived the *Challenger* disaster, changes of upper stage and trajectory, and the protests of antinuclear activists. A summary of the impacts on the Galileo mission that followed the *Challenger* tragedy is given in table 4.3.

Table 4.3. **Impacts of *Challenger* on the Galileo mission and spacecraft.**

CHANGE	COMMENTS
Centaur upper stage was eliminated.	A liquid-hydrogen-fueled rocket engine was considered too dangerous to carry in the Shuttle cargo bay.
Solid-fueled Inertial Upper Stage (IUS) was used instead.	Perceived as safer, but its lower thrust would lengthen travel time to Jupiter and necessitate a more complex trajectory.
VEEGA trajectory replaced direct trajectory to Jupiter.	Venus and Earth gravity-assists were needed to minimize fuel consumption.
Thermal blankets and Sun shades were added.	Necessary because on the VEEGA trajectory, Galileo would approach closer to the Sun.
Sun shields were added to shade the high-gain antenna and other equipment.	Galileo had to orient its axis toward the Sun so that Sun shields could shade sensitive parts of the spacecraft. Radio contact with Earth had to be maintained with two low-gain antennas instead of a high-gain antenna. Data transmission rates were affected.

[123] Ibid.

[124] "NASA Clears Atlantis for Liftoff Tuesday," *Washington Post* (15 October 1989): A18; "NASA Deploys Guards as Shuttle Launch Nears," *Washington Post* (16 October 1989): A20.

[125] Kathy Sawyer, "Galileo Launch Nears; Court Appeal rejected," *Washington Post* (17 October 1989): A3.

Table 4.3. **Impacts of *Challenger* on the Galileo mission and spacecraft.** (continued)

CHANGE	COMMENTS
Galileo spacecraft/upper stage interface was modified.	IUS connections required different hardware from that needed for Centaur.
Due to launch delay, encounter with Amphitrite asteroid was no longer possible.	Opportunities for flying by the asteroids Ausonia, Gaspra, and/or Ida could be exploited instead.
Attrition occurred among project staff.	Staff members retired, died, or changed programs.
Nuclear material aboard Galileo became a major public relations issue.	Public confidence in the safety of NASA spacecraft had already been damaged by the *Challenger* disaster. Widespread fear of plutonium releases that might occur if Galileo crashed led to protests and lawsuits.

The Launch

The new liftoff date, 17 October 1989, looked like a go. The Shuttle appeared to be operating as it should, with Galileo in its belly and the astronaut crew aboard. Then cloud cover began to build up. Countdown continued, but finally the Launch Director scrubbed liftoff once again. Visibility was poor enough that if problems arose and the Shuttle had to return to Earth shortly after takeoff, it would not be able to land at Kennedy. Launch was delayed until the following day.[126]

As the Galileo team prepared for the next day's launch, a magnitude 7.1 earthquake struck northern California, just miles south of the Inertial Upper Stage (IUS) control center, which was crucial to mission operations. The crew had to evacuate, and it looked like the mission would have to be pushed back once more. But the crew was able to return later in the night and go back online. The next morning, however, ominous clouds at Kennedy Space Center put the launch in jeopardy one more time. Fortunately, the skies finally cleared enough for the Shuttle to launch safely. At 12:54 p.m. eastern daylight time (EDT) on 18 October 1989, the Shuttle *Atlantis*, carrying its Galileo payload, took off into a sky filled with puffy white cumulus clouds. Galileo was finally on its way (see figure 4.5).[127]

Six hours and 21 minutes after launch, mission specialist Shannon Lucid initiated the deployment procedure for Galileo, ejecting it and the IUS rocket to which it was attached from *Atlantis*'s cargo bay. After deployment was completed, *Atlantis* commander Donald Williams declared, "Galileo is on its way to another world. It's in the hands of the best flight controllers in the world—fly safely." One hour later, when the Shuttle had flown to a safe distance, the IUS rocket fired, speeding the spacecraft into interplanetary space and on to its first objective—Venus.[128]

---

[126] "Launch to Landing," *Galileo Messenger* (November 1989): 1–2.

[127] Ibid.

[128] Ibid.

Figure 4.5. **Launch of Galileo: liftoff of the Space Shuttle Atlantis, carrying the Galileo spacecraft and its Inertial Upper Stage. (STS34(S)025)**

The theme of the Galileo mission was accomplishment in the face of adversity. Even after the successful launch of *Atlantis*, deployment of Galileo, and firing of its IUS, there was trouble. Heavy winds were forecast for *Atlantis*'s landing site at Edwards Air Force Base in the California desert. To minimize the danger, two orbits of Earth, which would have taken 3 hours, were cut from the Shuttle's flight plan so that it could get down to the ground as soon as possible. Even so, shortly before landing, the fog started rolling down off nearby hills into the dry lakebed where *Atlantis* was to touch down. Luckily, the skies cleared at the last moment, and the Shuttle made a perfect landing. In attendance at the landing site were reporters, Galileo flight team members, and other people, some of them now retired, who had helped to design and construct Galileo.[129]

R. J. Spehalski, who had recently taken over as Galileo Project Manager from John Casani, had this to say about the events of 18 October 1989:

> The Galileo launch, to many of us, represents a culmination of years of personal and team efforts and sacrifices. The emotions present within each of us at such an event defy description . . . I was fortunate to witness the coming together of such a diverse set of activities that it boggles the mind. The professionalism, teamwork, and dedication visible in this process was truly inspiring to behold.[130]

---

[129] Ibid.

[130] R. J. Spehalski, "From the Project Manager," *Galileo Messenger* (November 1989): 1.

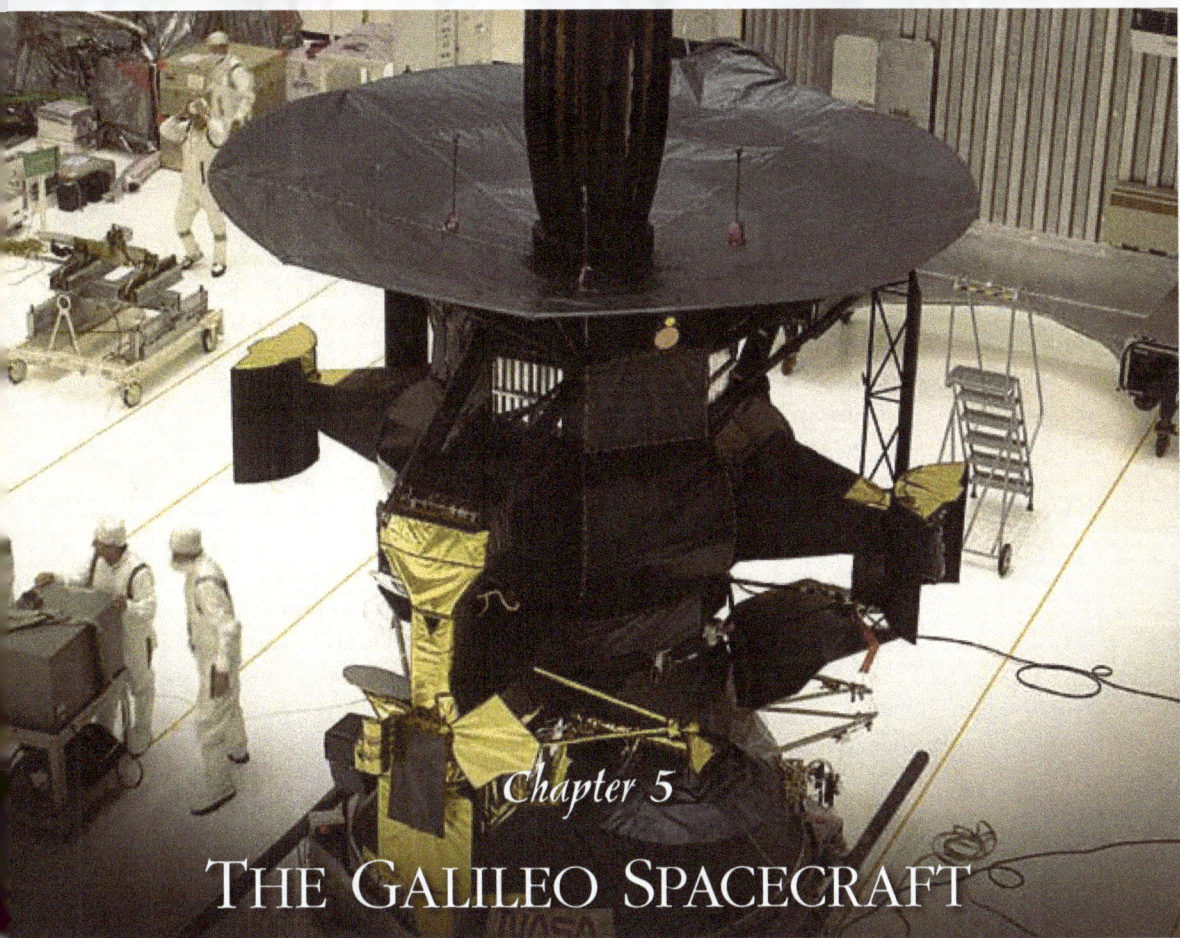

Chapter 5

THE GALILEO SPACECRAFT

THE GALILEO SPACECRAFT WAS DESIGNED TO SPEND YEARS TRAVELing through the inner solar system and asteroid belt before it reached its main objective: the Jovian planet-satellite system. NASA intended the spacecraft to take some observations of inner planets and asteroids but to focus on investigating Jupiter, its moons, and the fields and particles of its magnetosphere at a level of detail never before attempted. The overall mission would last more than a decade.

NASA intended that the Galileo project advance the goals of the Agency's Lunar and Planetary Exploration Program, which were to understand 1) the origin and evolution of the solar system; 2) Earth, through comparisons with other planetary bodies; and 3) the origin and evolution of life. To make progress toward these goals, the Galileo mission was to conduct a comprehensive investigation of the entire Jovian system, including both in situ and remote observations of the planet, its environment, and its satellites. In order to make so many different types of measurements, the Galileo spacecraft needed to have a wide range of capabilities. NASA constructed the spacecraft in three segments, each of which had a specific objective:

- **Analyze the composition and physical properties of Jupiter's atmosphere.** To accomplish this, a pod of instruments called the Atmospheric Probe detached from the spacecraft and descended into the planet's atmosphere. Its in situ measurements broadened our understanding of the events and processes that formed the planets approximately five billion years ago.

- **Explore the vast region of magnetic fields and ionized gas (plasma) that surrounds Jupiter.** To meet this objective, the main section of the Galileo Orbiter was given a spin, which allowed its instruments to sweep through and measure characteristics of fields and particles in all directions relative to the spacecraft. The spinning section of the Orbiter also carried communications antennas, the main computers, and most of Galileo's support systems.

- **Investigate the nature of the Jovian system's principal satellites.** Cameras and sensors capable of forming high-resolution images in the visible and other parts of the spectrum were needed to meet this objective, as well as a stable platform from which to take the pictures. Galileo's designers included a "despun" (nonspinning) section in the Orbiter that could be pointed and maintained at a given orientation. From this platform, Galileo's imaging system obtained pictures of Jovian satellites at resolutions from 20 to 1,000 times better than Voyager's best.[1]

The above objectives necessitated developing a spacecraft of great complexity and technological capability. Figure 5.1 provides an illustration of some of the Galileo spacecraft's many parts and instruments. The spacecraft made it possible for scientists to conduct a long-term, close-range study of the Jovian system. The Atmospheric Probe, which weighed 340 kilograms (750 pounds), carried six scientific instruments. The Orbiter carried 10 instruments and weighed 2,200 kilograms (4,900 pounds), including about 900 kilograms (2,000 pounds) of rocket propellant. The spacecraft radio link to Earth and the Probe-to-Orbiter radio link served as instruments for additional scientific investigations. Galileo communicated with its controllers and scientists through the Deep Space Network's tracking stations in California, Spain, and Australia.[2]

The Political and Economic Reasons for a Three-Segment Spacecraft

The complex Galileo spacecraft design arose not only to satisfy the scientific objectives discussed above, but also because of politics and cost factors involved in creating the mission. In the 1970s, the space science community debated at great length what types of projects should follow the Pioneer and Voyager flyby missions to the outer planets. As discussed in chapter 2, the space science community formed many committees and conducted numerous studies to help answer this question. After the successful Pioneer encounters with Jupiter in 1973 and 1974, more difficult spacecraft maneuvers seemed possible. Pioneer collected enough data about the Jovian system, for instance, that NASA engineers thought that they could now design a probe spacecraft to descend through Jupiter's atmosphere and stay operable long enough to make valuable observations.

---

[1] John R. Casani, testimony for Space Science and Applications Subcommittee, Committee on Science and Technology, United States House of Representatives, 1 March 1978, JPL Archives; "Galileo's Science Instruments," *Galileo: Journey to Jupiter, http://www.jpl.nasa.gov/galileo/instruments/* (accessed 11 April 2000); NASA Office of Space Science and Applications, "Mission Operation Report," Report No. E-829-34-89-89-01, 1989, p. 12.

[2] "Galileo Overview," *Galileo: Journey to Jupiter, http://www.jpl.nasa.gov/galileo/overview.html* (accessed 1 November 2000).

Magnetometer
Sensors

Plasma Wave
Search Coil Sensor

Energetic
Particles
Detector

Plasma Wave
E-Field Sensor

Spun Section

High Gain Antenna
(Earth Communications
and Radio Science)

Plasma Science

Radioisotope
Thermoelectric
Generator

Dust Detector

Ultraviolet Spectrometer

Solid State Imaging

Radiosotope Thermoelectric
Generator

Retropropulsion
Module

Near Infrared
Mapping Spectrometer

Probe Relay Antenna

Despun
Section

Atmospheric
Entry Probe

Photopolarimeter
Radiometer

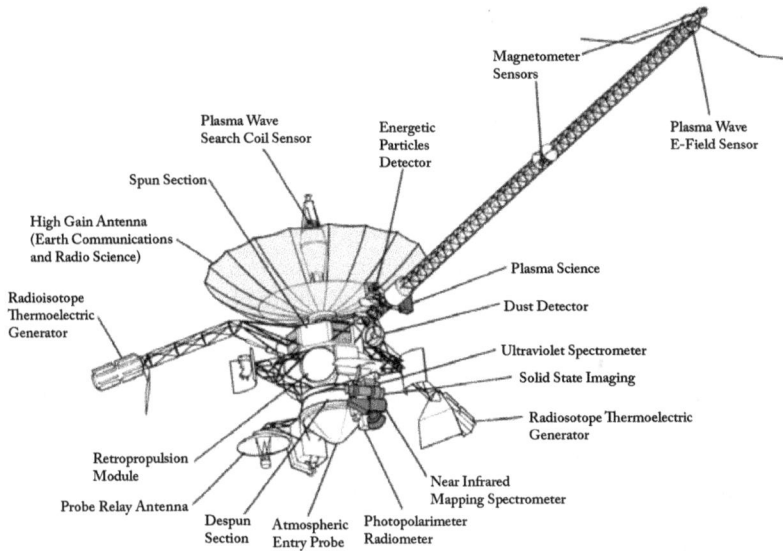

Figure 5.1. **The Galileo spacecraft consisted of three major parts: the Atmospheric Probe and the spun and despun sections of the Orbiter, each of which had specific tasks to perform. (Adapted from JPL image no. 230-1235)**

What made Jupiter mission planning complicated was that several different factions were involved in the debates, and each one envisioned a different kind of mission. Atmospheric scientists wanted a probe descent into the Jovian atmosphere. The "space physics" faction, which wanted to study the Jovian magnetosphere, favored a spinning spacecraft that would orbit Jupiter and take fields-and-particles data. And the "remote sensing" faction needed a stable, nonrotating orbiting platform on which to mount its imaging equipment. The trouble was, trying to do everything that the scientists wanted could require two or three trips to Jupiter, and this was economically untenable.

A spacecraft needed to be designed that would satisfy at least some objectives of the different scientific disciplines. In 1976, James Van Allen of the University of Iowa headed a NASA committee known as the Jupiter Orbiter Probe Science Working Group (JOPSWG), whose purpose was to "develop a follow-on Jupiter mission of more advanced capability than the Pioneers and the Voyagers, with a deep atmospheric entry probe and an orbiter having a useful lifetime of at least two years."[3] JOPSWG attempted to combine the best aspects of the different scientific factions' visions into one spacecraft concept. According to Torrence V. Johnson, who served on Van Allen's committee and was later appointed Galileo's Project Scientist, "What Galileo became was basically a massive compromise in trying to get all the various scientific disciplines on board . . . . Van Allen was actually the godfather of the mission as it currently exists. He played the role

[3] James A. Van Allen, "Planetary Exploration," section 11 in "What Is a Space Scientist? An Autobiographical Example," *University of Iowa, Department of Physics and Astronomy, http://www-pi.physics.uiowa.edu/java/*, 22 August 2002. This article originally appeared in *Annual Review of Earth and Planetary Sciences* 18 (1990): 1–26.

of bringing the [factions] together and getting them to agree to necessary compromises to get a mission that would do a good job for everybody. And he did it in a pretty effective and objective way."[4]

Selecting Science Instrumentation To Meet Mission Objectives

In the process of developing the mission concept, JOPSWG had to consider the "number, type, and capabilities of instruments that might be carried on the mission to make a credible case the mission would achieve its primary scientific objectives."[5] Torrence Johnson commented that this was

> . . . always a somewhat incestuous process since the people on the committee inevitably have their own axes to grind—involving conflicts between discipline areas like remote sensing versus space physics, cameras versus spectrometers, whose instrument worked best on the last mission, etc. Nevertheless, the (JOP)SWG had to come to consensus on a package that would have credibility with the rest of the science community. This involved putting together lists of prioritized objectives and how precise measurements had to be if they were to be useful, and comparing those with the capabilities of previously flown instruments and instruments under development.[6]

Numerous researchers and instrument developers gave presentations to JOPSWG during this time, trying to sell their current instrument concepts. JOPSWG eventually developed a "straw-man" payload, or, in other words, a set of instruments that could meet the minimum objectives of the mission and which were credible in terms of their costs, masses, power requirements, and data-collection rates.

In NASA's "Announcement of Opportunity for Outer-Planets Orbiter/Probe (Jupiter)," the Agency solicited proposals for instrumentation to meet primary Jupiter mission objectives. Proposers were given information about the straw-man payload but were not required to limit themselves to just those instruments—if a group felt that it could meet the objectives better with another approach, it was free to propose it. The straw-man instrument types did "have an edge" over other instruments in the selection process, however, having already been examined and discussed by JOPSWG and identified as being capable of meeting mission objectives.[7]

The specific instruments selected are discussed later in this chapter in the "Atmospheric Probe" and "Orbiter Scientific Experiments" sections.

---

[4] Torrence V. Johnson interview, tape-recorded telephone conversation, 31 July 2001; Craig B. Waff, "Jupiter Orbiter Probe: The Marketing of a NASA Planetary Spacecraft Mission" (paper presented in the "National Observatories: Origins and Functions (The American Setting)" session of the American Astronomical Society meeting, Washington, DC, 14 January 1990).

[5] Torrence V. Johnson, e-mail message, 24 October 2001.

[6] Ibid.

[7] Johnson e-mail message; "Announcement of Opportunity for Outer-Planets Orbiter/Probe (Jupiter)," 1 July 1976, A.O. No. OSS-3, folder 18522, NASA Historical Reference Collection, Washington, DC.

Spacecraft Features

NASA's lead Center for robotic exploration of the solar system, the Jet Propulsion Laboratory in Pasadena, California, fabricated the Galileo Orbiter and was given overall responsibility for the mission. NASA's Ames Research Center, one of whose specialties was atmospheric research, oversaw the design and development of the Jupiter Atmospheric Probe. The retropropulsion module (RPM), which was part of the Orbiter's spun section, was built by Messerschmitt-Bolkow-Blohm of the Federal Republic of Germany (West Germany). It contained the 400-newton engine used for trajectory maneuvers and Jupiter Orbit Insertion, and it also included two 10-newton thrusters for spin and altitude control.[8]

Figure 5.1 depicts a long extension jutting out from the spacecraft. This was Galileo's 11-meter (36-foot) science boom, which was attached to the Orbiter's spinning section and on which fields-and-particles instruments were mounted. These instruments were placed there because they happened to be extremely sensitive to local fields generated by Galileo's other equipment, and the distance that the science boom provided helped minimize interference. For Galileo's launch, the science boom was collapsed into a cylinder only 0.6 meter (2 feet) long. After Galileo separated from its Inertial Upper Stage, the boom was extended like a telescope to its full 11-meter length.

Figure 5.1 also depicts the plutonium-bearing RTGs, which, like the fields-and-particles instruments, were located a safe distance away from the main body of the Galileo spacecraft.[9]

Flight Cabling

Galileo was the most ambitious and complex interplanetary spacecraft that had yet been launched. If laid end to end, the electrical wire used in the Orbiter's many cables would have stretched 7,500 meters (25,000 feet), or nearly 5 miles. Over 700 connectors joined these cables to one another and to Galileo equipment. Every connector and cable had to be electrically shielded to protect equipment from Jupiter's strong magnetic fields and radio signals, as well as from electrostatic discharges on the spacecraft. Poorly shielded cables and connectors could have caused interference or damage to computers and sensitive instruments. Voyager, which spent only weeks in Jupiter's hostile environment, experienced some electrical discharge problems. Galileo would spend years in the region and would get much closer to Jupiter.[10]

Project staff spent four months installing the myriad cables in the Orbiter's despun section and half a year in the spun section. This was painstaking, labor-intensive work, sometimes performed with tweezers (see figure 5.2). The cabling team had to solder

---

[8] James W. Stultz, "Thermal Design of the Galileo Spun and Despun Science," *J. Spacecraft* 28 (March–April 1991): 139; "About JPL," *http://www.jpl.nasa.gov/about_JPL/about_JPL_index.html* (accessed 20 November 2001); John Casani interview, tape-recorded telephone conversation, 29 May 2001.

[9] "Galileo Frequently Asked Questions (FAQ)—General Spacecraft Anatomy," *http://www.jpl.nasa.gov/galileo/faqana. html* (accessed 4 December 2000); "Thermal Blanketing," *Galileo Messenger* (March 1983): 3.

[10] "Flight Cabling," *Galileo Messenger* (March 1983): 2.

tens of thousands of electrical contacts, and connectors had to be "potted"—filled with a molding compound that held individual wires in place and made the cables more durable and less likely to fail during the mission. Cables that would be directly exposed to space required fiberglass sleeving to protect them from micrometeoroids. The cabling crew consisted of 15 technicians who worked 60-hour weeks to deliver the completed cables on schedule and who had been intensively trained in cable fabrication, soldering, crimping (an alternate method to soldering of fastening wires or electrical shielding to connectors), potting, and various cleaning techniques. This training was critical—a carelessly soldered, "cold" joint or improperly crimped connection could impair an instrument's functioning or separate during the mission and disrupt the gathering of key data.[11]

Figure 5.2. **Galileo's despun section receives its cabling.**

Galileo's Dual-Spin Design

Unlike previous planetary spacecraft, Galileo was designed with a "dual-spin" feature. Part of its Orbiter rotated continuously at typically 3 rpm, which accommodated fields-and-particles experiments. The craft's magnetometer, for instance, built up data on a magnetic field by sweeping through it. The despun section of the Orbiter remained stationary, providing a fixed orientation for imaging equipment and other sensors (see figure 5.1).

During delicate propulsive maneuvers, as well as at the time of Probe release, spacecraft stability was critical. To maximize stability, the sections of the spacecraft were locked together and spun as one at a higher-than-usual rotation rate of 10 rpm. A stable spacecraft helped to precisely aim the Probe toward the Jovian atmosphere when it was released from Galileo. Galileo's higher spin rate also helped direct the Probe on its proper course in much the same way as a gun barrel's rifling spins a bullet to send it on a straighter line toward its target. Aiming the Probe accurately was absolutely essential

---

[11] Ibid.

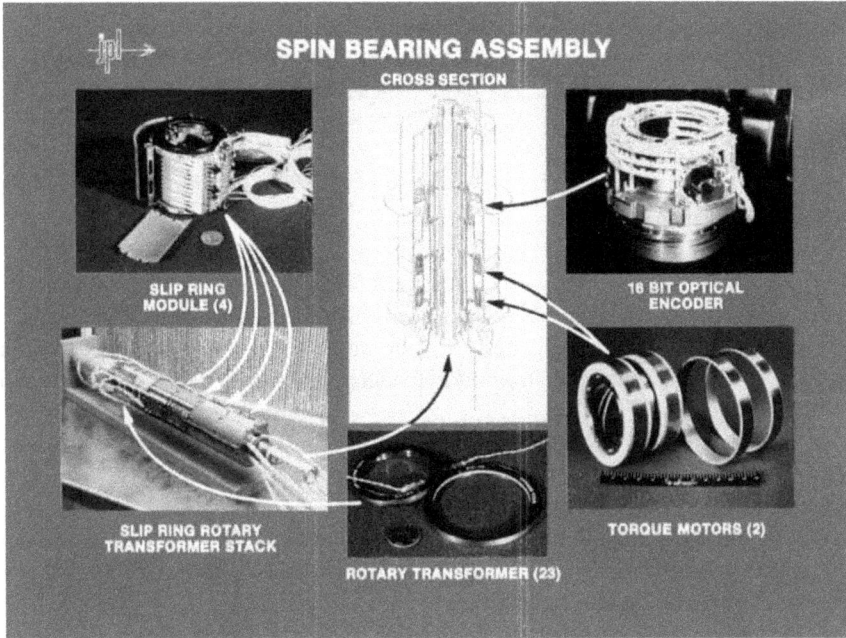

SPIN BEARING ASSEMBLY
CROSS SECTION

SLIP RING MODULE (4)

16 BIT OPTICAL ENCODER

SLIP RING ROTARY TRANSFORMER STACK

ROTARY TRANSFORMER (23)

TORQUE MOTORS (2)

Figure 5.3. **Cross section of Galileo's spin bearing assembly showing slip-ring and rotary transformer assemblies. (JPL image 230-957Bc)**

because the Probe had no propulsion system of its own to correct any mistakes. Once it was sent away from Galileo, its course could not be altered.[12]

Fabricating a dual-spin spacecraft required that a reliable electrical interface be developed to carry signals and power between its despun and spinning sections. Cables could not be used; they would soon have become twisted and wrapped around various parts of the spacecraft. Instead, the spun and despun sections of Galileo were mechanically connected by means of a "spin bearing assembly (SBA)," and it was via this that electricity was passed. Electrical power, as well as data at low rates, was sent across the SBA through the sliding contacts of 48 slip-rings. In these assemblies, a flexible brush contact rode over a large revolving ring that was electrically connected to the spun section of Galileo. Data were transmitted at high rates through 23 rotary transformers that consisted of pairs of coiled wires wrapped around a ferrite core. In each pair, one coiled wire was connected to and rotated with the spun section of Galileo; the other wire was connected to the despun section. The electric current passing through one coil created a magnetic field that induced current in the other coil. Data were transmitted without physical contact between the coils (see figure 5.3).

It took project staff considerable effort to achieve reliable data transmission through the spin bearing assembly. During extensive testing begun in 1984, electrical noise interfered with signal transmission. After laborious disassembly and inspection of

---

[12] "Galileo FAQ"; "WWWWWH," *Galileo Messenger* (March 1983): 4.

the assembly, the staff determined that chemical contamination accumulated during the manufacturing process was probably the cause of the interference. Ultraclean components replaced questionable ones, and redundancies were built into the system to make it less sensitive to electrical noise.[13]

Multipurpose Blanketing for Galileo

Galileo staff covered most of the spacecraft, except for its radioisotope thermal generators, antennas, and certain radiating areas, with nearly 300 thermal blankets. The blanketing design process began with paper patterns fit onto models of various Galileo components. The blankets were then fabricated from these patterns and laboriously laced together into place on the spacecraft.[14]

The blanketing consisted of multiple layers of insulation. The stratum closest to the spacecraft was a sort of "thermal underwear" made up of 10 to 20 layers of aluminized Mylar and Dacron net. Not only did the blankets help vital parts of Galileo to retain heat, but they also protected against impacts from micrometeoroids. The blanketing on most of Galileo was mounted with a space between it and the surface in order to better disperse the energy of a micrometeoroid impact.

A third function of the blanketing was to prevent electrical discharges. This was accomplished by covering Mylar reflecting layers with electrically conductive blanketing that was grounded to the rest of the spacecraft. In the past, spacecraft had been plagued by electrical arcing, which sometimes occurred when one surface amassed a large differential charge relative to another surface. Signals caused by this arcing had been wrongly interpreted as radio signals from a source external to the spacecraft. Galileo's blanketing electrically connected different regions of the spacecraft surface with each other so that there would be no more than 10 volts of potential between the regions at any time. This equalization of potential prevented arcing and spurious radio signals.[15]

Galileo's Main Power Source: Radioisotope Thermal Generators

Electrical power for most of Galileo's equipment was provided by its RTGs, in which heat produced by radioactive decay was converted to 500 watts of direct current. (The Probe's more modest needs were supplied by long-life lithium batteries.)[16]

Several NASA spacecraft had employed RTGs in the past, including Viking Mars landers, Apollo Moon landers, and the Voyager and Pioneer craft on missions to the outer planets. NASA's policy has been to use RTGs on robotic spacecraft traveling beyond the orbit of Mars, where the faint radiation from the Sun makes solar panel arrays impractical; they would have to be prohibitively large to collect enough energy for a spacecraft's

---

[13] "Spin Bearing Assembly," *Galileo Messenger* (September 1985): 2–3.

[14] "Thermal Blanketing," *Galileo Messenger* (March 1983): 3.

[15] Ibid., p. 3.

[16] "Mission Operation Report," p. 14.

system. RTGs were used on Mars and lunar landers as well because the long nights would have reduced the effectiveness of solar panels.[17]

The Atmospheric Probe

Attached to one end of the Galileo Orbiter was a pod of scientific instruments called the Jupiter Atmospheric Probe (or, as figure 5.1 calls it, the "Atmospheric Entry Probe"). Nearly five months before Galileo reached Jupiter, its two parts—Orbiter and Probe—separated from each other with the aid of explosive nuts. Three springs pushed the Probe away at just the right angle to send it on a collision course with the planet.[18]

Once the Probe was released, it drifted in freefall toward Jupiter, for it had no on-board propulsion system. The planet's strong gravitational force reeled the Probe in at ever-increasing speeds. The Probe then conducted the most challenging atmospheric entry ever attempted by NASA. By the time the Probe entered the Jovian atmosphere 147 days after release, it had attained a speed of 171,000 kilometers per hour (106,000 miles per hour) relative to the planet—a speed sufficient to fly from Washington, DC, to San Francisco in 100 seconds. When the fast-moving Probe hit the Jovian atmosphere, it experienced a braking force equal to 250 times the gravitational force at Earth's surface. Then it fell through a region where atmospheric pressure ranged from one-tenth to over 20 times Earth's surface pressure. The Probe had to be extremely durable to withstand such entry forces and pressures and still be able to collect and transmit useful data.[19]

Probe Components

In order to simulate the Probe's expected response to atmospheric entry stresses, NASA's Ames Research Center had built special equipment during Galileo's design process that included arcjet and laser facilities. Ames, working with NASA Langley Research Center and outside contractors, also wrote a sophisticated modeling program that predicted the Probe's ability to withstand expected entry temperatures. These simulations helped to identify the need for a Probe composed of two major segments, the deceleration module and the descent module (see figure 5.4).

The deceleration module provided an extremely tough outer shell for the Probe; it surrounded and protected its payload during the first 2 minutes of atmospheric entry, when temperatures would be the highest. The deceleration module consisted of fore and

---

[17] "Backgrounder: Shuttle Launches of Ulysses and Galileo," December 1984, John Casani Collection, folder 48, box 6 of 6, JPL 14, Galileo Discreet Correspondence, JPL Archives.

[18] Ames Research Center, "Probe Mission Time Line," in *Galileo Probe Background*, no date given, folder 18522, NASA Historical Reference Collection, Washington, DC.

[19] Ames Research Center, "Galileo Probe To Look for Secrets of Jupiter" and "The Probe Spacecraft," both in *Galileo Probe Background*, no date given, folder 18522, NASA Historical Reference Collection, Washington, DC; J. R. Casani, "Galileo Fact Sheet," 4 June 1981, JPL Interoffice Memo GLL-JRC-81, Galileo Correspondence 4/81–6/81, folder 28, box 3 of 6, JPL Archives; *Galileo to Jupiter: Probing the Planets and Mapping the Moons*, JPL Document No. 400-15 7/79 (Washington, DC: Government Printing Office (GPO) No. 1979-691-547, 1979); "Probe Mission Events," *http://spaceprojects.arc.nasa.gov/Space_Projects/galileo_probe/htmls/Probe_Mission.html* (accessed 17 October 2004), folder 18522, NASA Historical Reference Collection, Washington, DC.

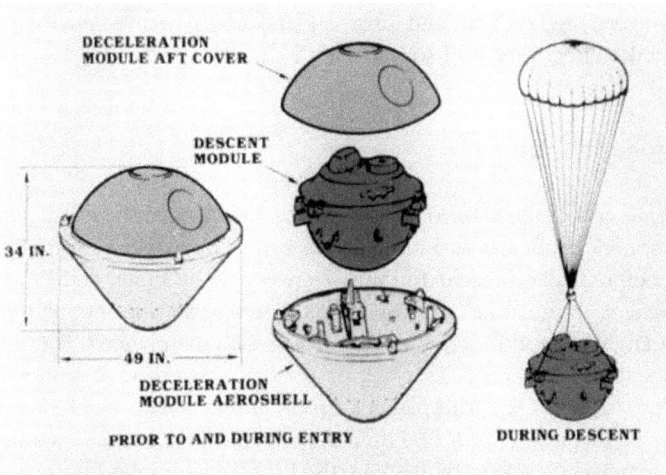

Figure 5.4. **The Probe deceleration and descent modules. (JSC digital image collection, NASA photo number S89-44175)**

aft heatshields, the supporting structure for the heatshields, and thermal-control hardware for the Probe's atmospheric entry. The fore and aft heatshields were made of carbon phenolic and phenolic nylon, respectively, which had been used often for Earth reentry vehicles but were subjected to conditions at Jupiter that the materials had never before experienced in flight. The deceleration module had to withstand the enormous stresses that occurred when the Probe hit the dense Jovian atmosphere at 171,000 kilometers per hour. The Probe slammed into and rapidly compressed the atmospheric gases ahead of it. It also experienced tremendous friction from gases streaming by it. The friction and atmospheric compression heated and ionized nearby atmospheric gases to temperatures twice as hot as the surface of the Sun, thereby generating an incandescent plasma envelope that surrounded the Probe as it fell.

The Probe's plunge through Jupiter's upper atmosphere has been compared to a trip through a nuclear fireball. The job of the heatshields was to disperse the enormous thermal energy to which they were subjected before the sensitive electronics within the Probe got cooked. The shields did this by means of an ablative process in which heat energy was first absorbed as it vaporized the shields' surface, then carried away by the gases produced. This process quickly eroded the shields (see figure 5.5). During the first 2 minutes after atmospheric entry, this erosion reduced the forward heatshield mass from 152 kilograms (335 pounds) to only 70 kilograms (154 pounds).[20]

---

[20] "Galileo Probe To Look for Secrets of Jupiter," "Probe Mission Time Line," and "The Probe Spacecraft," all in *Galileo Probe Background*; "The Galileo Probe Spacecraft," *http://spaceprojects.arc.nasa.gov/Space_Projects/ galileo_probe/htmls/probe_spacecraft.html* (accessed 23 March 2004), available in folder 11626, NASA Historical Reference Collection, Washington, DC; "Probe Mission Events," *http://spaceprojects.arc.nasa.gov/Space_Projects/ galileo_probe/Probe_Mission.html* (accessed 17 October 2004), available in folder 11626, NASA Historical Reference Collection, Washington, DC; "Galileo Probe Heat Shield Ablation," *http://spaceprojects.arc.nasa.gov/ Space_Projects/galileo_probe/htmls/Heat_Shield.html* (accessed 18 October 2004), available in folder 11626, NASA Historical Reference Collection, Washington, DC.

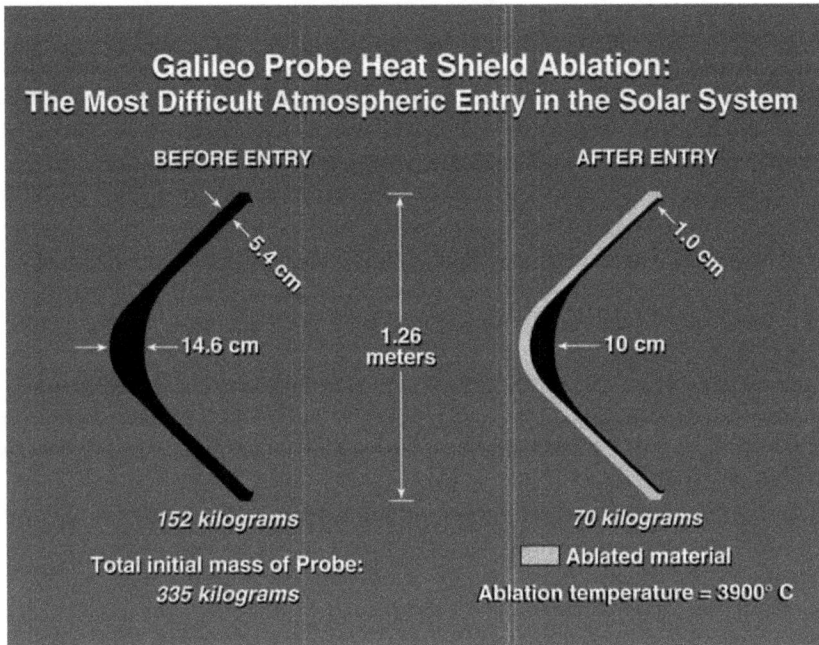

Figure 5.5. **Probe heatshield ablation. The forward heatshield, which was made of carbon-phenolic, accounted for about one-half of the Galileo Probe's total mass before entry (as shown in the left half of the figure). More than one-half of the heatshield's mass was ablated during the entry period. The right half of the figure shows the final heatshield shape (in black) as determined from sensors within the heatshield. The ablated material is shown in grey. (NASA Ames photo number ACD96-0313-13)**

Most of the Probe's reduction in speed took place within those first 2 minutes. By the end of that time, the braking action of Jupiter's atmospheric gases had cut the Probe's speed to about 1,600 kilometers per hour (1,000 miles per hour), slow enough that parachutes could be deployed without getting ripped apart or melted. As will be explained in chapter 8, the parachutes helped to separate the Probe's deceleration module components from its descent module. This descent module contained the payload—delicate scientific instruments and electronics that would study the Jovian atmosphere. The descent module fell slowly through the Jovian atmosphere, its speed controlled by a parachute, while its instruments gathered data. These data were then transmitted back to the over-flying Orbiter, which stored them for later transmission to Earth.[21]

---

[21] "The Probe Spacecraft"; "The Galileo Probe Spacecraft"; "Probe Mission Events"; Ames Research Center, "Artwork of Parachute Deployment," *http://spaceprojects.arc.nasa.gov/Space_Projects/galileo_probe/htmls/Parachute_deployment.html* (accessed 17 October 2004), available in folder 18522, NASA Historical Reference Collection, Washington, DC.

Power Source for the Probe

The Galileo Probe needed an electrical power source that would retain a high energy density during the years of interplanetary travel and be able to deliver its power quickly during the final phase of the Probe's life as it approached, then plunged down through Jupiter's ever-thickening atmosphere. To achieve such a power source, three parallel battery modules were developed, each containing 13 lithium–sulfur dioxide cells connected in series. NASA staff had chosen this design in 1976, 13 years before the Galileo launch, during the early development of the Probe. In making the selection, the discharge capabilities of sample cells and modules were evaluated while the batteries were subjected to rigorous high-g simulations of the launch and the plunge into Jupiter's atmosphere. Life testing was an important part of the evaluation. The lithium–sulfur dioxide cells had to demonstrate that their voltage delay at turn-on, as well as their ampere-hour capacity, would remain within acceptable limits even after the multiyear journey through space.

Because the Galileo launch was repeatedly delayed, numerous battery-cell lots were fabricated. For maximum battery life, it was vital that the Probe carry the freshest cells possible. The 10th lot of cells, manufactured in 1988, became the lot used on the spacecraft. All of the lithium–sulfur dioxide cells were manufactured at Honeywell's Power Sources Center in Pennsylvania. Each cell was designed to be the size of a D battery so that the company's existing tooling and assembly experience could be exploited.[22]

Probe Entry Trajectory

The Probe's trajectory was designed to intersect Jupiter's atmosphere at an angle of only 8 degrees below horizontal. The angle needed to be shallow so that the Probe would slow gradually and be able to withstand the forces of its deceleration, as well as the frictional heat it generated. But an angle of 8 degrees was also just steep enough to prevent the Probe from bouncing back out of the atmosphere like a stone skipping along the surface of a pond.[23]

Probe Scientific Experiments and Instrumentation

The Probe's atmospheric exploration focused on advancing our understanding of Jupiter beyond what was known from Earth-based and previous spacecraft studies. These earlier investigations had to rely on remote observations of the Jovian atmosphere. Analysis of the planet's complex, three-dimensional atmospheric dynamics had to be gleaned from two-dimensional images of cloud tops. But to fully understand the colorful swirls and rapid movements visible at the cloud tops, scientists needed to know about the dynamics and composition deep within the atmosphere and the nature of the powerful forces that drive Jovian weather patterns. The Probe had the advantage of being able to provide

---

[22] L. M. Hofland, E. J. Stofel, and R. K. Taenaka, "Galileo Probe Lithium–Sulfur Dioxide Cell Life Testing," in *IEEE Proceedings of the Eleventh Annual Battery Conference on Applications and Advances* (held in Long Beach, CA, 9–12 January 1996), pp. 9–14.

[23] "Galileo Probe To Look for Secrets of Jupiter."

key in situ measurements of atmospheric characteristics as a function of depth. What the Probe found shed light not only on Jupiter's atmosphere, but also on those of the other gas giant planets—Saturn, Uranus, and Neptune. However, the Probe did have the limitation that, at a given moment, it could take observations of only one place in the atmosphere. An important question that arose during the interpretation of Probe data was how typical it was of the rest of the Jovian atmosphere.

James Van Allen's JOPSWG team, whose activities were discussed above, had furnished NASA in 1976 with a "rationale for the mission's objectives and measurement requirements," which helped NASA staff to develop its Announcement of Opportunity soliciting proposals for instrumentation to meet mission objectives. The announcement identified a series of specific objectives that an Atmospheric Entry Probe was to fulfill. These objectives were as follows:

- Characterize structure (temperature, pressure, and density) of the Jovian atmosphere to a pressure depth of at least 10 bars (10 times the atmospheric pressure at sea level on Earth).

- Determine the chemical composition of the Jovian atmosphere.

- Determine the location and structure of the Jovian clouds.

- Measure the vertical energy flux to determine the local radiative energy balance.

- Characterize the upper atmosphere.

- Determine the nature and extent of cloud particles.

Furthermore, the Probe's payload of science instruments was to weigh no more than 30 kilograms (66 pounds).[24]

Most of the scientists on Van Allen's JOPSWG team who advocated an atmospheric entry probe either had served on the Pioneer Venus mission or had connections to it. Not surprisingly, their instrumentation formed the basis for the Probe's straw-man payload design.[25]

Critical to the Probe's success was a method of analyzing the Jovian atmosphere's chemical composition. In selecting the right instrument for this, weight considerations could never be forgotten, since the total Probe payload weight could not exceed a very

---

[24] T. V. Johnson, C. M. Yeates, and R. Young, "Space Science Reviews Volume on Galileo Mission Overview," *Space Science Reviews* 60 (1992): 7–8; *Galileo to Jupiter*; "Galileo Project Information," *http://nssdc.gsfc.nasa.gov/planetary/galileo.html* (accessed 15 March 2000); Hofland, Stofel, and Taenaka, "Galileo Probe Lithium–Sulfur Dioxide Cell Life Testing," p. 9; "Galileo Probe To Look for Secrets of Jupiter."

[25] Johnson e-mail message.

modest 66 pounds. The neutral mass spectrometer (NMS)[26] was a tried-and-true instrument for analyzing neutral-particle atmospheric chemical compositions that had been flown on the Pioneer Venus and other missions. It also had a modest weight. The version flown on Pioneer Venus's large atmospheric probe weighed 9 kilograms, or about 20 pounds.[27]

Another common instrument used for chemical analysis is the gas chromatograph (GC).[28] Although GCs and NMSs were used by the Pioneer Venus mission, a gas chromatograph was not considered an adequate replacement for a mass spectrometer aboard the Galileo Probe. According to Joel Sperans, former Probe Project Manager, "You get more information with a mass spec than with a gas chromatograph."[29] The issue before JOPSWG was more "how massive and costly [the mass spectrometer] could afford to be," rather than whether or not to use a mass spectrometer.[30] What the JOPSWG team did consider was using a combination gas chromatograph–mass spectrometer (GC/MS), a very powerful analytical tool. But JOPSWG was "mindful that the Viking [Mars] lander GC/MS at that time was the most expensive thing ever flown and had created huge programmatic problems for Viking."[31] JOPSWG concluded that "while a GC/MS would be nice for Jupiter, it was not required to meet our minimum objectives and would probably prove to be too costly and massive to be affordable."[32]

NASA's instrument-selection staff ended up agreeing with JOPSWG. Following the Galileo Announcement of Opportunity, NASA chose a proposal from Goddard Space Flight Center to build a basic mass spectrometer, "but with some clever additions of gas adsorption cells" that added GC/MS-like capabilities to the instrument.[33]

A critical question that the Probe addressed was, how much helium did the Jovian atmosphere contain? Precise measurements of helium quantities and of the ratio of hydrogen to helium would give important clues "as to the origins of the atmosphere and to the origins of the planet." Although neutral mass spectrometer data would shed some light on helium abundance, scientists planning the mission thought they could "use a little more horsepower" in addressing the helium question. They decided to add another instrument, a helium abundance detector (HAD), to enhance the Probe's helium analysis.[34]

---

[26] A mass spectrometer is an instrument that performs chemical analyses by converting molecules of a gas sample into ions and then separating the ions according to their mass-to-charge ratio. The combination of different masses from a substance, called its mass spectrum, provides a "fingerprint" of the substance. The Probe used its mass spectrometer to identify the different chemical constituents of Jupiter's atmosphere. "Mass Spectrometer," *Microsoft® Encarta® Online Encyclopedia*, 2001, *http://encarta.msn.com*, © 1997–2000 Microsoft Corporation, all rights reserved.

[27] "Neutral Particle Mass Spectrometer," *National Space Science Data Center (NSSDC) Master Catalog: Experiment, http://nssdc.gsfc.nasa.gov/nmc/tmp/1978-078D-6.html* (accessed 7 September 2000).

[28] The gas chromatograph is a device that performs chemical analysis by separating the volatile constituents of a substance. "Gas Chromatograph," *Encarta World English Dictionary* (North American Edition), *http://dictionary.msn.com/find/entry.asp?refid=1861694680*, © and (P) 2001 Microsoft Corporation, all rights reserved.

[29] Joel Sperans, former Probe Project Manager, interview, tape-recorded telephone conversation, 26 October 2001.

[30] Johnson e-mail message.

[31] Ibid.

[32] Ibid.

[33] "The Pioneer Venus Multiprobe," *Pioneer Venus Project Information, http://nssdc.gsfc.nasa.gov/planetary/pioneer_venus.html, 5 January 2001*; Johnson e-mail message; "Probe Mass Spectrometer," *Galileo Messenger* (August 1982): 2.

[34] Joel Sperans interview.

Other key measurements that the Probe had to perform were of atmospheric temperature, pressure, density, and molecular weight as functions of depth. These parameters were measured by sensors in the atmospheric structure instrument (ASI). The parameter values also served as reference scales for other Probe experiments and as critical constraints in computer models of atmospheric composition and dynamics.

The Probe's nephelometer (NEP) and NMS studied Jovian cloud particle states (liquid versus solid) and the composition, structure, and location of cloud layers, while the lightning and radio emissions detector (LRD) searched for any lightning activity in the clouds. Sunlight penetrating into the atmosphere and infrared radiation upwelling from the planet were characterized using the Probe's net flux radiometer (NFR).

One reason for adding a lightning detector to the suite of Probe instruments was that Voyager imagery and plasma wave data had indicated the possible existence of lightning phenomena in the Jovian atmosphere, and in situ measurements were needed to verify the phenomenon. In addition, scientists were interested in the planet's radio-frequency emissions, which had been received for many years by stations on Earth. Some scientists thought that Jovian lightning could provide an energy source for these emissions.[35]

An energetic particles instrument (EPI) was also added to the Probe to study ionized magnetospheric particles from an altitude of 5 Jupiter radii down to the atmosphere.[36] A camera was not selected as a Probe instrument. The biggest problem with including a camera would have been its data generation rate. Digital pictures require a considerable amount of data, but the Probe was only capable of transmitting data at the rate of 128 bits per second. The total quantity of data that the Probe transmitted during its entire atmospheric descent was about one million bits, enough for only one or two images. According to Charlie Sobeck, who served as the Probe Deputy Project Manager and Systems Engineering Manager, "We just didn't have the data [transmission] rate to support a camera."[37]

Characteristics of the Probe's instruments are summarized in table 5.1. Note the wide range of institutions involved in managing the Probe's experiments. Photos of Probe instruments are included in figure 5.6. The findings obtained from the various Probe instruments' measurements are discussed in detail in chapter 8, "Jupiter Approach and Arrival."

Neutral Mass Spectrometer. The NMS was a quadrupole mass spectrometer, developed by Goddard Space Flight Center in Maryland, that repeatedly measured chemical and isotope compositions of the gases in Jupiter's atmosphere and recorded vertical variations in the compositions. The NMS weighed 27.8 pounds (12.6 kilograms) and drew only 29 watts of power (see table 5.1). At the time the instrument was built, it was considered to be very light and economical, although new generations of the NMS are even smaller and draw less power.[38]

---

[35] "Galileo Project Information"; "Pioneer Venus Multiprobe"; "The Probe Science Instruments," *Galileo Messenger* (September 1995).

[36] "Probe Instruments," in *Galileo Probe Background*; "Probe Science Instruments."

[37] Charlie Sobeck, former Probe Deputy Project Manager and Systems Engineering Manager, interview, tape-recorded telephone conversation, 26 October 2001.

[38] Mass and power data were supplied by Charlie Sobeck, NASA Ames Research Center Probe Systems Engineering Manager at the conclusion of the Probe project. Sobeck's data were drawn from the "Galileo Probe Mass Properties Report" of 2 April 1984 and Hughes space and communications group instrument fact sheets.

Table 5.1. **Atmospheric Probe instrumentation.**

INSTRUMENT	FUNCTION	MASS IN KILOGRAMS (POUNDS)	POWER REQUIREMENT IN WATTS	PRINCIPAL INVESTIGATOR(S) AND INSTITUTION(S)
Neutral mass spectrometer (NMS)	Analyzed gas composition.	13 (28)	29	Hasso Niemann, NASA Goddard Space Flight Center
Helium abundance detector (HAD)	Determined atmospheric hydrogen/helium ratios.	1.4 (3.1)	1.1	Ulf von Zahn, Bonn Universitat & Institut fur Atmosparenphysik an der Universitat Rostock
Atmospheric structure instrument (ASI)	Recorded temperature, pressure, density, and molecular weight.	4.1 (8.9)	6.3	Alvin Seiff, NASA Ames Research Center and San Jose State University Foundation
Nephelometer (NEP)	Located cloud layers and analyzed characteristics of cloud particles.	4.8 (11)	14	Boris Ragent, NASA Ames Research Center and San Jose State University Foundation
Lightning and radio emissions detector (LRD) and energetic particles instrument (EPI)	LRD recorded radio bursts and optical flashes. EPI measured fluxes of protons, electrons, alpha particles, and heavy ions orbiting Jupiter's magnetosphere.	2.7 (6.0)*	2.3**	Louis Lanzerotti, Bell Laboratories, University of Florida, and Federal Republic of Germany
Net flux radiometer (NFR)	Determined differences between light and heat being radiated downward versus upward at each altitude.	3.0 (6.7)	7.0	L. Sromovsky, University of Wisconsin
Radio equipment	Besides transmitting data up to the Orbiter, the Probe's radio equipment served double duty as an aid in measuring Jovian wind speeds and atmospheric absorption.			David Atkinson, University of Idaho

* Combined mass of LRD and EPI.
** Combined power draw of LRD and EPI.

Data sources:

[1.] Mass and power data supplied by Charlie Sobeck, NASA Ames Research Center Probe Systems Engineering Manager at the conclusion of the Probe project. Sobeck's data were drawn from the "Galileo Probe Mass Properties Report" of 2 April 1984 and Hughes space and communications group instrument fact sheets.

[2.] "The Atmosphere Structure Instrument," *Galileo Messenger* (April 1981).

[3.] "The Probe Science Instruments," *Galileo Messenger* (September 1995).

[4.] "Probe Mass Spectrometer," *Galileo Messenger* (August 1982).

[5.] "Net Flux Radiometer—Studying the Atmosphere," *Galileo Messenger* (April 1987).

[6.] "Probe Nephelometer," *Galileo Messenger* (March 1983).

[7.] Ames Research Center, "*Galileo Probe Background*," no date given. Copy available in folder 11626, NASA Headquarters Historical Reference Collection.

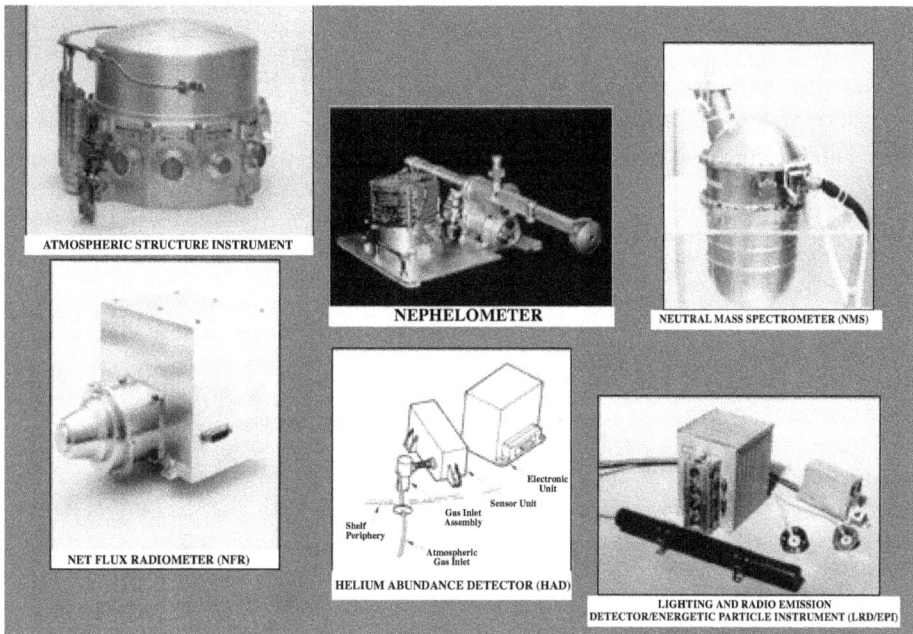

Figure 5.6. **Probe science instruments. (Adapted from JPL image number 230-900)**

A quadrupole mass spectrometer typically identifies gases by bombarding a neutral gas species with an electron beam to create ions. These ions are then passed through a four-pole "mass filter" that uses specific voltages between its poles to allow only ions of a certain mass-to-charge ratio to pass through the array and strike a detector. Other ions are deflected away before they reach the detector. The voltages between poles of the mass filter are varied rapidly, so that the mass-to-charge ratios allowed to pass through to the detector are also constantly varying. An ion's mass-to-charge ratio can be surmised from the time that the ion strikes the detector.[39]

A mass spectrometer is calibrated and programmed to record the mass and intensity of ions (that is, the rate at which ions hit the detector) for a wide range of atomic masses. The Probe's NMS measured all species with atomic masses from 1 through 52, as well as selected species of higher atomic masses, including krypton and xenon. In addition, the NMS also occasionally performed sweeps from 1 up to 150 atomic masses. The sampling range of the NMS was designed so that almost all of the Jovian atmospheric gases that entered the instrument could be analyzed. NMS measurements gave a cross section of Jovian atmospheric composition, and this aided greatly in understanding the dynamic processes that created the planet's multihued, complex cloud formations.[40]

---

[39] "Neutral Mass Spectrometer (NMS)," *NSSDC Master Catalog: Experiment*, NSSDC ID:1989-084E-3, *http://nssdc.gsfc. nasa.gov/database/MasterCatalog?sc=1989-084E&ex=3* (accessed 8 August 2003); Hugh Gregg, chemist and mass spectrometer specialist, Lawrence Livermore National Laboratory, CA, interview, 2001; "Probe Mass Spectrometer," *Galileo Messenger* (August 1982), *http://www.jpl.nasa.gov/galileo/messenger/oldmess/3Probe.html*; Atmospheric Experiment Branch in the Laboratory for Atmospheres at NASA Goddard Space Flight Center, "Introduction to the Galileo Mission," *Introduction to the Branch Activities, http://webserver.gsfc.nasa.gov/* (accessed 8 August 2003).

[40] "Neutral Mass Spectrometer"; "Probe Mass Spectrometer"; Gregg interview.

Gases entered the NMS through two inlet ports at the Probe's apex. For the protection of the instrument, these ports remained sealed by metal-ceramic panels until after the Probe entered Jupiter's atmosphere, at which time pyrotechnic devices blew the panels away.[41]

Jupiter's atmosphere is starlike in composition, with hydrogen and helium composing approximately 90 percent and 10 percent of it, respectively. The atmosphere also contains minor amounts of methane, water vapor, ammonia, hydrogen sulfide, acetylene, and ethane, as well as the inert gases neon, argon, krypton, and xenon. Inert gases are so named because they do not generally combine with other elements to form compounds. They also do not settle out of a planetary atmosphere by liquefying or freezing, but remain suspended and in their gaseous state. Because of their stability, they are expected to occur in the same abundances as they did at the beginning of the solar system's existence. Thus, their concentrations in the Jovian atmosphere give us information about our early planetary system. Cosmic abundances of the inert gases krypton and xenon are not well known, and so measurements taken by the Probe's NMS instrument were especially important, for they offered the first opportunity to measure these elements in an undisturbed reservoir. Results obtained were being used to calibrate a large amount of data on noble gas abundances on Earth, the other inner planets, and meteorites.[42]

Voyager's discovery of Jovian lightning raised the possibility that organic compounds exist on the planet. Organic compounds, which are the basis of life on Earth, can be formed when sparks arc across various mixtures of gases such as may be found on Jupiter. The Probe's neutral mass spectrometer analyzed the Jovian atmosphere closely for various organic compounds such as hydrogen cyanide and acetonitrile.[43]

Helium Abundance Detector. The HAD, which was developed at Bonn University in Germany, determined the abundance ratio of helium to hydrogen in Jupiter's atmosphere at pressures from 3 to 8 bars. The ratio was also measured by the NMS, but accurate measurements of this parameter were considered so important that the HAD, which was able to make the measurements with an uncertainty one-tenth that of NMS, was included in the Probe as well. When the instruments for the Probe were first chosen in 1977, the predominant scientific opinion was that Jupiter's helium abundance was a relic of the distant past, in that it was the same as the helium abundance created during the Big Bang, and the same as that present in the solar nebula from which our Sun and planets were formed. Accurate measurements of Jupiter's helium abundance would thus, it was believed, provide critical data on conditions at the instant of the universe's creation.[44]

The prevailing theories were called into question when Voyager data from Saturn and Uranus, as well as more detailed information about Jupiter and the Sun, suggested that Jovian evolutionary processes could have altered the original helium abundance. HAD measurements are now seen as providing valuable clues about the origin and evolution of Jupiter itself.[45]

---

[41] "Neutral Mass Spectrometer"; "Probe Mass Spectrometer."

[42] "Neutral Mass Spectrometer"; "Probe Mass Spectrometer."

[43] "Neutral Mass Spectrometer"; "Probe Mass Spectrometer."

[44] "The Probe Science Instruments," *Galileo Messenger* (September 1995).

[45] Ibid.

To make these measurements, the HAD used an optical sensor called a Jamin-Mascart two-arm interferometer, with an infrared light source. One arm of the interferometer was vented to Jupiter's atmosphere, while the other arm was connected to a reference gas reservoir. The interferometer determined relative helium abundance by comparing the difference in refractive index between the atmosphere and the reference gas. (A medium's index of refraction is calculated by dividing light waves' speed through a vacuum by their speed through the medium.) When the light beam from the infrared source is split and directed through the gases in each of the interferometer's arms, then merged back into one beam, an interference pattern of light and dark bands is formed. The particular pattern is dependent on the refractive index of the Jovian atmospheric gas, which is, in part, a function of the percentage of helium present. The HAD experimental apparatus was small compared to the NMS, weighing only 3.1 pounds (1.4 kilograms) and drawing only 1.1 watts of power (see table 5.1).[46]

Atmospheric Structure Instrument. The ASI, developed by San Jose State University Foundation in California, measured physical properties of the atmosphere—temperature, pressure, density, and molecular weight—over an altitude range from about 1,000 kilometers (600 miles) above the cloud deck down to the atmospheric depth at which the Probe ceased to operate, about 150 kilometers (90 miles). From the instrument's measurements, Probe altitude and velocity of descent were calculated and then used in other experiments' calculations as well as in the ASI experiment. Mission scientists were especially interested in the stability of the atmosphere and the levels at which various cloud layers occur. Stability observations were important for determining whether the Jovian atmosphere was turning over (like Earth's lower atmosphere, the troposphere, does during a storm) or stagnantly layered (similar to Earth's stratosphere, or to the troposphere in regions of temperature inversions).[47]

Nephelometer. The word nephele is Greek for cloud.[48] The purpose of the Probe's nephelometer was to determine the physical structure of Jupiter's clouds, including cloud particle shapes, sizes, and concentrations, as well as locations of cloud layers. The shapes of the particles provided an indication of their state—solid (ice) or liquid.[49]

Understanding the characteristics of Jupiter's clouds is essential to understanding the planet's energy balance. Jupiter is very different from Earth. It is a giant planet, composed almost completely of hydrogen and helium, and it resembles a star perhaps more than it does a terrestrial-type planet. Like a star, Jupiter gives off more energy than it takes in. Jupiter radiates twice as much energy as it receives from the Sun, and it is this internal

---

[46] "The Probe Instruments," in *Galileo Probe Background*; "Helium Abundance Detector (HAD)," *NSSDC Master Catalog Display: Experiment,* http://nssdc.gsfc.nasa.gov/cgi-bin/database/www-nmc?89-084E-01 (accessed 15 March 2000); Charles Sobeck, NASA Ames Research Center Probe Systems Engineering Manager, at the conclusion of the Probe project, interview, tape-recorded telephone conversation, 26 October 2001.

[47] "Atmospheric Structure Instrument (ASI)," *NSSDC Master Catalog Display: Experiment,* http://nssdc.gsfc.nasa.gov/cgi-bin/database/www-nmc?89-084E-02 (accessed 15 March 2000); "The Probe Instruments"; "The Atmosphere Structure Experiment," *Galileo Messenger* (April 1981); "Galileo Probe Mission Events." Also, mass and power data supplied by Charlie Sobeck, NASA Ames Research Center Probe Systems Engineering Manager, at the conclusion of the Probe project. Sobeck's data were drawn from the "Galileo Probe Mass Properties Report" of 2 April 1984 and Hughes space and communications group instrument fact sheets.

[48] *Webster's New World Dictionary of the American Language—College Edition* (Cleveland and New York: World Publishing Company, 1962), p. 984.

[49] "Probe Nephelometer," *Galileo Messenger* (March 1983).

heat that drives its complex weather patterns. Greater understanding of Jovian weather and of its cloud and atmospheric structure provides clues regarding Jupiter's basic dynamics and internal processes.[50]

As the spinning Probe fell through the clouds, the nephelometer analyzed cloud particles by shining an infrared laser through them and measuring how the light beam scattered. Cloud particle sizes were determined by the intensity of scattered light intercepted by special sensing mirrors positioned at 5.8 degrees, 16 degrees, 40 degrees, 70 degrees, and 178 degrees off the direction of the laser beam.[51]

The nephelometer weighed 11 pounds (4.8 kilograms) and operated on 14 watts. The San Jose State University Foundation in California oversaw this experiment.[52]

Lightning and Radio Emissions Detector. Jupiter was named for the Roman god of the sky, who was said to have kept a stock of lightning bolts on hand. Before the Voyager 1 flyby of Jupiter, scientists hypothesized that Jovian lightning served as an energy source for the planet's nonthermal radio emissions, which had been readily detected on Earth. Photos that Voyager took of Jupiter's night side appeared to show the existence of lightning, as did the craft's plasma wave experiment, which detected "whistlers" (signals thought to be caused by electrical discharges propagating through the atmosphere). But to verify the existence of Jovian lightning and to delve deeper into its physical characteristics and understand how it is generated, how frequently it occurs, and how intense it is, in situ rather than brief flyby measurements needed to be made. Galileo's LRD was well suited to this task. It was built to take into account large uncertainties regarding the nature of Jovian lightning. It also had the capacity to carry out the second part of the experiment, the measurement of radio-frequency emissions and magnetic field characteristics near Jupiter.[53]

The LRD was funded and built by the Federal Republic of Germany, one of several international partners on the Galileo project. The LRD consisted of two types of sensors: photodiodes to take optical observations and an antenna to make radio-frequency (RF) and magnetic measurements. The LRD shared its electrical system with the energetic particles instrument (EPI) described below.[54]

While Galileo sped through space on its way to Jupiter, the LRD experiment's principal investigator, Louis Lanzerotti of Bell Laboratories and the University of Florida, and his team made numerous studies of lightning on Earth in order to calibrate a duplicate

---

[50] Ibid.

[51] Ibid.

[52] "Nephelometer," *NSSDC Master Catalog Display: Experiment, http://nssdc.gsfc.nasa.gov/cgi-bin/database/www-nmc?89-084E-05* (accessed 15 March 2000); "Probe Nephelometer." Also, mass and power data supplied by Charlie Sobeck, NASA Ames Research Center Probe Systems Engineering Manager, at the conclusion of the Probe project. Sobeck's data were drawn from the "Galileo Probe Mass Properties Report" of 2 April 1984 and the Hughes space and communications group instrument fact sheets.

[53] "The Probe Science Instruments," *Galileo Messenger* (September 1995).

[54] "Lightning and Radio Emission Detector (LRD)," *NSSDC Master Catalog Display: Experiment, http://nssdc.gsfc.nasa. gov/database/MasterCatalog?sc=1989-084E&ex=6* (accessed 15 March 2000); "Probe Instruments," in *Galileo Probe Background*; "Probe Science Instruments." Also, mass and power data supplied by Charlie Sobeck, NASA Ames Research Center Probe Systems Engineering Manager, at the conclusion of the Probe project. Sobeck's data were drawn from the "Galileo Probe Mass Properties Report" of 2 April 1984 and the Hughes space and communications group instrument fact sheets.

LRD. Doing so would help them better understand the data they would receive from Galileo, as well as aid in the ongoing interpretation of Earth's electrical phenomena.[55]

Energetic Particles Instrument. The goal of the EPI experiment was to study fluxes of protons, electrons, alpha particles, and heavy ions orbiting in Jupiter's magnetosphere at speeds of up to tens of thousands of miles per second. The EPI was the only Probe instrument to start recording data before the Probe entered the Jovian atmosphere. The EPI analyzed energies and angular distributions of charged particles at altitudes that ranged from 5 Jovian radii down to the top of the atmosphere. At 5 planetary radii, the study took place in the vicinity of a torus-shaped (donut-shaped) field of plasma that is thought to have originated from particles spewed out during the moon Io's frequent volcanic eruptions. This torus of plasma and the magnetic field associated with it corotate with Jupiter, which means that they travel four times faster than Io and repeatedly overtake it.[56]

Between Io and Jupiter, four tiny moons orbit. At about 2 Jovian radii is a dust ring, and scientists believe that this ring and the four small moons influence the energetic particle populations in the region, sweeping up particles in their paths. The Probe's EPI experiment was the first to directly measure the region's energetic particles, which travel at "relativistic" velocities (comparable to the speed of light). Prior to Galileo, the only means of "seeing" into Jupiter's inner magnetosphere was to observe electromagnetic radiation that the fast-moving charged particles emitted.[57]

The EPI instrument employed two silicon-disc particle detectors with a brass particle absorber inserted between them. These parts were contained in a cylindrical tungsten shield that was open in the front, admitting particles as a telescope admits light. The EPI instrument was mounted under the aft heatshield of the Probe, which meant that only particles with enough energy to pass through the heatshield (at least 2.6 million electron volts, or "MeV") would be counted. The energies of the particles were further determined by the parts of the instrument that they were able to penetrate. For instance, a particle that struck only the first silicon detector disc had to pass through the Probe's heatshield but didn't have enough energy to get through the brass absorber to the second detector. Such a particle had to have at least 2.6, but not more than 7, MeV of energy. Particles that could get through the brass absorber and hit the second silicon detector had energies of at least 7 MeV, and particles that could get through the tungsten shield had at least 20 MeV of energy.[58]

Net Flux Radiometer. The NFR, whose operation was overseen by the University of Wisconsin, measured heat and light energy, at different levels in the atmosphere, that was radiated both from below by Jupiter and from above by the Sun. The NFR's optical system performed upward and downward observations in order to determine the difference in energy emitted between Jupiter's internal heat source and that coming from the Sun (see figure 5.7). Because the Probe was spinning and the Sun was near the horizon, the NFR was also able to locate and measure the opacities of substantial cloud layers around the Probe that scattered sunlight. Note from figure 5.7 the sudden increase in solar net flux at

---

[55] "Probe Science Instruments."

[56] "Probe Instruments," in *Galileo Probe Background*; "Probe Science Instruments"; Ed Tischler, former experiments manager for Probe instruments, interview, telephone conversation, 26 October 2001.

[57] "Probe Science Instruments."

[58] "Probe Instruments."

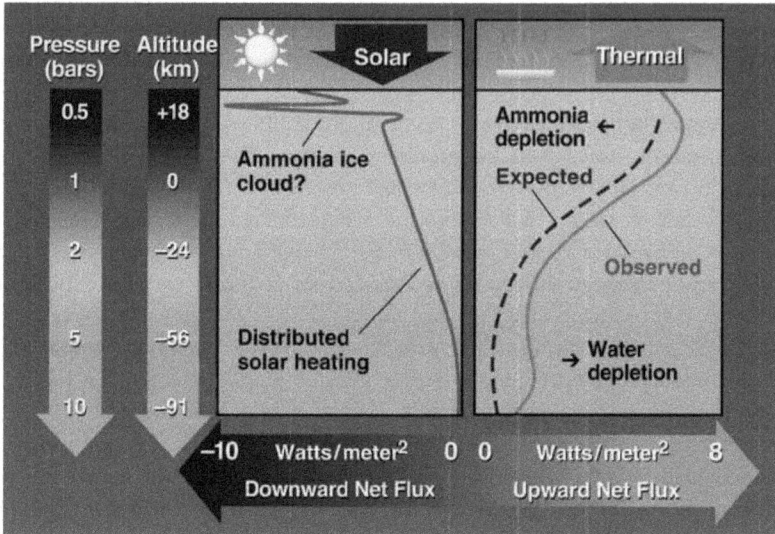

Figure 5.7. **Jupiter's net radiation fluxes. The net flux radiometer experiment measured the difference between upward and downward energy flow ("net flux") at the wavelengths of visible sunlight, as well as at the wavelengths of thermal infrared radiation, as a function of depth in Jupiter's atmosphere. The sudden rise in solar net flux followed by an abrupt dropoff is probably caused by an ammonia ice cloud layer. (NASA Ames photo number ACD96-0313-6)**

0.5 bar pressure, immediately followed by a sharp decrease at 0.6 bar. This phenomenon is thought to be caused by scattering from an ammonia ice cloud layer.[59]

Radio-Science Experiment: Investigations Using Relay Radio Equipment. Dave Atkinson of the University of Idaho oversaw the use of radio equipment on the Probe and Orbiter to carry out two analyses: Doppler wind determination and atmospheric absorption. During the Probe's descent through the Jovian atmosphere, its position and velocity were partly functions of the planet's 28,000-mph rotational speed and its massive gravity field. These effects could be predicted. But atmospheric winds also had a significant effect. To study this effect, the Probe's transmitter and antenna worked in conjunction with the Orbiter's Probe relay radio hardware to conduct the Doppler wind experiment. The Probe's transmitted frequencies were monitored during the entire descent by the Orbiter, and their Doppler shifts were noted.[60] From the Doppler shifts, mission scientists were able to reconstruct the Probe's trajectory during descent. After separating out the Doppler shifts expected to be caused by gravitational and rotational forces, they were able to surmise the shifts caused by Jovian winds and develop a profile of those winds at the time of descent. These data gave them clues as to the forces behind Jupiter's global circulation patterns. They also provided confirmation, or "ground truth," for the wind

---

[59] "Probe Instruments"; "Jupiter's Net Fluxes," *http://spaceprojects.arc.nasa.gov/Space_Projects/galileo_probe/ htmls/NFR_results.html* (accessed 19 October 2004), available in folder 18522, NASA Historical Reference Collection, Washington, DC.

[60] These are the shifts in frequency due to changes in velocity. They are similar to the changes in pitch one hears in an ambulance siren as it approaches, rushes past, and speeds away.

speeds determined by the Orbiter's imaging team. The Doppler wind experiment will help future researchers understand the dynamics, not only of Jupiter, but also of Saturn, Uranus, and Neptune, the other gas-giant planets.[61]

The atmospheric absorption experiment was also carried out using the radio signal from Probe to Orbiter. This signal was affected by density variations in the atmosphere. In particular, electron densities in the planet's ionosphere had a significant effect on the radio signal. Atmospheric absorption of the signal's energy was determined by comparing the Probe's transmitter output with the actual signal strength received by the Orbiter. Variation of atmospheric absorption with changes in the Probe's position provided data on atmospheric structure as well as ionospheric electron densities.[62]

Orbiter Scientific Experiments

The Orbiter's tasks, after relaying data from Probe to Earth, were to conduct its own investigations of Jupiter's atmosphere and magnetosphere and to perform repeated flybys of Jovian satellites, collecting a wide range of data. The mission's Announcement of Opportunity listed these objectives for the Orbiter's scientific experiments:

- Define the topology and dynamics of the outer magnetosphere, magnetosheath, and bow shock.

- Describe the nature of magnetospheric particle emission.

- Determine the distribution and stability of trapped radiation.

- Study the magnetosphere-satellite interaction.

- Determine the surface composition of the satellites.

- Identify the physical state of the satellite surfaces and characterize their surface morphology.

- Measure the satellite magnetic, gravitational, and thermal properties and thereby obtain their geophysical characteristics.

- Study the satellite ionospheres and atmospheres, as well as the emission of gas.

- Conduct a synoptic study of the Jovian atmosphere.[63]

---

[61] "Probe Instruments."

[62] Ibid.

[63] "Announcement of Opportunity for Outer-Planets Orbiter/Probe (Jupiter)." A synoptic climate analysis attempts to characterize an entire weather situation that exists in a given area at a given time, and involves meteorological analysis over a large region of the atmosphere, of a scale typically in excess of 2,000 kilometers. See Synoptic Climatology Laboratory, Department of Geography, University of Delaware, March 2003, http://www.udel.edu/SynClim/; University Corporation for Atmospheric Research, National Center for Atmospheric Research, "Definition of the Mesoscale," *MetEd Meteorology and Training*, 31 July 2003, *http://meted.ucar.edu/mesoprim/mesodefn/print.htm.*

To meet these objectives, the Orbiter had to visit different parts of Jupiter's magnetosphere, view the parent planet from many different angles, and fly by some of its satellites. This required that the spacecraft's trajectory "be adjusted from time-to-time by powered maneuvers and by satellite flybys (involving gravity assists) in order to optimize observing conditions for diverse scientific purposes."[64] In their review of the Galileo mission plan, Johnson, Yeates, and Young compared the Probe's and Orbiter's tasks by saying that "while the Probe's success is keyed to its ability to penetrate the Jovian atmosphere, the Orbiter's success depends on its unique trajectory, which provides for unprecedented new measurements."[65]

The instruments chosen to meet the Orbiter's objectives had to be of several types. Some had to assess magnetic and electric fields, others needed to characterize particles of various energy levels, and those in a third group were to investigate Jovian satellites remotely. Figure 5.8 depicts the science instruments used for these ends. Instruments designed to take fields-and-particles data were mounted on the Orbiter's spun section, where they swept through and measured the surrounding environment in all directions as the spacecraft flew through it. The Orbiter's spun section carried not only these instruments, but also Galileo's power supply, propulsion module, most of its computers and control electronics, its high-gain antenna, and one of its low-gain antennas. Instruments that remotely sensed satellite and planetary characteristics and that required stable, high-accuracy pointing were mounted on the Orbiter's despun section.[66]

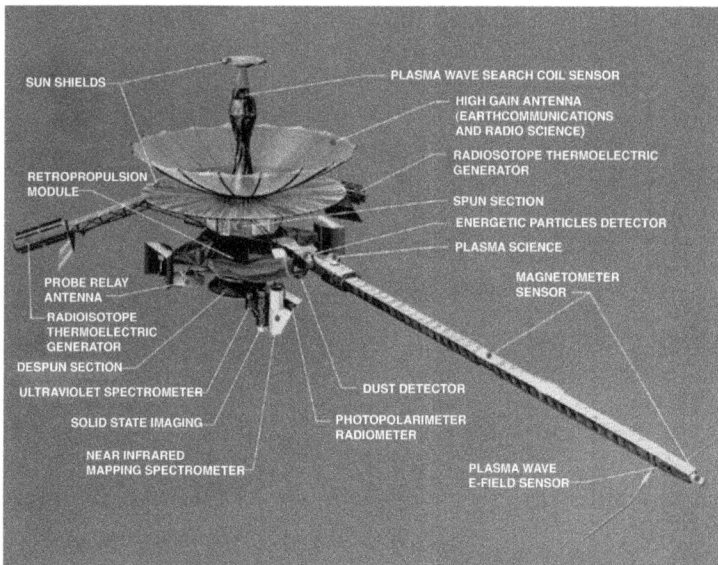

Figure 5.8. **The Galileo Orbiter. Fields-and-particles instruments were mounted on the spun section (including the long science boom); remote sensing experiments were mounted on the Orbiter's stable despun section (at the bottom of the spacecraft drawing). (Adapted from JPL image number P-31284)**

---

[64] "Announcement of Opportunity for Outer-Planets Orbiter/Probe (Jupiter)."

[65] Johnson et al., pp. 19–20.

[66] Ibid., pp. 12–13.

Instruments Designed To Measure Fields and Particles

The Orbiter required instruments that could measure the Jovian magnetosphere's magnetic fields, as well as its electromagnetic and electrostatic wave characteristics over a range of frequencies. The instruments also needed to investigate the magnetic fields of Jupiter's satellites. Furthermore, the Orbiter needed equipment to characterize the energies, angular distributions, and compositions of the particles that the magnetosphere contained. Mission planners had a good idea of what instruments would best accomplish these goals because similar experiments to those that Galileo performed had been conducted on the earlier Jupiter flybys of Pioneer and Voyager spacecraft, as well as on missions that had orbited Earth. These other missions had used magnetometers to characterize magnetic fields; particle detectors, plasma analyzers, and other instruments to study charged- and neutral-particle populations; and a plasma wave system to investigate wave phenomena. On-board communications equipment also had been utilized for radio-science experiments that helped determine satellite and atmospheric properties. According to Torrence Johnson, "The only major [Galileo] proposal issues were which universities and labs could produce the best instruments at the lowest cost and mass. Groups with instruments already on other missions usually have an advantage in this area, of course."[67] Table 5.2 lists the fields-and-particles instruments selected for the Orbiter and the organizations that supplied them.

Table 5.2. **Orbiter fields-and-particles instruments.**

INSTRUMENT	FUNCTION	MASS IN KILOGRAMS (POUNDS)	POWER DRAW IN WATTS	PRINCIPAL INVESTIGATOR(S) AND INSTITUTION(S)
Dust detection system (DDS)	Measured mass and speed of small Jovian dust particles and studied the distribution of interplanetary dust.	4.1 (9.0)	1.8	Eberhard Grun, Max Planck Institut fur Kernphysik, Heidelberg, Germany
Energetic particles detector (EPD)	Determined angular distributions, temporal fluctuations, intensities, and compositions of energetic charged particles in Jupiter's magnetosphere and investigated the processes responsible for replenishing particles that escape into interplanetary space.	10 (23)	6	D. J. Williams, Johns Hopkins University
Magnetometer (MAG)	Measured magnetic fields of Jupiter's magnetosphere and satellites. Characterized interplanetary magnetic fields and those of asteroids.	7 (15)	4	Margaret Kivelson, UCLA

[67] Johnson et al., pp. 12, 19–20; Johnson e-mail; "Pioneer 10," *NSSDC Master Catalog: Spacecraft,* last updated 28 February 2003, *http://nssdc.gsfc.nasa.gov/nmc/tmp/1972-012A.html*; "Pioneer 11," *NSSDC Master Catalog: Spacecraft,* last updated 7 September 2000, *http://nssdc.gsfc.nasa.gov/nmc/tmp/1973-019A.html*; "Voyager Project Information," *NSSDC: Planetary Sciences,* version 2.4, 20 May 2003, *http://nssdc.gsfc.nasa.gov/planetary/voyager.html*.

Table 5.2. **Orbiter fields-and-particles instruments.** (continued)

INSTRUMENT	FUNCTION	MASS IN KILOGRAMS (POUNDS)	POWER DRAW IN WATTS	PRINCIPAL INVESTIGATOR(S) AND INSTITUTION(S)
Plasma subsystem (PLS)	Measured densities, temperatures, bulk velocities, and composition of low-energy plasmas (1.2 electron volts, or eV, to 50 keV), which make up the bulk of Io's plasma torus.	13 (29)	11	Lou Frank, University of Iowa
Plasma wave subsystem (PWS)	Determined intensities of plasma waves in Jovian magnetosphere, as well as radio waves emitted by Jupiter, Earth, and the Sun.	7 (16)	10	Donald Gurnett, University of Iowa
Radio science	Employed the Orbiter's radio telecommunication subsystem and Earth-based instruments to perform celestial mechanics and relativity experiments, as well as atmospheric studies.			John Anderson, JPL, and H. Taylor Howard, Stanford University

Sources:

1. "Galileo's Science Instruments," *Galileo Home Page,* *http://www.jpl.nasa.gov/galileo/instruments/index.html* (accessed 19 February 2001).

2. "Galileo Magnetometer Project Overview," *Galileo Magnetometer Team's Homepage,* Institute of Geophysics and Planetary Physics, UCLA, *http://www.igpp.ucla.edu/galileo/framego.htm* (accessed 2 January 2001).

3. "Dust Detection Experiment," *Galileo Messenger* (December 1983).

4. "PLS–Plasma Subsystem," *Galileo's Science Instruments, Galileo Home Page, http://www.jpl.nasa.gov/galileo/instruments–pls.html>,* accessed 15 March 2000.

5. "PWS–Plasma Wave Subsystem," *Galileo's Science Instruments, Galileo Home Page, http://www.jpl.nasa.gov/galileo/instruments/pws.html* (accessed 15 March 2000); "Plasma Wave Investigation," *Galileo Messenger* (July 1986).

6. "Energetic Particle Detector (EPD)," *Galileo Messenger* (February 1986).

Dust Detection System. The exquisite ring system surrounding Saturn has been seen by millions of people on Earth—many looking through fairly modest telescopes. Uranus's rings have been viewed by far fewer people, although they too can be seen using more sophisticated Earth-based instruments. In contrast, Jupiter's rings are too faint to spot with a telescope sitting on Earth. Their discovery, in fact, came from one fortuitous photograph taken by Voyager 1, in which the Sun happened to backlight the ring's particles in just the right way for Voyager's cameras to pick them up. Jupiter's tenuous ring system is believed to be composed entirely of dust particles, making them quite different from Saturn's rings, where constituents range from dust-size to the size of a house.[68]

Orbiter's dust detection system (DDS) was designed to measure the motions of dust streams near Jupiter, as well as electrical charges on the larger particles. The DDS could analyze as many as 100 dust impacts per second. These measurements cast light on the particles' sources and transport mechanisms and on how dust streams interact with plasmas. Observations of Jupiter's electrostatically charged dust were of particular interest. Scientists wanted to determine whether charges on dust particles led to particular ring structures, such as the radial spoke features that have been observed in Saturn's rings. The

---

[68] "Dust Detection Experiment," *Galileo Messenger* (December 1983).

theory was that fine, charged dust particles interact with a planet's magnetic fields in such a way as to get forced out of the ring plane and leave voids in the rings.

Scientists also wanted to study "fluffy" aggregates of small particles. If such aggregates get strongly charged, mutually repulsive electrostatic forces can break them apart into small chunks, creating a swarm of micrometeoroids. Such swarms have been observed in Earth's atmosphere by a detector on a Highly Eccentric Orbit Satellite (HEOS).[69]

The DDS was a modified micrometeoroid detector that had been used successfully on the HEOS-2 satellite. As particles flew through the instrument's entrance grid, they induced currents in it from which their charges were surmised. The particles then struck a gold target and were detected by the plasma (ionized gas) generated from the impact. An electric field separated the ions and electrons of this plasma by their charges and directed each stream toward separate amplifiers, which produced current pulses whose rise times were functions of the particles' speeds and whose amplitudes were functions of both the particles' masses and speeds. From pulse rise time and amplitude, particle mass and speed were determined.[70]

Energetic Particles Detector (EPD). Jupiter's magnetosphere is the region in which the planet's intense magnetic field and charged particles are confined by the solar wind (see figures 5.9a and b). A planet's magnetosphere can be thought of as a "bubble" in the solar wind. Charged particles in the Jovian magnetosphere have so much kinetic energy that radiation damage to the Orbiter was of great concern to project engineers—after all, the spacecraft would have to operate in the area for years. These energetic particles continuously escape into interplanetary space but are quickly replaced with new particles. One of the EPD's main objectives was to study how this replacement occurred; the other was to analyze the dynamics of the Jovian magnetosphere.

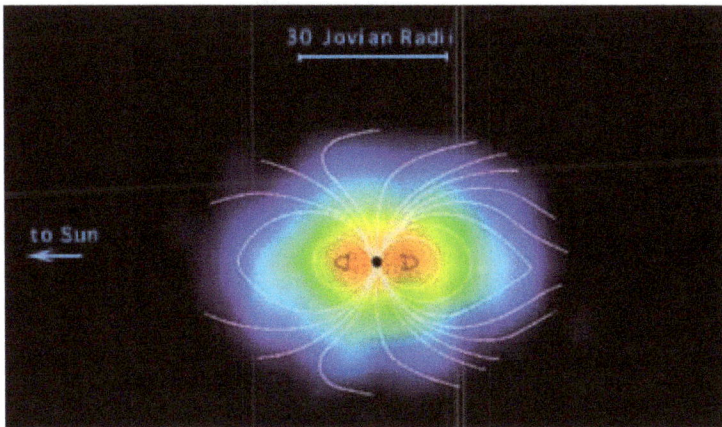

Figure 5.9a. **Jupiter's magnetosphere, a region of intense magnetic fields and rapidly moving charged particles. The small black circle at the center shows the relative size of Jupiter. Also depicted are lines of Jupiter's magnetic field. (Adapted from JPL image number PIA-03476)**

---

[69] Ibid.; "Heos 2," *NSSDC Master Catalog: Spacecraft*, NSSDC ID:1972-005A, 17 May 2000, *http://nssdc.gsfc.nasa. gov/database/MasterCatalog?sc=1972-005A.*

[70] "Dust Detection Experiment."

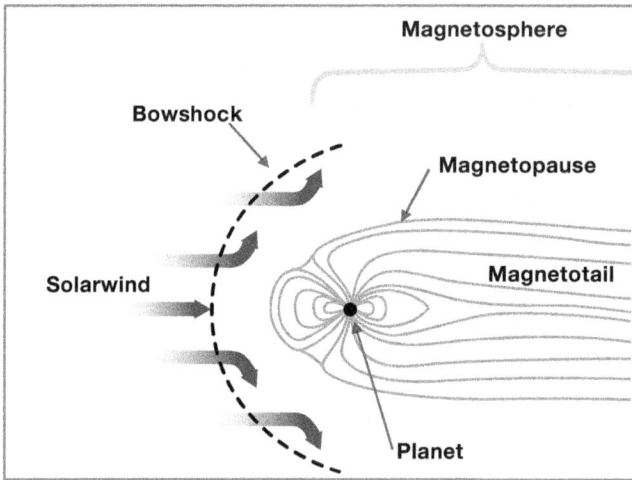

Figure 5.9b. **Parts of a planet's magnetosphere. Deflected solar-wind particles flow around a magnetosphere. Note the magnetosphere's typically elongated, rather than spherical, shape. The planet's bow shock is the interface between the region of space influenced by the planet's magnetic field and the region of the undisturbed, interplanetary solar wind. (Figure drawn by NASA History Office from the Web site *http://www2.jpl.nasa.gov/galileo/ messenger/oldmess/Earth3.html*)**

Voyager's observations identified three possible sources for energetic-particle replenishment: the Sun, Jupiter's ionosphere, and its moons. The Sun's energetic-particle streams and its solar wind were deemed the most likely source for the helium, carbon, nitrogen, oxygen, neon, magnesium, silicon, and iron-bearing particles detected in Jupiter's outer magnetosphere. The satellite Io and its plasma torus were suspected of furnishing the particles containing sulfur, sodium, and oxygen that had been seen closer

Figure 5.10. **The energetic particles detector (EPD) included two telescope systems: the low-energy magnetospheric measurements system (LEMMS) and the composition measurement system. These determined particle charges, energies, and species. (JPL image no. 230-1027A)**

to the planet. Molecular-sized ions of the hydrogen isotope H_3 were thought to have originated in Jupiter's ionosphere. The relative contributions of these sources had been shown by Voyager data to vary considerably over time, and so years of observations by Galileo would be crucial to obtaining an accurate picture of Jovian magnetospheric processes.[71]

The EPD (figure 5.10) used magnetic deflection and various absorber materials to distinguish between incoming particle charges and types. The instrument consisted of two different telescope systems that together studied electrons with energies from 15,000 electron volts (15 keV) to 11 million electron volts (11 MeV) and ions with energies from 10 keV to 55 MeV. The EPD was able to identify ion species that ranged from hydrogen to iron. It determined particle energy, composition, and location in the vicinity of Jupiter and its satellites, and it also generated a three-dimensional map of particle distribution. The EPD principal investigator was from Johns Hopkins University, although the science team for the instrument included staff from Germany's Max Planck Institut fur Aeronomie, the University of Alaska, the University of Kansas, and Bell Laboratories.[72]

Magnetometer. Within a magnetosphere, the magnetic field of the planet, rather than that of the solar wind, is dominant and controls the behavior of charged particles.[73] The size of a magnetosphere is related to the strength of the planet's field. Jupiter has a very strong magnetic field and an enormous magnetosphere. The Jovian magnetosphere is the largest single object in the solar system, with an average diameter of about 9 million miles (15 million kilometers). If our eyes could see its magnetic fields, it would appear from Earth to be larger than either our Moon or the Sun. It would occupy 1.5 degrees of sky, compared to the Sun's 0.5 degrees of sky. And it would take up this space in our sky even though it is far more distant from Earth—about 710 million kilometers (440 million miles)—than is the Sun at 150 million kilometers (93 million miles).[74]

The mammoth scale of Jupiter's magnetosphere arises from several factors. The solar wind which envelopes it has a low dynamic pressure at Jupiter's distance from the Sun, down by a factor of more than 25 from its pressure at Earth. This is not surprising; the solar wind has hundreds of millions more miles to spread out and become increasingly diffuse as it streams outward through the solar system and eventually reaches Jupiter. Jupiter's large size, with its radius 11 times that of Earth, also contributes to the scale of its magnetosphere, as does its hefty magnetic moment (a measure of the strength and orientation of a magnetic field), which is four orders of magnitude larger than that of Earth. Finally, the Jovian system fills its magnetosphere with a relatively dense plasma that is centrifugally accelerated by Jupiter's rapid rotation and strong field. The result is that this plasma pushes outward and "inflates" the magnetosphere from within.[75]

A planet's magnetosphere is not actually shaped like a sphere. Jupiter's magnetosphere resembles a giant wind sock, with its rounded head facing toward the Sun (and

---

[71] "Energetic Particle Detector (EPD)," *Galileo Messenger* (February 1986).

[72] Ibid.; "EPD—Energetic Particles Detector," *Galileo's Science Instruments, Galileo Home Page, http://www.jpl.nasa. gov/galileo/instruments/epd.html* (accessed 15 March 2000); "Energetic Particles Detector (EPD)," *NSSDC Master Catalog Display: Experiment, http://nssdc.gsfc.nasa.gov/cgi-bin/database/www-nmc?89-084B-06* (accessed 15 March 2000).

[73] "Magnetometers," *Galileo Messenger* (May 1982).

[74] Ibid.; "Plasma Investigation," *Galileo Messenger* (September 1985).

[75] M. G. Kivelson, K. K. Khurana, J. D. Means, C. T. Russell, and R. C. Snare, "The Galileo Magnetic Field Investigation," *Space Science Reviews* 60 (1992): 357–383.

the solar wind) and its long, flapping "magnetotail" stretched out away from the Sun. The tail is so long that it may extend as far as Saturn's orbit, roughly 650 million kilometers (400 million miles) away from Jupiter. This is more than four times the distance from Earth to the Sun and nearly the distance from Jupiter to the Sun. Galileo was able to travel only about 11 million kilometers (7 million miles) down the tail.

The Orbiter's magnetometer investigated magnetic fields throughout Jupiter's magnetosphere. Pioneer and Voyager missions also made observations of the Jovian magnetosphere's size, shape, and structure as they flew by the planet, but they did not stay in the area long enough to record significant temporal changes. In its years of residence in the Jovian system, Orbiter's magnetometer was able to study the dynamic nature of the magnetosphere more thoroughly.[76]

Jupiter's four largest moons—Io, Europa, Ganymede, and Callisto, the "Galilean" satellites that were discovered by the Italian astronomer—all orbit within Jupiter's magnetosphere. As Jupiter rotates, its magnetic field sets the charged particles in its magnetosphere in motion. These interact with the Galilean moons in ways that differ depending on how good an electrical conductor the moon is, how strong its magnetic field is, and whether it has an ionosphere. A focus of the Orbiter's tour through the Jovian system was to study relationships among the planet, its magnetospheric particles, and its satellites.

The Orbiter also characterized the electrical and magnetic properties of the satellites themselves. Evidence of a strong magnetic field suggested that a satellite might have a molten, electrically conducting metallic core such as is found inside Earth.[77]

During Galileo's long voyage to Jupiter, its magnetometer studied the fields of interplanetary space and the solar wind. One aspect of this study was to correlate changes in solar activity with long-term changes in the solar wind's field at great distances from the Sun. Also observed were the interactions of the solar wind with Venus, Earth, and the asteroids Gaspra and Ida.[78]

The magnetometer consisted of two clusters of three sensors each, mounted on the Orbiter's 11-meter (36-foot) science boom. The sensors had to be placed a distance away from the rest of the spacecraft in order to minimize the effect of magnetic fields generated by electronic circuitry in the Orbiter and Probe, which could be confused with magnetic fields from outside the craft. The cluster that measured the faint fields originating in the distant Jovian magnetotail, as well as from the solar wind, was located at the end of the 11-meter boom. The cluster designed to measure the much stronger fields of Jupiter's inner magnetosphere, and also to provide redundancy with the other cluster's measurements, was mounted about 7 meters down the boom from the spacecraft. (Figure 5.8 shows the two locations of the magnetometer sensor clusters.)[79]

Procedures were also carried out during data processing to reduce the effects of spacecraft-generated fields. Rotation of the spacecraft was used to distinguish between the natural magnetic fields that NASA wanted to measure and "engineering-induced" fields produced by and moving with the spacecraft. These latter fields were quantified and separated from the magnetospheric data before transmission to Earth.[80]

---

[76] "Magnetometers."

[77] Ibid.

[78] Ibid.; "Magnetometer (MAG)," *NSSDC Master Catalog Display: Experiment, http://nssdc.gsfc.nasa.gov/cgi-bin/ database/www-nmc?89-084B-03* (accessed 15 March 2000).

[79] "Magnetometers"; "Magnetometer (MAG)."

[80] "Magnetometers"; "MAG—Magnetometer," *Galileo's Science Instruments, http://www.jpl.nasa.gov/galileo/instruments/ epd.html* (accessed 24 January 2000).

Galileo's magnetometer system was designed by UCLA's Earth and Space Sciences Department. The basic device that measured magnetic fields was called a ring-core sensor, manufactured by the Naval Surface Weapons Center in Silver Spring, Maryland. This sensor was chosen because of its proven stability and low noise.[81] The magnetometer was calibrated in flight using a specially designed coil that generated a known magnetic field at a set frequency. This calibration enabled mission staff to eliminate the effects of yet another source of error in measurement—the bending and twisting of the long boom on which the magnetometer was mounted. These deformations altered the magnetometer's orientation, and a calibration scheme was needed in order to determine its orientation accurately at any given time.[82]

Plasma Subsystem. A plasma is a collection of charged particles, usually made up of about equal numbers of ions and electrons. In some respects, plasmas act like gases. But unlike gases, they are good electrical conductors and are affected by magnetic fields. Plasmas are the most common form of matter, making up more than 99 percent of the visible universe.[83]

Galileo's PLS measured the composition, energy, temperature, motion, and three-dimensional distribution of low-energy plasma ions passing by the spacecraft. Its objectives were to 1) determine the sources of magnetospheric plasma; 2) investigate the plasma's interaction with Jupiter's moons; 3) analyze the plasma's role in generating energetic charged particles; 4) determine the nature of a sheet of current in Jupiter's equatorial plane; and 5) evaluate the effects that rotational forces, electric currents aligned with magnetic fields, and other phenomena have on Jovian magnetospheric dynamics.[84]

Observations by the Pioneer and Voyager flybys had shown Jupiter's magnetosphere to be an enormous reservoir of charged particles. Galileo added to the data collected by these missions but had the advantage of spending years, rather than weeks, in the vicinity of Jupiter. Galileo's PLS was also far more advanced than the instruments of the other spacecraft. The Galileo PLS had the ability to identify chemical compositions of ion populations in Jupiter's magnetosphere.

Galileo's plasma instrument used two electrostatic analyzers to measure the energy per unit charge for electrons and positive ions (see figure 5.11). These analyzers also measured the particles' directions of flow. In addition, three miniature mass spectrometers measured the mass per unit charge of the plasmas' positive ions, which allowed identification of ions' species.[85]

Plasma Wave Subsystem. Plasma waves are oscillations that involve both the charged particles of a plasma and the electric and magnetic fields associated with the plasma. There are two basic types of plasma waves: electrostatic waves, which resemble

---

[81] "Magnetometers"; "Galileo Magnetic Field Investigation"; "MAG—Magnetometer."

[82] "MAG—Magnetometer."

[83] "Perspectives on Plasmas—Basics," *Plasmas International*, 1999, *http://www.plasmas.org/basics.htm*; "What Is Plasma?" Coalition for Plasma Science, 2000, *http://www.plasmacoalition.org/what.htm*; "Plasma Investigation," *Galileo Messenger* (September 1985).

[84] "PLS—Plasma Subsystem," *Galileo's Science Instruments*, *http://www.jpl.nasa.gov/galileo/instruments/pls.html* (accessed 15 March 2000).

[85] "Plasma Investigation."

sound waves, and electromagnetic waves, which are more similar to light waves. The PWS was designed to analyze the properties of varying fields in these waves. The PWS measured electric field oscillations in the plasma waves over frequencies ranging from 5 hertz (cycles per second) to 5.6 megahertz, as well as magnetic field oscillations over the range of 5 hertz to 160 kilohertz. Plasma electric fields were measured with an electric dipole antenna, a simple design often used on Earth for radio reception. The magnetic fields were analyzed using two "search coil" magnetic antennas.[86]

When Voyager passed by Jupiter, it detected several kinds of plasma wave phenomena. The Galileo mission had a chance to study these phenomena in more detail and sought to understand how the waves were generated. The spacecraft made three-dimensional measurements of the areas where the waves occurred, which included Jupiter's equatorial plane and the plasma torus that encircles the planet and encloses the satellite Io. Jupiter and its nearest moon, Io, are strongly coupled in various ways that do not occur in our Earth-Moon system. By studying Jovian plasma wave phenomena, Galileo was able to improve our understanding of this relationship between Jupiter and Io. In particular, Galileo data shed light on how energy is transferred among Io, the plasma torus, and Jupiter's ionosphere. The rate of energy transfer is sizable—10^{12} watts of power flows from Io to Jupiter. Galileo also collected data on the nature of certain Jovian polar auroras, which occur along magnetic field lines that pass through both the planet and Io.[87]

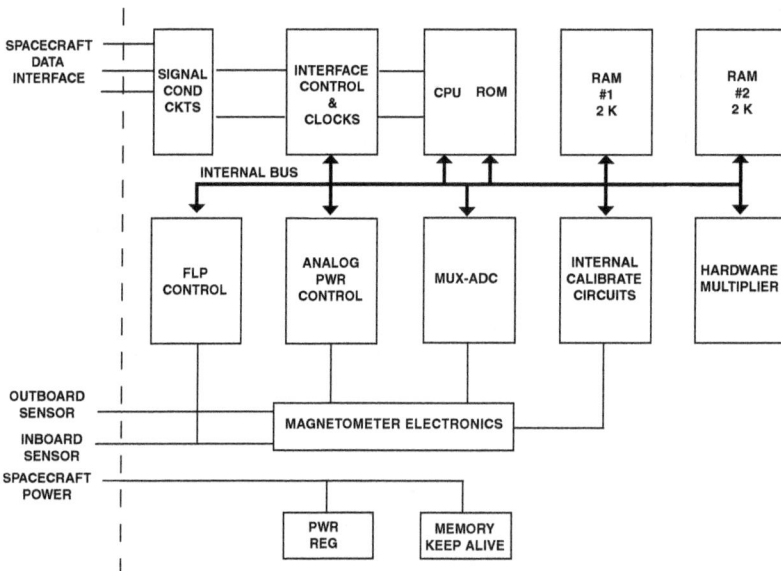

Functional block diagram of the experiment.

Figure 5.11. **The plasma subsystem measured energy and mass per unit charge of plasma particles and was able to identify particular chemical species.**

---

[86] "PWS—Plasma Wave Subsystem," *Galileo's Science Instruments*, http://www.jpl.nasa.gov/galileo/instruments/pws. html (accessed 15 March 2000); "Plasma Wave Investigation," *Galileo Messenger* (July 1986).

[87] "Plasma Wave Investigation."

As was mentioned above, all four of Jupiter's Galilean satellites—Io, Europa, Ganymede, and Callisto—lie within the planet's magnetosphere. This situation gives the plasma within this magnetosphere the opportunity to interact with each of the Galilean moons. Jupiter's rapid rotation and strong magnetic field serve to accelerate plasmas in the magnetosphere to great speeds. Wakes and waves are formed as the plasma is swept past slower-moving satellites. Factors that influence plasma-moon interactions include a moon's size, magnetic moment, and surface properties, as well as the nature of its atmosphere (or lack of an atmosphere). The flow velocity and magnetic field strength of the plasma are also important. The data collected by Galileo's PWS were critical for understanding the complex interactions in the magnetospheric region.[88]

Radio Science. Three separate radio-science experiments were planned for Galileo that used the Orbiter's radio equipment for functions other than communication with Earth. The experiments were to study celestial mechanics, search for gravity waves, and measure radio-wave propagation characteristics. Unlike the mission's other investigations, approximately 80 percent of the measuring equipment for radio-science experiments was located on Earth, at our Deep Space Network (DSN) stations, rather than on the spacecraft. This equipment included arrays of ultrasensitive antennas and receiving systems.[89]

The celestial mechanics studies sought to detect tiny changes in Galileo's trajectory caused by gravitational fields. These trajectory changes could be measured by the small increments they caused in the frequency of the radio signal transmitted by the spacecraft and received on Earth. The amount of frequency increment was dependent on the magnitude of the spacecraft's velocity change. From the velocity changes caused by Galileo's close passage to Jupiter or one of its satellites, scientists were able to determine details regarding the planet's gravitational field and mass. Also, features regarding the internal distribution of that mass could be surmised. For instance, were there distinct layers of rock and ice within a satellite, or was it homogeneous throughout? Small increments in signal frequencies shed light on such issues. In addition, the data enabled Galileo scientists to determine more exactly the orbits of Jupiter and its satellites.[90]

Another radio-science experiment searched for gravitational waves. These are extremely faint perturbations in the interstellar medium caused by catastrophic events such as the collapse of a star. Normally, there is a constant level of gravitational-wave energy in space. This is occasionally overwhelmed by some momentous stellar or galactic event, which generates very long-wavelength, low-frequency oscillations with periods of about 20 minutes.

Gravity waves perturb electromagnetic radiation that they encounter, and it is this property that was used by Galileo to search for such waves. If two black holes were to collide, the event would generate gravitational waves rippling through interstellar space. These waves would eventually encounter and interfere with Galileo's radio signal to Earth, and receiving stations on Earth would pick up low-frequency "pops" indicative of this interference. The Galileo mission presented a rare opportunity because the radio

---

[88] Ibid.

[89] "Radio Science," *Galileo Messenger* (December 1986).

[90] "Radio Science," *Galileo's Science Instruments, http://www.jpl.nasa.gov/galileo/instruments/rs.html* (accessed 15 March 2000); "Radio Science," *Galileo Messenger.*

signal stretching from spacecraft to Earth constituted a gravitational-wave detector that was hundreds of millions of kilometers long. A detector of such enormous scale was critical for picking up gravity-wave phenomena. The detection of a gravity-wave signal could be confirmed by other spacecraft in the vicinity or by finding visual confirmation of the causal event using Earth-based telescopes. It would take a very large event, however, to generate gravitational waves capable of detection. John Anderson, JPL's team leader for the celestial mechanics experiment, estimated that the collapse of a star with a mass 10 million times larger than the Sun might be necessary to produce a measurable phenomenon.[91]

Galileo's radio-wave propagation studies measured the minute changes in frequency, power, time of flight to Earth, and polarization of Galileo's radio signals that resulted from their passages through the atmospheres and ionospheres of Jupiter and its satellites. Atmospheres refract as well as scatter the radio signals that pass through them. The process of refraction bends and slows a radio signal, while scattering diffuses the electromagnetic waves of the signal. When the signal was received by a DSN station on Earth, these modifications to the signal showed up as amplitude and frequency changes. Analyses of the changes yielded Jovian system atmospheric information that included refractive indices and pressure and temperature profiles. Ionospheric electron densities were also estimated from these measurements.[92]

Galileo conducted a radio-wave experiment that studied the solar corona. When Galileo was on the opposite side of the Sun from Earth, a radio signal sent by the spacecraft had to pass through the solar corona in order to reach DSN antennas. Changes in the signal as it passed through the corona gave scientists data on its nature.

Another radio-science experiment was conducted with the aid of the Probe. By monitoring its radio signal during its descent into the Jovian atmosphere, the Galileo team was able to collect information on wind-speed changes with respect to altitude and on the direction of heat flow—downward or upward—through the atmosphere.[93]

Partially in preparation for radio transmissions from Galileo, NASA built several sensitive new antennas and receiving systems at its DSN sites. These were able to detect the faint signal from Galileo continuously, which the old equipment would have had trouble doing. By the time it reached Earth, Galileo's signal was so minuscule as to be over one billion times smaller than the sound of a transistor radio playing in New York—as heard from Los Angeles. Moreover, the phenomena that the radio-science experiment observed were also minute. Detecting these phenomena was comparable to measuring the distance from New York to Los Angeles with an error no greater than the diameter of a human hair.[94]

Orbiter's Remote Sensing Instruments

In order to satisfy the objectives stated in NASA's Announcement of Opportunity for Outer Planets, the Orbiter's remote sensing instruments (see table 5.3) needed to determine

---

[91] "Radio Science," *Galileo Messenger.*

[92] "Radio Science," *Galileo's Science Instruments*; "Radio Science," *Galileo Messenger.*

[93] "Radio Science," *Galileo Messenger.*

[94] Ibid.

Jovian satellite compositions and surface morphologies, as well as making observations of Jupiter's atmospheric characteristics. Central to these investigations was a means of capturing high-resolution optical images. As NASA had successfully done on many of its previous missions, it decided to keep control of the design and fabrication of the Orbiter's imaging system rather than soliciting proposals for its development. The Agency instead solicited proposals from individual investigators who wanted to join the team that would use the imaging system. The selected investigator team did, however, get to refine the requirements for the imaging system's color capability and some of its other features.[95]

The major instrument-selection issue with which NASA dealt for remote sensing experiments was what types of spectrometers were most appropriate to include on the mission. Spectrometers analyze reflected and emitted light, both in visible and nonvisible (infrared and ultraviolet) parts of the spectrum. Spectral data can give valuable information on the chemical compositions of planetary surface features and atmospheres.

NASA considered developing an in-house spectrometer for the Orbiter, as it was doing for the imaging system. NASA had in mind a variant of the spectrometer flown on Voyager. This instrument, designed primarily to analyze atmospheric gases, had high resolution in the "far" infrared range (so named because its wavelengths were significantly longer than those of visible light). This part of the light spectrum was useful for studying thermal emissions, such as those from a planetary atmosphere.

The scientific community also made cases for ultraviolet and infrared reflection instruments. Those interested in satellite surface analysis thought that a reflection spectrometer operating in the near-infrared region (the section of the spectrum that begins with wavelengths just slightly longer than those of visible red light) was needed to achieve the Announcement of Opportunity's objective to "determine the surface composition of the satellites." Torrence Johnson remembers at least two proposals for a near-infrared instrument being submitted, one of which was from a team that he headed. His team's concept for a near infrared mapping spectrometer (NIMS) won out over other proposals and convinced NASA to develop it rather than supplying a Voyager-type spectrometer for the mission.[96, 97]

The mission still needed a far-infrared instrument to study thermal emissions from the Jovian system. What NASA selected was a radiometer, an instrument that quantitatively measured electromagnetic radiation, and included it as one of the instruments in the photopolarimeter radiometer (PPR) experiment. The radiometer specialized in analyzing the intensity of radiant thermal energy, and that was very useful for "sounding" the atmosphere in several infrared bands of the spectrum, as well as for measuring temperatures of satellite surfaces. A radiometer did not have the strong chemical composition analysis capabilities that a spectrometer would have had, but NASA mission planners did not perceive such capabilities as necessary. NASA's thinking was that the Probe's in situ compositional experiments would provide in-depth analyses of Jupiter's atmospheric composition and that the NIMS instrument would perform major

---

[95] Announcement of Opportunity; Johnson e-mail.

[96] Announcement of Opportunity; Johnson e-mail.

[97] Spectroscopy, the study of spectra, is based on the idea that each chemical substance has its own characteristic spectrum. See "Spectroscopy," *Microsoft Encarta Online Encyclopedia*, 2001, *http://encarta.msn.com*. Infrared light consists of electromagnetic waves whose wavelengths are too long for the human eye to see, whereas ultraviolet light has wavelengths to short to be detected by our eyes. See "Infrared Radiation" and "Ultraviolet Radiation," both in *Microsoft Encarta Online Encyclopedia*, 2001, *http://encarta.msn.com*.

compositional studies of the satellites. (NIMS actually ended up supplying valuable atmospheric compositional information as well.)[98]

Mission planners also selected an ultraviolet spectrometer (UVS) because of its capability for making measurements of Jupiter's high atmosphere, analyzing the ionized gas regimes surrounding the planet, and determining whether the satellites have atmospheres.[99]

Table 5.3. **Orbiter remote sensing equipment.**

INSTRUMENT	FUNCTION	MASS IN KILOGRAMS (POUNDS)	POWER DRAW IN WATTS	PRINCIPAL INVESTIGATOR(S) AND INSTITUTION(S)
Near infrared mapping spectrometer (NIMS)	Determined satellite chemical compositions. Also analyzed Jovian atmospheric compositions.	18 (40)	12	Robert Carlson, JPL
Photopolarimeter Radiometer (PPR)	Measured radiant thermal energy intensities. Determined distribution of Jupiter's cloud and haze particles. Analyzed Jupiter's energy budget.	5 (11)	11	J. E. Hansen, Goddard Institute for Space Studies, NY
Solid state imaging (SSI) camera	Galileo's main imaging device. Especially useful for studying satellite geology.	30 (65)	15	Michael Belton, National Optical Astronomy Observatories
Ultraviolet spectrometer (UVS)	Located cloud layers and analyzed characteristics of cloud particles. Analyzed Jupiter's high atmosphere and surrounding ionized gas regions. Searched for satellite atmospheres.	4.2 (9.2)	4.5	C. W. Hord and Ian Stewart, University of Colorado

Sources:

1. "Galileo's Science Instruments," *Galileo Home Page*, *http://www.jpl.nasa.gov/galileo/instruments/index.html* (accessed 19 February 2001).

2. Institute of Geophysics and Planetary Physics, UCLA, "Galileo Magnetometer Project Overview," *Galileo Magnetometer Team's Homepage*, *http://www.igpp.ucla.edu/galileo/framego.htm* (accessed 2 January 2001).

3. "Ultraviolet Spectrometer," *Galileo Messenger* (June 1983).

4. "The PPR: Finding More Than Meets the Eye," *Galileo Messenger* (August 1989).

Near Infrared Mapping Spectrometer. NIMS analyzed Jovian cloud strata and temporal and spatial atmospheric composition variations. It determined temperature

---

[98] Johnson e-mail; NASA Earth Observatory, "Radiometer," Glossary, *http://earthobservatory.nasa.gov:81/Library/ glossary.php3?xref=radiometer* (accessed 9 August 2003); "Radiometer," *Microsoft Encarta Online Encyclopedia*, 2001, *http://encarta.msn.com*.

[99] Johnson e-mail.

versus pressure in the region between 1 and 5 bars of pressure. (As noted earlier, a bar is equal to Earth's atmospheric pressure at sea level.) NIMS also tried to answer the question, what are the Jovian moons made of? NIMS was able to recognize many common minerals, including silicates, nitrates, and carbonates. One product of the observations was a detailed geological map of each satellite. These data are expected to shed light on the processes governing solar system evolution.[100]

Galileo's NIMS (figure 5.12) was a brand-new instrument that had never flown in space before. It was designed and built to be sensitive to the part of the electromagnetic

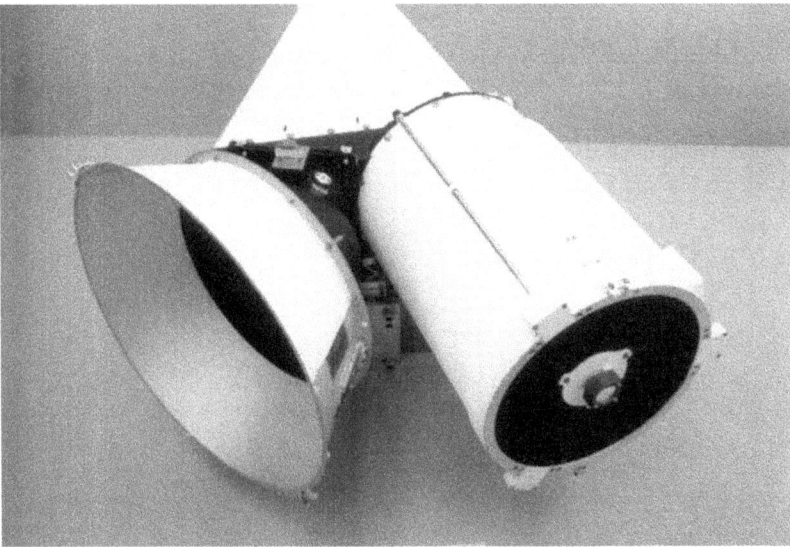

Figure 5.12. **NIMS was especially useful for studying gaseous chemical species and reflections from solid matter. The cone on the left was a radiative cooler. (JPL image 382-2165A)**

spectrum that begins with the deepest red that we can see (whose wavelength is about 0.7 micrometers) and then moves out of the visible to wavelengths as long as 5 micrometers. The near-infrared spectrum is quite useful for analyzing gaseous chemical species, as well as light reflected off solid matter. NIMS was thus a powerful diagnostic tool for both Jupiter's gaseous atmosphere and the solid surfaces of its satellites.[101]

NIMS's main components were a telescope and a spectrometer. The telescope focused light on the spectrometer, which acted like a prism and divided the near-infrared component of the light up into individual wavelengths, or "colors." Most were not, however, colors that our eyes could see—their wavelengths were too long. An array of detectors measured the intensity of light at each individual wavelength. The relative intensities of different wavelengths of light reflected or emitted from a particular material constitute a unique "signature" of that substance. Large databases exist of characteristic spectral

[100] "Near Infrared Mapping Spectrometer," *Galileo Messenger* (July 1981).

[101] Ibid.

signatures from many thousands of different materials. These were used to interpret the NIMS data observed by Galileo and identify various constituents of the Jovian atmosphere and satellites. The overall amplitude of a signal received by NIMS was also of use, for it gave an indication of the quantity of the substance present.[102]

NIMS was built at JPL, with scientists on the team from the United States, England, and France. Because warm surfaces in the instrument would have emitted infrared (thermal) radiation that might have interfered with infrared signals from the objects being observed, NIMS's detector array was lowered to a cryogenic temperature of 64 K (the equivalent of 64 Celsius degrees above absolute zero) using a radiative cooler. This action minimized background infrared radiation.[103]

Photopolarimeter Radiometer. The PPR was designed to measure intensity and polarization of light and was used during the Galileo mission to determine the following:

- Variations in intensity of reflected sunlight and emitted thermal radiation from both Jupiter and its satellites.

- Vertical and horizontal distribution of Jupiter's cloud and haze particles, as well as their shapes, sizes, and refractive indices.

- Jupiter's energy budget.

- The thermal structure of the Jovian atmosphere and the vertical distribution of absorbed solar energy within it.

- Photometric, polarimetric, and radiometric properties of the Galilean moons (Io, Europa, Ganymede, and Callisto).[104]

The PPR (see figure 5.13) was an instrument system with several functions. Its photometer measured the intensity of incoming light and used various colored filters to pass only light in the desired spectral band. The signal intensity that the PPR observed could be correlated with the abundance of a particular material present. Of the seven bands of light that the photometer observed, one was in a region of the spectrum in which methane strongly absorbs light, and another was in a region where ammonia strongly absorbs light.[105]

The PPR's polarimetric measurements determined the polarization characteristics of reflected sunlight. Polarization is the suppression of light-wave vibrations in certain directions, coupled with the allowance of light-wave vibrations in other directions (this is what polarized sunglasses do). Polarization is also related to the times at which vibrations

---

[102] Ibid.

[103] "Near Infrared Mapping Spectrometer"; "NIMS—Near Infrared Mapping Spectrometer," *Galileo's Science Instruments*, *http://www.jpl.nasa.gov/galileo/instruments/nims.html* (accessed 3 January 2001).

[104] "Photopolarimeter-Radiometer (PPR)," *NSSDC Master Catalog Display, Experiment*, *http://nssdc.gsfc.nasa.gov/cgi-bin/database/www-nmc?89-084B-08* (accessed 15 March 2000).

[105] "PPR—Photopolarimeter-Radiometer," Galileo's Science Instruments, *http://www.jpl.nasa.gov/galileo/instruments/ppr.html* (accessed 15 March 2000); "The PPR: Finding More Than Meets the Eye," *Galileo Messenger* (August 1989).

in different directions reach their maxima and minima. For instance, in a particular beam of light, vertical vibrations may be "in phase" with those in the horizontal direction, meaning that they both reach their maxima at the same time. Alternatively, vertical vibrations may peak a fraction of a cycle before or after horizontal ones. A "phase angle" describes the amount by which vibrations in one direction are out of phase with those in another.

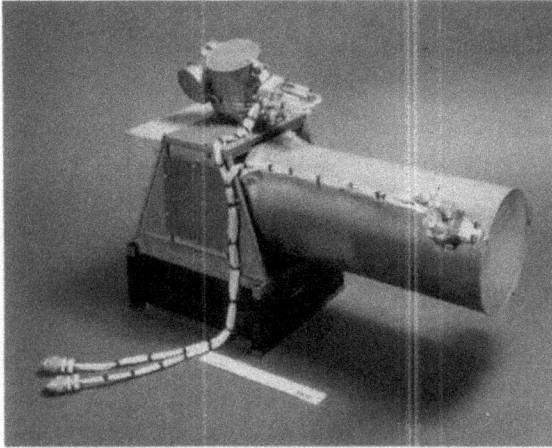

Figure 5.13. **Photopolarimeter radiometer. (Image number 23-1158A)**

To measure polarization, the PPR passed incoming light through a prism, which separated it into beams of vertically and horizontally polarized components. These beams were directed onto a pair of photodiode detectors, which measured the intensities of the two polarized components by converting their light into electric current.[106]

In the Jovian atmosphere, as in that of Earth, cloud particles and atmospheric molecules play important roles in scattering and polarizing sunlight, as well as in determining the color of the atmosphere. On Earth, the blue of the sky is caused by the scattering of the short-wavelength blue components of white sunlight by particles suspended in the atmosphere. The scattering of these blue light waves is more pronounced than the scattering of waves with longer visible wavelengths, such as those of red light. This is why Earth's sky does not appear red. The scattering of light by Jupiter's atmospheric molecules also favors the blue wavelengths, as well as the violet wavelengths. This type of frequency-dependent scattering by small particles is called Rayleigh scattering; it occurs only when the atmospheric particles doing the scattering are very small compared to the wavelength of the light.[107]

One of the PPR's science goals was to measure the relative importance of Rayleigh scattering in the Jovian atmosphere above cloud-top level, from which an estimate of the amount of gas in that region could be made, as well as identification of cloud-top height variations.

The PPR's radiometry function was to measure thermal infrared emissions. The PPR observed in seven radiometry bands, which corresponded to different regions of the infrared spectrum that are absorbed, to varying extents, by the atmosphere. Emissions in

---

[106] "The PPR: Finding More Than Meets the Eye."

[107] Georgia State University, "Rayleigh Scattering," *HyperPhysics* Web site, 2000, *http://hyperphysics.phy-astr.gsu.edu/hbase/atmos/blusky.html*; "Color," *Microsoft Encarta Online Encyclopedia*, 2001, *http://encarta.msn.com*; "Rayleigh Scattering," *Encarta World English Dictionary* (North American Edition), 2001, *http://dictionary.msn.com/find/print.asp?refid=1861727369&search=&wwi=8274*; "The PPR: Finding More Than Meets the Eye."

those parts of the spectrum that are absorbed poorly could originate deeper down in the atmosphere and still be observed by the Orbiter. Thus, depending on which emissions it was observing, the PPR was able to detect signals from various atmospheric depths. This capability allowed the PPR to analyze a range of different levels of the Jovian atmosphere. From the characteristics of the thermal radiation observed, atmospheric temperatures at each depth could be estimated. The PPR could also make direct temperature measurements of the surfaces of Jupiter's satellites. It was the only one of Galileo's instruments with this capability.[108]

Solid State Imaging (SSI) Camera. The SSI camera (figures 5.14 and 5.15) was used for a wide range of science objectives, including the following:

- Analyzing Jupiter's atmospheric dynamics and cloud structure.

- Measuring sizes, shapes, and features of Jupiter's moons.

- Determining the geological processes that formed the satellites' surfaces. To do this, the SSI mapped those surfaces at resolutions finer than 1 kilometer (0.6 mile) over a range of viewing angles and lighting conditions.

- Identifying and mapping ice and mineral distributions on the satellites.

- Searching for auroral and other atmospheric emission phenomena on Jupiter's night side, in the region around the planet, and on its satellite surfaces.

The SSI camera needed to study both atmospheric motion and geological formations, which required a high-resolution, large-format design that could pick out and identify very small features. But the SSI was also used for composition analyses of satellite surfaces, which meant that it needed to employ a range of spectral filters to help in mineral identification. Analyses of low-light phenomena such as auroras, lightning, and the Jovian rings required a camera with an extremely sensitive detector and an optical system with very little scattering of light so as not to obscure faint images.

To prevent damage to the SSI from Jupiter's intense radiation, the instrument needed to be heavily shielded. It also had to be protected from propellant byproducts that might contaminate it during launch or in flight. Finally, the SSI had to be made as light as possible and require only limited electrical power to operate.[109]

To meet the above requirements, the Orbiter used a 1,500-millimeter (60-inch) focal-length Cassegrain reflecting telescope with a 176-millimeter (7-inch) aperture that had operated successfully on Voyager. Light from an image was collected by the telescope's primary mirror and directed to a smaller, secondary mirror. This mirror then channeled the light through a hole in the center of the primary mirror and onto a solid-state silicon image-sensor array called a "charge-coupled device" (CCD). The CCD had 640,000 individual sensors, or "pixels," to provide high-resolution images. Except where light entered, the CCD was surrounded by a 1-centimeter-thick (0.4-inch) tantalum shield to protect it from the

---

[108] "The PPR: Finding More Than Meets the Eye"; "PPR—Photopolarimeter-Radiometer."

[109] "Solid-State Imaging," *Galileo Messenger* (December 1984).

intense radiation fields within Jupiter's magnetosphere. Data collected by the CCD were digitized and sent to the Orbiter's tape recorder for storage, eventually to be played back, edited, and compressed by the spacecraft's computers, then sent to Earth. After the data were received by tracking stations on Earth, they were sent to JPL's image-reconstruction equipment.[110]

The SSI used an eight-position filter wheel to obtain images of the same scenes in different spectral bands. These images were combined electronically on Earth to produce color images. The spectral response of the SSI ranged from light with wavelengths of about 0.4 micrometers to light of 1.1 micrometers (a micrometer is one-millionth of a meter). This range covered both the visible light spectrum (0.4 to 0.7 micrometers) and part of the near-infrared region.[111]

Texas Instruments provided the CCD for the SSI camera. The SSI used RCA microprocessors and contained 600 integrated circuits. The telescope, shutter, and filter designs were inherited from Voyager, but NASA improved its design to better reject scattered light, which could interfere with the imaging of faint objects. The electronics chassis was machined at JPL.[112]

The SSI and three other optical remote sensing instruments (the near infrared mapping spectrometer, ultraviolet spectrometer, and photopolarimeter radiometer) were mounted on a movable "scan platform" connected to the despun section of the Orbiter. The scan platform could be oriented so as to point the instruments toward the object being studied. The optical axes of all four instruments were aligned so that they all looked toward the same areas at the same time. The simultaneous outputs from the four instruments could then be correlated so as to interpret various phenomena better.[113]

Ultraviolet Spectrometer. The UVS studied ultraviolet light from the Jovian system, which was especially helpful in understanding properties of Jupiter's high atmosphere, the compositions of the planet's satellites, and the nature of Io's plasma torus (the donut-shaped cloud of ionized gas originating from the moon and encircling Jupiter). Ultraviolet (UV) light includes wavelengths shorter than those of visible light but longer than those of x rays. The shorter the wavelength is, the higher the light wave's frequency and energy are. As an illustration of this principle, UV light from the Sun is able to burn our skin. X rays, which have shorter wavelengths and are more energetic than UV light, can penetrate our bodies and do serious damage. Ultraviolet wavelengths begin at the wavelength of violet light, 4,000 angstroms (an angstrom is one ten-millionth of a millimeter), and range to as short as 150 angstroms, where the x-ray spectrum begins. The Orbiter's UVS had detectors that could observe light in all but the shortest wavelengths of the ultraviolet spectrum.[114]

---

[110] "Solid-State Imaging"; "SSI—Solid-State Imaging," *Galileo's Science Instruments, http://www.jpl.nasa.gov/galileo/ instruments/ssi.html* (accessed 9 January 2001).

[111] "Solid-State Imaging"; "SSI—Solid-State Imaging."

[112] "Solid-State Imaging"; "SSI—Solid-State Imaging."

[113] "Solid-State Imaging".

[114] "UVS—Ultraviolet Spectrometer," *Galileo's Science Instruments, http://www.jpl.nasa.gov/galileo/instruments/uvs. html* (accessed 15 March 2000); "Ultraviolet Spectrometer," *Galileo Messenger* (June 1983); "Ultraviolet Radiation," *Microsoft Encarta Online Encyclopedia*, 2000, *http://encarta.msn.com*.

Figure 5.14. **The Orbiter's SSI. (JPL image number 352-8284)**

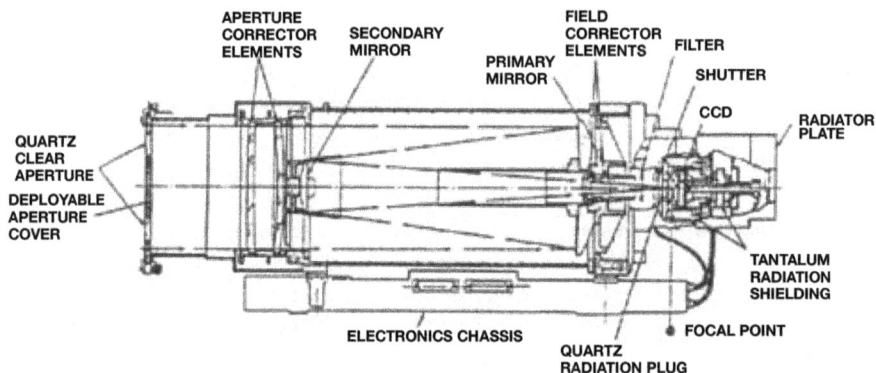

Figure 5.15. **Cutaway schematic of the SSI. Shows principal optical components, charge-coupled device detector, particle radiation shield, and front aperture. (JPL 230-537)**

The UVS searched for evidence of certain complex hydrocarbon molecules in the Jovian atmosphere that, on Earth, were "building blocks" for life. In its investigation of the Galilean satellites Io, Europa, Ganymede, and Callisto, the UVS searched for evidence of atmospheres. Such evidence could indicate that volatiles were escaping from the satellites and that their compositions were still evolving. Reflected light from the satellites was picked up by the UVS and analyzed for signs of ammonia, ozone, and sulfur dioxide.

Particles from Io's volcanic eruptions are believed to be the source of its ionized plasma torus. UVS measurements, as well as observations by Orbiter's fields-and-particles instruments, of electrons and ions in the torus provided comprehensive data on Io's evolution and its relationship with Jupiter's magnetic field.

The UVS was developed and built by the University of Colorado's Laboratory for Atmospheric and Space Physics. The UVS team was led by principal investigator Charles Hord. Although the experiment's main focus was to study the Jovian system, UVS measurements were also carried out during Galileo's six-year voyage to Jupiter in order to obtain data at Venus, Earth and its Moon, and two asteroids.[115]

---

[115] "Ultraviolet Spectrometer"; "Galileo," *Laboratory for Atmospheric and Space Physics, http://lasp.colorado.edu/ programs_missions/present/galileo/* (accessed 11 January 2001).

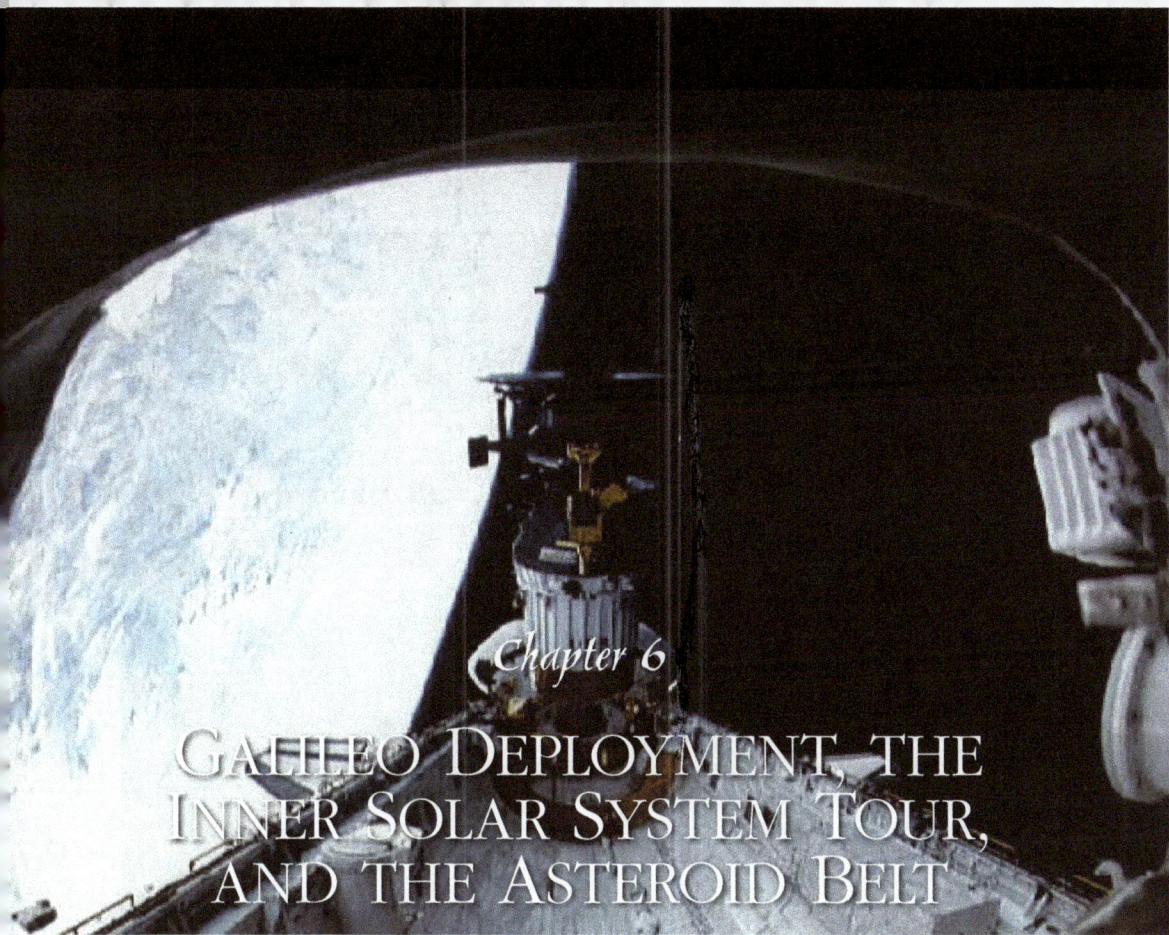

Chapter 6

GALILEO DEPLOYMENT, THE INNER SOLAR SYSTEM TOUR, AND THE ASTEROID BELT

W H E N THE SPACE SHUTTLE *ATLANTIS* LIFTED OFF FROM Kennedy Space Center at 12:53 p.m. EDT on 18 October 1989, it began the Galileo Prime Mission, which explored Venus, two asteroids, a comet, and the Jovian system. The Shuttle carried in its cargo bay both the Galileo spacecraft and the Inertial Upper Stage (IUS) rocket to which the spacecraft was connected, bringing them up to a low-Earth orbit. At 7:15 p.m. EDT, the Shuttle crew deployed the 38,500-pound package that included the Galileo spacecraft and its IUS booster (see figure 6.1). As this spacecraft-rocket combination emerged from the Shuttle with its booms and antennas still folded, it "resembled a butterfly emerging from its cocoon, with all its appendages."[1]

Once the spacecraft and IUS were deployed, the Shuttle flew away to a safe distance. An hour after deployment, the IUS fired, slowing Galileo's orbital speed around the Sun by 10,000 mph. As a result, Galileo began falling toward the inner solar system and its first destination: Venus.[2]

The IUS booster rocket, which was supplied by the U.S. Air Force, consisted of two solid-fueled sections. The first, containing 21,400 pounds (about 9,700 kilograms) of

---

[1] C. T. Russell, foreword to "The Galileo Mission," *Space Science Reviews* 60, nos. 1–4 (1992); Everett Booth, "Galileo: The Jupiter Orbiter/Space Probe Mission Report," December 1999, p. 2, JPL internal document, Galileo—Meltzer Sources, folder 18522, NASA Historical Reference Collection, Washington, DC; Kathy Sawyer, "Galileo's 6-Year Flight to Jupiter Begins," *Washington Post* (19 October 1989): A4.

[2] Russell, foreword to "The Galileo Mission"; Booth, "Galileo: The Jupiter Orbiter/Space Probe"; Sawyer, "Galileo's 6-Year Flight."

Figure 6.1. **The deployed Galileo spacecraft and Inertial Upper Stage booster rocket. Note the two circular sun shields installed to protect the spacecraft's sensitive equipment. (Image number S89-42940)**

propellant, lifted Galileo out of low-Earth orbit while the smaller second stage, containing 6,000 pounds (about 2,700 kilograms) of propellant, "injected" Galileo into its interplanetary Venus-Earth-Earth gravity-assist (VEEGA) trajectory. After deployment from the Shuttle, the IUS received its commands from Onizuka Air Force Base, located south of San Francisco and very close to NASA Ames Research Center. In the event of an IUS malfunction, Onizuka could issue corrective commands.[3]

Checkout of Galileo Equipment and First Observations

Shortly after launch, the Galileo flight team, whose members were located at Kennedy Space Center, Ames Research Center, and JPL, began remotely verifying the functionality of each of the spacecraft's systems and subsystems. They feared that vibrations during

---

[3] NASA Office of Space Science and Applications, Mission Operation Report (Report No. E-829-34-89-89-01, 1989), p. 1; Booth, "Galileo: The Jupiter Orbiter/Space Probe"; Sawyer, "Galileo's 6-Year Flight."

launch and the IUS burn might have "flipped" some of the craft's electrical relays, which would need to be reset. Such an occurrence would have been especially problematic for the Probe, so it was the first system examined. The Probe ran on nonrechargeable lithium batteries, and a flipped relay could have resulted in power being drained from the batteries. If the Probe's batteries were depleted, it would not be able to function at Jupiter, and a major part of the mission would be ruined. Fortunately, none of the Probe's relays had been flipped. Future tests were scheduled annually, with the next one to occur during the first Earth flyby in December 1990.[4]

A day after Galileo's 18 October launch, it had already journeyed almost 300,000 miles (480,000 kilometers) from its home planet and was speeding toward Venus at more than 9,000 miles per hour (14,500 kilometers per hour).[5] Several days after launch, Galileo's mission operations team began sending commands to Galileo. In general, the spacecraft was operated by command sequences sent from Earth that were then stored for later execution. "Real-time" commands could also be sent as needed, however, for quicker response. By 27 October 1989, the mission operations team had sent over 250 real-time commands.[6]

In the first few days of the mission, the spacecraft's heavy ion counter was powered on and began detecting high-energy ions from solar flares. Emissions from solar flares were of immediate concern because they were energetic enough to disrupt sensitive equipment aboard the spacecraft, causing what was termed a single event upset (SEU). Fortunately, no SEUs were detected by Galileo. SEU incidents that had plagued earlier missions had been thoroughly studied, and special shielding and hardening of electronic components had been included in Galileo's design.[7]

In late October, Galileo's solid state imaging and near infrared mapping spectrometer instruments were successfully checked out. Data from these tests were stored on Galileo's data memory subsystems tape recorder, to be played back later and transmitted to Earth as the spacecraft's telemetry link capabilities allowed. The tape recorder would be used extensively during each planetary, satellite, and asteroid encounter in order to record and preserve valuable observation data until they could be forwarded to Earth. Galileo was being tracked by the Deep Space Network (DSN) system of antennas in California, Australia, and Spain.[8]

Testing to be performed on Galileo's flight to Venus included the characterization of spacecraft equipment. One of the first pieces of equipment to be tested was the attitude-control subsystem, which would be needed during trajectory correction maneuvers (TCMs) in early November. The remaining scientific instruments would all be checked out and turned on in time for the Venus flyby.[9]

---

[4] "Checking Out the Probe," *Galileo Messenger* (November 1989): 3.

[5] Kathy Sawyer, "Galileo Travels 292,500 Miles Toward Venus," *Washington Post* (20 October 1989): A24.

[6] "Galileo: Up to Date," *Galileo Messenger* (November 1989): 3.

[7] Ibid., p. 3.

[8] Ibid., p. 3; Douglas J. Mudgway, *Uplink-Downlink: A History of the Deep Space Network, 1957–1997* (Washington, DC: NASA SP-2001-4227, 2001).

[9] "Galileo: Up to Date," p. 3.

Trajectory Correction Maneuvers

Included in the mission were opportunities for several TCMs that would optimize the spacecraft's course and minimize propellant requirements. The first TCM after launch, called TCM-1, was initiated on 9 November 1989 and corrected for IUS trajectory-injection errors. The IUS burn had been extremely accurate so only a small correction to Galileo's speed was required—less than a 2-meter-per-second change. TCM-1 also included a preventative maneuver. When traveling inside Earth's orbit, Galileo had to orient its axis directly toward the Sun in order for the spacecraft's Sun shields to shade the equipment and keep it from overheating. This orientation introduced a potential problem: orbital corrections that involved a thrust component directly *away* from the Sun would be inefficient to perform, requiring excessive fuel consumption. TCM-1 was designed to reduce to less than 1 percent the probability that such an "anti-Sun" maneuver would be needed.[10]

TCM-2, performed on 22 December 1989, removed the very small maneuver-execution errors of TCM-1 and aimed Galileo at the desired Venus flyby location. The third scheduled maneuver (TCM-3) on the Earth-Venus trajectory leg turned out not to be needed because the aimpoint at Venus achieved by TCM-2 had been "virtually perfect."[11]

The Venus Encounter

Galileo reached and flew by Venus on 10 February 1990, less than four months after launch. The spacecraft instruments had been designed for a broad range of planetary investigations, and they took many useful measurements at Venus. But because Galileo was closer to the Sun than Earth orbit during this part of its trajectory, steps had to be taken to protect its equipment—in particular, its high-gain antenna (HGA)—from possible thermal damage. The HGA's parabolic-shaped dish had to be kept folded up like an umbrella during this time. The HGA also had to stay pointed toward the Sun so that a sun shield mounted at its tip remained in the right position to keep the HGA shaded and cool. This configuration prevented the spacecraft from rotating to optimum orientations for data collection. In addition, keeping the HGA dish closed made it impossible to send data back to Earth at the normal rate for planetary missions. Two small low-gain antennas (LGAs) had to be used instead, transmitting data at rates so slow that most of the data collected at Venus had to be stored on the spacecraft's tape recorder until Galileo flew close to Earth months later and could transmit more quickly. But the volume of data that could be stored on the tape recorder was limited, and this restricted the amount of data collected at Venus. In particular, it restricted the number of images collected because they required large amounts of digital storage space.[12]

---

[10] Louis A. D'Amario, Larry E. Bright, and Aron A. Wolf, "Galileo Trajectory Design," *Space Science Reviews* 60, nos. 1–4 (1992): 26–28.

[11] Ibid., p. 28.

[12] Torrence V. Johnson, Clayne M. Yeates, Richard Young, and James Dunne, "The Galileo Venus Encounter," *Science* 253 (27 September 1991): 1516; Michael J. S. Belton, Peter J. Gierasch, Michael D. Smith, Paul Helfenstein, Paul J. Schinder, James B. Pollack, Kathy A. Rages, Andrew P. Ingersoll, Kenneth P. Klaasen, Joseph Veverka, Clifford D. Anger, Michael H. Carr, Clark R. Chapman, Merton E. Davies, Fraser P. Fanale, Ronald Greeley, Richard Greenberg, James W. Head III, David Morrison, Gerhard Neukum, and Carl B. Pilcher, "Images from Galileo of the Venus Cloud Deck," *Science* 253 (27 September 1991): 1531–1536.

Venus had already been visited by many U.S. and Soviet craft, which had performed flybys and sent atmospheric probes, balloons, and landers toward and onto the planet's surface. Despite Galileo's limitations and the previous data that had been collected, however, the spacecraft was able to perform an interesting set of observations. Some of Galileo's instruments, such as its near-infrared mapping spectrometer (NIMS), had never been employed on a Venus-bound spacecraft. Others, such as Galileo's solid state imaging system, constituted major improvements over instruments that had been used before. Galileo extended observations that had previously been made of particle, field, and plasma wave phenomena and of the planet's limb region. As Galileo encountered Venus, the spacecraft's instruments worked in concert with ground-based observations from Earth, yielding more accurate information than had been available about Venus's plasma fields, cloud patterns, and lightning.[13]

Infrared Observations

Galileo's NIMS revealed the structure not only of Venus's clouds and dense atmosphere, but also of the surface beneath. NIMS was able to probe from a height of about 100 kilometers downward. For the first time, space scientists were able to map small-scale features in mid-latitude and equatorial regions. The NIMS experiment was particularly useful in studying the night side of Venus. Recent telescope observations had discovered that the Venusian atmosphere was transparent to limited parts of the infrared spectrum. Galileo's observations at these wavelengths were able to collect thermal radiation from deep in the planet's atmosphere, far below the visible clouds. As a result, Galileo was able to make night-side maps of Venus with resolutions three to six times better than those of Earth-based telescopes.[14]

Galileo was able to take night-side infrared images with high spatial resolution. The images indicated substantial lower cloud deck opacity variations that were centered at an altitude of 50 kilometers. The Galileo project team estimated that variations of 25 percent in the cloud deck opacity were caused by a narrow cloud band between an altitude of 48 kilometers (the bottom of the Venusian cloud deck) and roughly 52 kilometers. The fact that this region had such a marked effect on the opacity of the entire cloud deck suggested dramatically different conditions inside this narrow band from those of the rest of the cloud deck.[15]

In order to support the Galileo mission's investigation of Venus, Earth-based observations were carried out before, during, and after the spacecraft's flyby. They included thermal maps taken of Venus by the NASA Infrared Telescope Facility on Hawaii's Mauna Kea, as well as near-infrared images obtained by a global network of ground-based observatories. Astronomers on Earth saw a large dark cloud at low latitudes (within 40 degrees of the equator) that extended halfway around Venus and persisted throughout the

---

[13] Johnson et al., "The Galileo Venus Encounter," pp. 1516–1517.

[14] "Galileo's First Encounter: Venus is in View," *Galileo Messenger* (February 1990): 1, 4; Johnson et al., "The Galileo Venus Encounter," p. 1517; R. W. Carlson, K. H. Baines, Th. Encrenaz, F. W. Taylor, P. Drossart, L. W. Camp, J. B. Pollack, E. Lellouch, A. D. Collard, S. B. Calcutt, D. Grinspoon, P. R. Weissman, W. D. Smythe, A. C. Ocampo, G. E. Danielson, F. P. Fanale, T. V. Johnson, H. H. Kieffer, D. L. Matson, T. B. McCord, and L. A. Soderblom, "Galileo Infrared Imaging Spectroscopy Measurements at Venus," *Science* 253 (27 September 1991): 1541.

[15] Carlson et al., "Galileo Infrared Imaging," pp. 1541–1544.

observing period. Mid-latitudes (40 to 60 degrees) were characterized by bright east-west cloud bands. The highest latitudes observable, between 60 and 70 degrees, remained dark and featureless, indicating greater cloud opacity. The dynamics of sulfuric acid droplets in the planetwide Venusian cloud layer are believed to affect opacity significantly. For instance, the greater transparency of the atmosphere in some areas may be caused by downwelling currents that transport sulfuric acid droplets to warmer regions below the cloud base, where they evaporate.[16]

Solid State Imaging System Observations

Another Galileo instrument well-suited to investigating Venus was the solid state imaging (SSI) system, which provided high-resolution tracking of small ultraviolet features and also conducted an optical search for lightning. These observations addressed fundamental questions about Venusian meteorology, chemical composition, and atmospheric evolution. [17]

Solar Wind–Bow Shock Interactions

On its approach, Galileo repeatedly traversed Venus's magnetospheric bow shock, mapping its structure and that of the solar wind upstream of the shock. A planet's bow shock is the interface between the region of space influenced by the planet's magnetic field and the undisturbed, interplanetary solar wind (a current of ions, or plasma, streaming outward from the Sun). A planet's bow shock is very similar to the wave which appears in front of a boat as it passes through the water, or in front of a plane traveling at supersonic speeds. Planetary bow shocks slow the solar wind and help force it to go around the planet's magnetosphere (see figure 6.2).

The first hint that Galileo had crossed the Venusian bow shock was given by the craft's magnetometer, which registered sudden changes in magnetic field strengths. The plasma detector reported heating and compression of plasma flows, with electron temperatures rising by up to 20 percent and densities increasing by as high as a factor of three.[18]

The solar wind interacts very differently with Venus from how it interacts with Earth. Our planet's strong magnetic field deflects the solar wind while it is still far away from us, resulting in a bow shock that occurs at a distance of about 10 Earth radii (65,000

---

[16] D. Crisp, S. McMuldroch, S. K. Stephens, W. M. Sinton, B. Ragent, K.-W. Hodapp, R. G. Probst, L. R. Doyle, D. A. Allen, and J. Elias, "Ground-Based Near-Infrared Imaging Observations of Venus During the Galileo Encounter," *Science* 253 (27 September 1991): 1538–1541; Glenn S. Orton, John Caldwell, A. James Friedson, and Terry Z. Martin, "Middle Infrared Thermal Maps of Venus at the Time of the Galileo Encounter," *Science* 253 (27 September 1991): 1536–1538.

[17] Johnson et al., "The Galileo Venus Encounter," p. 1517.

[18] Margaret Galland Kivelson, "Serendipitous Science from Flyby of Secondary Targets: Galileo at Venus, Earth, and Asteroids; Ulysses at Jupiter," *Reviews of Geophysics,* supplement (July 1995): 565–566; Mona Kessel, author and curator, NASA Goddard Space Flight Center, "Collaborative Study of the Earth's Bow Shock," *NASA SpaceLink, http://spdf.gsfc.nasa.gov/bowshock/* (accessed 11 November 2000) (available in Galileo—Meltzer Sources, folder 18522, NASA Historical Reference Collection, Washington, DC); "Bow Shock," *http://www.windows.ucar.edu/cgi-bin/tour_def/glossary/bow_shock.html,* and "Magnetosphere," *http://www.windows.ucar.edu/tour/link=/glossary/magnetosphere.html&edu=high,* both from *Windows to the Universe,* copyright 1995–1999, The Regents of the University of Michigan, 2000 (accessed 15 November 2004), (available in Galileo—Meltzer Sources, folder 18522, NASA Historical Reference Collection, Washington, DC).

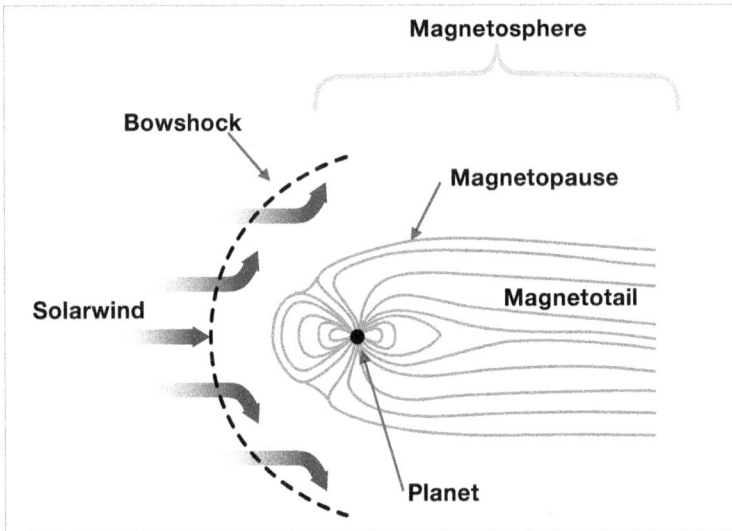

Figure 6.2. **Solar wind-bow shock-magnetosphere interaction for a planet such as Earth. Venus's bow shock occurs nearly on the planet's surface.**

kilometers or 40,000 miles) from the planet's center. Venus's weaker magnetic field cannot keep the solar wind so far away, and this results in a bow shock that occurs nearly on Venus's surface, or in other words, one Venus radius from the planet's center. (The radii of Venus and Earth are approximately the same.) The solar wind thus interacts directly with the Venusian ionosphere and dense atmosphere.[19]

The Search for Lightning

The spacecraft's plasma wave instrument searched for radio signals characteristic of lightning and recorded nine events in frequency ranges that suggested lightning was the source. The occurrence of lightning in Venus's atmosphere had been previously reported by others but was still a subject of controversy. The Soviet Venera spacecraft reported lightning at Venus, but there has always been a concern that the signals could actually have been electrostatic discharges from the spacecraft itself as it descended through the atmosphere. The U.S. Pioneer-Venus orbiter picked up very low-frequency "whistler" signals[20] on the order of 100 hertz (cycles per second) that were interpreted as being generated by a lightning event deep in the Venusian atmosphere. But skeptics believed that the whistlers could also have been locally generated plasma waves that had nothing to do with a lightning event.

[19] Kivelson, "Flybys of Planets and Asteroids," p. 565; L. A. Frank, W. R. Paterson, K. L. Ackerson, F. V. Coroniti, and V. M. Vasyliunas, "Plasma Observations at Venus with Galileo, *Science* 253 (27 September 1991): 1528; D. J. Williams, R. W. McEntire, S. M. Krimigis, E. C. Roelof, S. Jaskulek, B. Tossman, B. Wilken, W. Studemann, T. P. Armstrong, T. A. Fritz, L. J. Lanzerotti, and J. G. Roederer, "Energetic Particles at Venus: Galileo Results," *Science* 253 (27 September 1991): 1526.

[20] Whistlers are packets of electromagnetic waves in which the higher frequencies travel faster than the lower frequencies, reaching the receiving antenna first. When a whistler signal is played back at audible frequencies, it sounds like a whistle descending in pitch from high to low. Earth lightning is known to generate whistlers (Kivelson, "Flybys of Planets and Asteroids," p. 566).

A confirmation of lightning would be very important information because it would support the existence of convective storms in the Venusian atmosphere. It could also indicate active volcanism on the Venusian surface. On Earth, for instance, particles that rub together inside turbulent plumes of volcanic ash sometimes generate lightning; this might be happening on Venus as well. The Magellan spacecraft saw what appeared to be ash strewn downwind of volcanoes.[21]

Galileo's plasma wave detector was able to measure electric fields up to 5.6 megahertz (1 megahertz equals 10^6 cycles per second). This was well above the frequency range of locally generated plasma waves that could give a false positive indication of lightning. The nine plasma wave impulses that Galileo detected were in a frequency band that was able to propagate upward through Venus's atmosphere and ionosphere. Galileo's plasma team concluded that "there was no reasonable hypothesis other than lightning-generated whistler noise that could explain the observations."[22]

Due to the considerable interest in whether there is lightning activity in the Venusian atmosphere and what it would imply regarding atmospheric dynamics and possible volcanism, 10 pictures taken by Galileo's SSI camera were devoted to this search. No optical lightning images were obtained, but this result was far from conclusive. If Venus lightning flashes had the power ranges and frequency of occurrence characteristics of terrestrial lightning, then it was only "marginally possible" that the phenomenon would be detected by the SSI. The SSI observations serendipitously captured a star, κ-Geminorum, whose brightness was well known. This brightness was used to estimate that the minimum optical energy a lightning flash would need for the SSI to detect it was about three orders of magnitude greater than the energy of a typical terrestrial flash.[23]

Venus Cloud Deck

Galileo's SSI camera recorded useful images of the ever-changing states of Venus's cloud tops. Very little was known about small-spatial-scale, short-timescale dynamics of the Venusian atmosphere. The SSI's ability to image in the infrared allowed it to probe down to different cloud layers and follow the evolution of small dynamic phenomena in the atmosphere. Changes in cloud morphology were observed to occur far slower there than on Earth. Major changes in upper-level clouds observed by Galileo, as well as by the Pioneer spacecraft, evolved on timescales that ranged from days to years.[24]

---

[21] Johnson et al., "The Galileo Venus Encounter," p. 1517; M. G. Kivelson, C. F. Kennel, R. L. McPherron, C. T. Russell, D. J. Southwood, R. J. Walker, C. M. Hammond, K. K. Khurana, R. J. Strangeway, and P. J. Coleman, "Magnetic Field Studies of the Solar Wind Interaction with Venus from the Galileo Flyby," *Science* 253 (27 September 1991): 1518–1522; D. A. Gurnett, W. S. Kurth, A. Roux, R. Gendrin, C. F. Kennel, and S. J. Bolton, "Lightning and Plasma Wave Observations from the Galileo Flyby of Venus," *Science* 253 (27 September 1991): 1522; Richard A. Kerr, "Lightning Found on Venus At Last?" *Science* 253 (27 September 1991): 1492.

[22] Kivelson, "Serendipitous Science from Flybys of Secondary Targets," pp. 566–567.

[23] Belton et al., "Images from Galileo."

[24] Ibid., pp. 1531–1536.

Summary of Venus Tour

Important observations made by Galileo during its Venus flyby included the following:

- Probable detection of whistler phenomena that indicated the existence of lightning in the Venusian atmosphere.

- Interactions of Venus with the solar wind, which shed light on the nature of the planet's bow shock and plasma environment.

- Mapping of small-scale midlatitude and equatorial features of the Venusian surface.

- Collection of high-resolution night-side Venus data that indicated existence of a high-opacity cloud deck centered at a 50-kilometer altitude, as well as substantial differences between northern- and southern-hemisphere clouds.

- Cloud morphology observations that implied that significant changes in upper levels of clouds occur on timescales of years, far slower than on Earth.

Return to Earth

Galileo's post-Venus flyby of Earth in December 1990 marked the first time that a spacecraft had returned from interplanetary space, even if briefly. But to get to Earth from Venus, Galileo had to expend some of its fuel to perform another trajectory correction maneuver (TCM4). When JPL's Roger Diehl first discovered the VEEGA trajectory that would take Galileo to Jupiter, it appeared that the only time the craft could launch was in November 1989. Otherwise, the planets would not be lined up correctly. But continued research by Louis D'Amario and Dennis Byrnes revealed that a course correction between Venus and Earth widened Galileo's launch window to include most of October 1989 as well.[25]

TCM4 was the largest course correction that Galileo would have to perform to get to Jupiter, shifting the craft's velocity by 35 meters per second. It was performed in two parts: first from 9 through 12 April, then from 11 to 12 May. This and smaller TCMs put Galileo on the precise path for an Earth flyby and gravity-assist. The craft flew by Earth on 8 December at an altitude of 595 miles, only 5 miles higher than had been predicted. This meant that the predicted aiming point for the flyby was 99 percent accurate. The time of closest approach was predicted to be 12:34:34:00 Pacific standard time (PST) and proved to be only half a second off.[26]

---

[25] Peter N. Spotts, "NASA's Galileo Mission Clears Hurdles for Jupiter Voyage," *Christian Science Monitor* (3 December 1987): 5; Bill O'Neil, "From the Project Manager," *Galileo Messenger* (July 1990): 1.

[26] Bill O'Neil, "From the Project Manager," *Galileo Messenger* (July 1990): 1; "A Closer Look at the Earth and Moon," *Galileo Messenger* (April 1991): 1

Earth's Magnetic Field and the Solar Wind

Besides receiving its needed gravity-assist from Earth, Galileo also collected valuable data during its flyby of our planet and Moon. The first area studied was Earth's interplanetary environment, which included our magnetosphere's interaction with the solar wind. This plasma "wind" is extremely diffuse, with a density of only a few particles per cubic centimeter. While plasmas similar to the solar wind are difficult to maintain and analyze in Earth-based laboratories, the region of space near our planet can serve as a superb field lab for plasma study.[27]

The solar wind travels at speeds from 200 to 800 kilometers per second (450,000 to almost 2,000,000 mph) and carries magnetic field lines from the Sun along with it. Earth's magnetic field deflects the solar wind, and the interface between the two forms the boundary of Earth's magnetosphere (see figure 6.2). The bow shock forms on Earth's day side because it is this side that faces the Sun and the oncoming solar wind. The solar wind drags Earth's magnetic field on its night side back into a tail (the magnetotail) whose length is over a thousand times greater than Earth's radius.[28]

Galileo approached Earth from its night side, entering its magnetotail 560,000 kilometers (350,000 miles) behind the planet. The magnetosphere happened to be in a very active state at the time, and Galileo observed several magnetic storms. Many whistlers, generated by lightning strokes, were also detected.[29]

Moon Observations

Although our Moon has been studied for many years from both ground-based observatories and spacecraft, Galileo was able to provide some unusual observations that gave us a new view of parts of the satellite. In its orbit around Earth, the Moon always shows us the same face. It wasn't until the late 1960s and early 1970s that lunar orbiters showed us the Moon's far side. Apollo 15 made the most extensive survey of the far side, and its data led scientists to postulate the existence of a large basin on the far side's southern section. The new images that Galileo collected of the material covering this region led to the conclusion that the suspected impact basin did exist and was huge—1,940 kilometers (1,200 miles) across. If it was located on Earth, it would stretch from Mexico to Canada. Scientists now think that a 100-mile-diameter (160-kilometer-diameter) meteor formed the basin when it struck the Moon's surface "like a small state coming at you from space."[30] It probably hit so hard that it made the whole Moon wobble and lurch while it blew a large chunk out of the satellite's bottom half.[31]

---

[27] "A Closer Look," p. 2.

[28] Ibid., pp. 2–3.

[29] Ibid., pp. 3–4.

[30] Ibid., p. 4.

[31] Kathy Sawyer, "Galileo Probe Reveals Giant Basin on Moon," *Washington Post* (20 December 1990): A3.

Seeing Earth in a New Light

Because of Galileo's excellent instrumentation and unusual trajectory, it was able to carry out investigations that brought a new understanding of Earth's ozone hole, auroras, and global weather patterns. The spacecraft's near infrared mapping spectrometer (NIMS) was able to investigate very high clouds in Earth's mesosphere, an extremely cold atmospheric region above the stratosphere. These mesospheric clouds, seen only since 1885, are believed to be caused by methane generated from industrial processes. The methane affects global air flows, causing increased warm air to blow into polar regions, incrementing the amount of ice melted each spring and releasing more water vapor into the atmosphere. This water vapor forms the mesospheric clouds and reacts with the ozone, breaking it down and causing an ozone hole.[32]

These mesospheric clouds are sometimes seen in September or October. Galileo saw them in December, which was a rare event that might indicate that Earth's ozone population is indeed changing, as many scientists fear. The ozone layer helps shield our planet from harmful parts of the solar spectrum. A decrease in ozone could lead to a dangerous feedback loop in which increasing amounts of polar ice are melted, releasing more water vapor, which results in the destruction of ever-growing quantities of ozone. To develop reliable models of ozone depletion, atmospheric scientists have needed hard-to-acquire data on mesospheric water quantities. Galileo's investigations were important in filling this data gap.[33]

Can Life on Earth Be Surmised from Galileo's Observations?

Carl Sagan and other scientists asked this question: was it possible, from Galileo's observations, to deduce the existence of life on our planet? If Galileo were sending its data back to aliens on another world, what would they conclude about Earth?

Aliens receiving Galileo data would know that Earth's atmosphere had high oxygen and low carbon dioxide levels, in proportions that were not in equilibrium. There had to be a driving factor keeping these gases in disequilibrium, and the presence of life was one explanation. Galileo also detected narrow-band, pulsed, amplitude-modulated radio signals, as well as atmospheric methane in extreme thermodynamic disequilibrium. Methane is quickly oxidized to carbon dioxide and water. At thermodynamic equilibrium, not a single methane molecule should exist in our atmosphere. What mechanism, the aliens might wonder, could pump methane into our atmosphere so quickly that it outpaced oxidative processes? Decomposition of nonbiological organic matter would produce methane, but probably not at a sufficient rate. If the alien planet were similar to Earth, its scientists might know that the atmospheric methane levels could be caused by methane-generating bacteria, as well as activities such as agriculture, livestock-raising, and combustion processes.[34]

---

[32] "A Closer Look," p. 4.

[33] Ibid., pp. 4–5.

[34] Carl Sagan, W. Reid Thompson, Robert Carlson, Donald Gurnett, and Charles Hord, "A Search for Life on Earth from the Galileo Spacecraft," *Nature* 365 (21 October 1993): 715–721.

The existence of nitrous oxide (N_2O) in our atmosphere at high disequilibrium levels is another indicator of possible biological processes. Although certain nonbiological mechanisms such as lightning are able to produce N_2O, other sources such as nitrogen-fixing bacteria and algae would provide more credible explanations for the gas's abundance. Nitrogen-fixing bacteria and algae convert soil and oceanic nitrate (NO_3) to nitrogen gas and nitrous oxide.[35]

Spectral characteristics of Earth's surface also provided interesting information on the question of life. Project engineers took data from the six spectral bands that Galileo's SSI system observed and combined them in various ways to visualize spectral contrasts on the surface. Landmasses not obscured by clouds showed extreme albedo (reflectivity) contrasts. The colors of the lighter, more reflective areas were consistent with those of the rocks and soils that might be expected on a terrestrial-type planet with iron silicate surface composition. But there were widespread darker areas with a greenish tinge that, to an alien observer, could be puzzling. If photosynthetic processes occurred on the alien's home planet, it might reason that Earth's low albedo areas contained light-harvesting pigments. Widespread plant photosynthesis could explain the large atmospheric oxygen abundance on Earth.[36]

Galileo's high-resolution visible and near-infrared imaging of Earth's morphology, in which each pixel represented 1 to 2 square kilometers of surface, was not conclusive in establishing the existence of life. No geometric, artificial-appearing surface features or other compelling "artifacts of a technical civilization" were discerned. Previous studies of orbiting spacecraft imagery with resolutions finer than Galileo's have indicated that, when surface features as small as 0.1 kilometer can be resolved, the chances of finding convincing artifacts of life in any random image is only about one in a hundred. Galileo's poorer resolution would require that nearly all of Earth's surface be imaged in order to detect a civilization such as ours. Galileo's high-resolution imaging was mainly of Australia and Antarctica, representing a total of only 6 percent of Earth's surface. In addition, these areas are among the most sparsely populated land regions on the planet.[37]

The atmospheric composition and spectral data received by Galileo were suggestive of some kind of life on Earth, but they were far from conclusive. The strongest indication that Galileo received of intelligent, technologically sophisticated life was from radio transmissions. These were detected only on Galileo's night-side pass of Earth, when the signals were able to escape through the night-side ionosphere. Solar ultraviolet radiation provides a strong ionizing force when it shines on the day-side atmosphere, and it renders the ionosphere more opaque to radio waves.[38]

The radio transmissions that Galileo detected from ground-based antennas were not the only radio signals it received. Solar and auroral-related radio bursts were also recorded. Nevertheless, the ground-based signals had characteristics that suggested that the signals were being intentionally generated rather than naturally occurring. Their central frequencies remained constant over periods of hours. Naturally generated emissions

[35] Ibid., p. 718.

[36] Ibid., pp. 718–719.

[37] Ibid., pp. 718–719.

[38] Ibid., p. 720.

almost always drift in frequency. The artificial signals also displayed amplitude modulation patterns, a characteristic not seen in naturally occurring emissions and one that might be interpreted as intentionally designed into the signals to transmit information.[39]

Satellites equipped differently from Galileo have clearly traced radio signals back to the surface transmitters that generated them. Galileo, however, was not able to do this. A technologically advanced alien civilization might be able to deduce the existence of surface radio transmitters from Galileo's flyby data and make a strong case for an intelligent life-form on Earth, but such a conclusion would be far from certain.[40]

From the Galileo flyby, an alien observer previously unfamiliar with Earth could draw the following conclusions:

- Water is abundant on Earth in vaporous, liquid, and solid forms. If life does exist, it is likely water-based.

- The high abundance of oxygen, methane, and nitrous oxide gases in the atmosphere cannot readily be explained by nonbiological processes.

- Spectral data from Earth's land regions cannot be explained by geochemical processes but are suggestive of widespread photosynthetic processes. These could explain the atmospheric oxygen abundance.

- No clear sign of a technologically advanced civilization was obtained from the spacecraft's imaging activities.

- Amplitude-modulated, stable-frequency radio signals strongly suggest intelligent life.[41]

Out into the Asteroid Belt: The Gaspra Encounter

In planning the first spacecraft encounter with an asteroid, Galileo project staff set forth several objectives that included determining its size, shape, cratering characteristics, and composition, as well as surveying the surrounding environment. On 29 October 1991, a little over two years after Galileo had lifted off, it flew by the asteroid 951 Gaspra at a relative velocity of 8.0 kilometers per second (18,000 mph). The spacecraft was operating beautifully, with the exception of its high-gain antenna, which had failed to unfold into its intended dish shape. But there would be an opportunity to work on that problem, and the craft's low-gain antennas were adequate, for the time being, to transmit the data back to Earth. The transmission was slow, however, and most of the data were not received until November 1992, when Galileo was once again nearing Earth.[42]

---

[39] Ibid., p. 720.

[40] Ibid., p. 720.

[41] Ibid., pp. 715–721.

[42] "An Encounter with Gaspra," and Bill O'Neil, "From the Project Manager," both from *Galileo Messenger* (February 1992): 1, 8; "Gaspra's Story Continues," *Galileo Messenger* (February 1993).

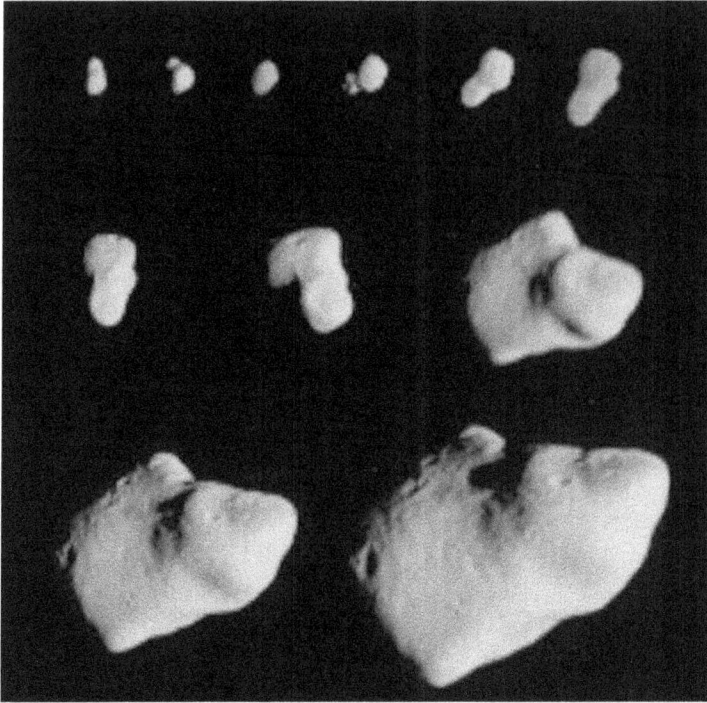

Figure 6.3. **The Gaspra asteroid as seen by the approaching Galileo spacecraft. The images show Gaspra growing progressively larger in the field of view of Galileo's SSI camera. The earliest view (upper left) was taken 164,000 kilometers (102,000 miles) from Gaspra, while the latest (lower right) was taken 16,000 kilometers (10,000 miles) away. (NASA image number PIA00079)**

Grigori Neujmin discovered Gaspra in 1916 at the Simeis Observatory in the Ukraine, naming the asteroid for a scientists' resort on the Crimean Peninsula. Gaspra, believed to be made up of both rocky and metallic minerals that included iron, nickel, olivine, and pyroxene, is one of the 5,000 bodies composing the "main asteroid belt," a donut-shaped area midway between Mars and Jupiter.[43]

The spacecraft's images revealed Gaspra to have a cratered surface and an irregular shape. It is about 20 kilometers (12 miles) long, 12 kilometers (7.4 miles) wide, and 11 kilometers (6.8 miles) thick, tiny compared to the largest asteroids, which range up to 800 kilometers (500 miles) in diameter. Figure 6.3 depicts a series of progressively larger Gaspra images, taken as the spacecraft approached the asteroid. These images capture almost one full rotation of the asteroid. Gaspra's irregular shape suggests that it was broken off a larger body through catastrophic collisions. Consistent with such a history is the prominence of groove-like linear features, believed to be related to fractures.

---

[43] "An Encounter with Gaspra"; "Gaspra's Story Continues."

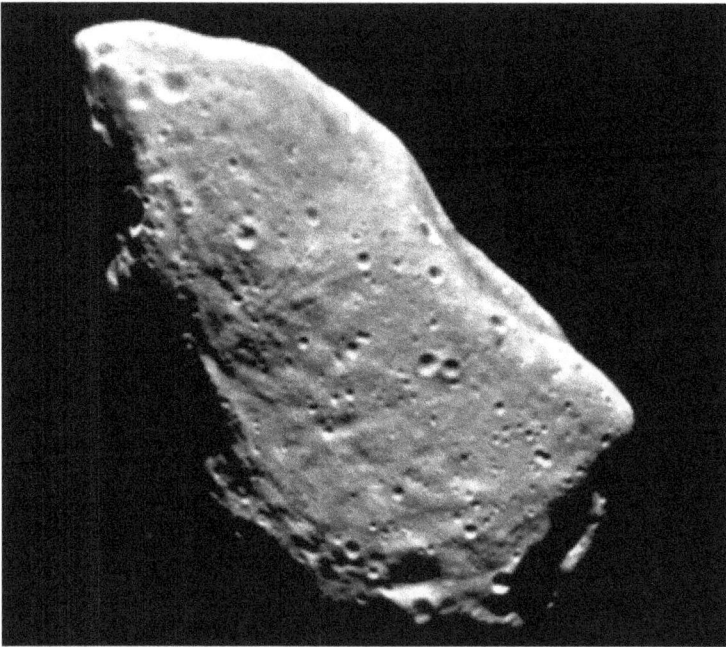

Figure 6.4. **Highest resolution picture of Gaspra, constructed as a mosaic of two images taken by the Galileo spacecraft from 5,300 kilometers (3,300 miles) away, 10 minutes before closest approach on 29 October 1991. The large concavity on the lower right limb is about 6 kilometers (3.7 miles) across, while the prominent crater on the left is about 1.5 kilometers (1 mile) in diameter. A striking feature of Gaspra's surface is the presence of more than 600 small craters, 100 to 500 meters (330 to 1,650 feet) in diameter. (NASA image number PIA00118)**

Figure 6.4, the highest resolution picture taken during the flyby, was formed from two images that Galileo took on 29 October 1991 from a distance of 5,300 kilometers (3,300 miles). This figure depicts an abundance of small craters—more than 600 of them with diameters ranging from 100 to 500 meters. Although many craters are visible on Gaspra, none approach the scale of the asteroid's radius. Gaspra apparently lacks the large craters commonly seen on many planetary satellite surfaces. This characteristic supports the theory that Gaspra is of comparatively recent origin. Gaspra is probably relatively young and has not been around long enough to pick up the major scarring that many other bodies have. The abundance of small craters on Gaspra is not surprising, however, since there are far more small asteroids and rocks that could hit another asteroid than large ones.[44]

---

[44] "An Encounter with Gaspra"; "Gaspra's Story Continues"; "Gaspra Approach Sequence," *NASA's Planetary Photojournal Catalog Page, http://photojournal.jpl.nasa.gov/cgi-bin/PIAGenCatalogPage.pl?PIA00079* (available in Galileo—Meltzer Sources, folder 18522, NASA Historical Reference Collection, Washington, DC); "Gaspra—Highest Resolution Mosaic," *NASA's Planetary Photojournal Catalog Page, http://photojournal.jpl.nasa.gov/cgi-bin/PIAGenCatalogPage.pl?PIA00118* (available in Galileo—Meltzer Sources, folder 18522, NASA Historical Reference Collection, Washington, DC); "Galileo Mission to Jupiter: Educators' Slide Set Vol. 1," *http://www2.jpl.nasa.gov/galileo/slides/* (accessed 15 November 2004) (available in Galileo—Meltzer Sources, folder 18522, NASA Historical Reference Collection, Washington, DC).

Galileo's observations showed that the solar wind changed direction a few hundred kilometers from the asteroid. This behavior of the ionized, electrically conducting gas that composes the solar wind is typical of what happens when it encounters a magnetic region and is called a "field rotation." Although this result suggested that Gaspra might have a magnetic field, it was not conclusive. Dr. Margaret Kivelson, a UCLA scientist who was the principal investigator for Galileo's magnetometer studies, emphasized that the field rotation might not have been caused by an interaction with Gaspra. The solar wind frequently undergoes field rotations, and it might simply have been coincidental that the observed one took place near the asteroid.[45]

One of the big questions that interest space scientists is, what were the processes that formed the asteroids and gave them their properties? The scientific community believes that asteroids' chemical compositions are related to those of meteorites, interplanetary fragments that have impacted Earth. If Gaspra indeed has a magnetic field, this finding could shed light on how it was formed and at what temperatures. Since asteroids are believed to have formed during the earliest stages of the solar system, the existence of a Gaspra magnetic field might also provide information on the magnetic environment of the ancient solar system. In addition, the strength of an asteroid's magnetic field could give an indication of how rich it is in iron and iron-nickel alloys. One day, asteroids may be mined for their ores, and magnetic field strength may provide a clue as to the body's economic value.[46]

Back to Earth

On 8 December 1992, two years after its last Earth encounter, Galileo once more flew by our planet, passing within 300 kilometers (190 miles) above the South Atlantic and using Earth's gravity to swing it onto a direct trajectory to Jupiter. The accuracy of Galileo's flyby maneuver was phenomenal. The spacecraft passed within 1 kilometer of its aimpoint, and as a result, a post-Earth trajectory correction maneuver (TCM-18) was canceled. This saved approximately 5 kilograms (11 pounds) of propellant that could be used later in the mission.[47]

Once again, the Galileo team was able to make valuable observations of our planet. The ultraviolet spectrometer (UVS) group, working with the near infrared mapping spectrometer (NIMS) group, discovered "a wealth of information" about the geocorona,[48] the outermost part of Earth's atmosphere (see figure 6.5). It is composed of diffuse hydrogen gas that has escaped from the lower reaches of the atmosphere. This hydrogen first absorbs, then reemits a certain part of the solar spectrum called Lyman-α radiation. Galileo was able to observe some of this reemitted radiation. The absorption and reemission process results in a force on the geocorona in the anti-Sun direction, creating a "geotail" feature of the geocorona (see figure 6.6).

---

[45] "Gaspra's Story Continues."

[46] "An Encounter with Gaspra"; "Gaspra's Story Continues."

[47] Kathy Sawyer, "Galileo Swings by Earth, Speeds Toward Jupiter," *Washington Post* (9 December 1992): A3; "A Farewell to Earth" and Bill O'Neil, "From the Project Manager," both from *Galileo Messenger* (February 1993): 1, 3.

[48] "A Farewell to Earth," *Galileo Messenger* (February 1993): 4.

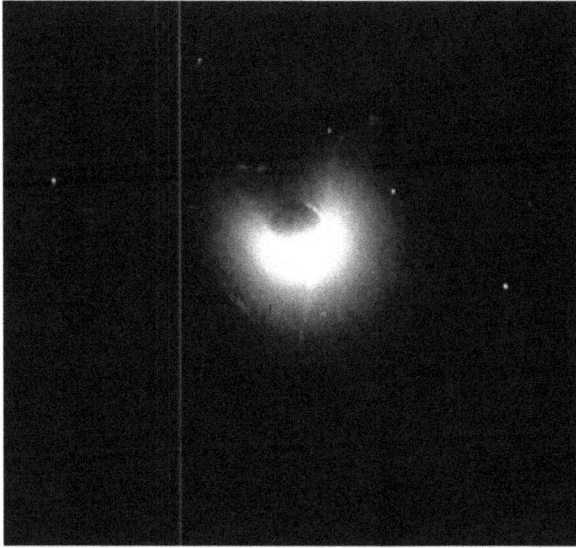

Figure 6.5. **The geocorona, shown here in a photograph taken with an ultraviolet camera, is a halo of low-density hydrogen around Earth that absorbs sunlight and emits "Lyman-α" radiation. (NASA image number AS16-123-19650)**

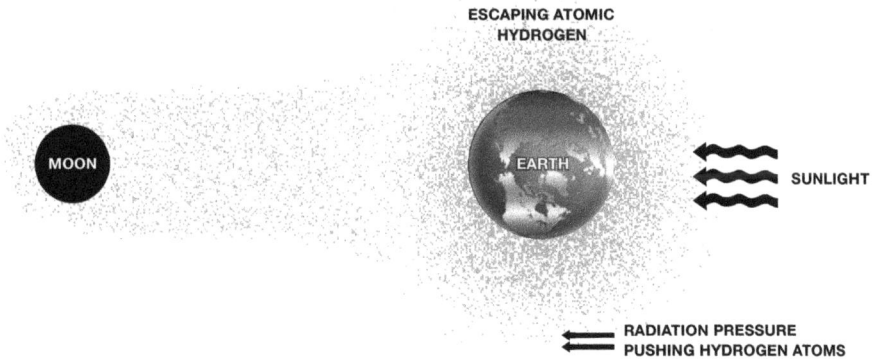

Figure 6.6. **The geocorona and its geotail. As the atomic hydrogen in the geocorona is exposed to sunlight, the hydrogen emits radiation. This process results in a force that sweeps much of the hydrogen in the direction opposite to that of the Sun, creating a geotail.**

The geocorona had previously been measured out to distances of 95,000 kilometers (59,000 miles) in the region of its geotail. During its second Earth flyby, Galileo detected an enormous geocorona bulge that extended 400,000 kilometers (250,000 miles) away from Earth in the geotail region—nearly to the Moon's orbit and four times the thickness of the traditional geocorona model—and even observed very diffuse atomic hydrogen (about 1 atom per cubic centimeter) in the vicinity of the Moon. The Galileo team believed that all or most of this hydrogen was associated with the geotail, rather than being an aspect of the Moon's tenuous atmosphere.

During the first Earth flyby in December 1990, Galileo investigated a series of high stratospheric clouds over Antarctica. Atmospheric scientists now know that these clouds are involved in the complex chemical reactions leading to the growing ozone hole over the South Pole. Due to the low temperatures of the Antarctic stratosphere, these clouds are composed of ice crystals, rather than water droplets, as is the case in warmer regions of the atmosphere. On the clouds' ice-crystal surfaces, reactions occur that result in the release of ozone-destroying chlorine. At the time of Galileo's first Earth flyby, scientists were surprised by the occurrence of these stratospheric ice clouds and hoped that they were an anomaly. Galileo's NIMS instrument found the clouds during the second flyby as well, however. Dr. Robert Carlson, NIMS principal investigator, thought that they might actually be a frequent occurrence over the South Pole.[49]

Galileo was able to fly by the northern polar region of the Moon that had been photographed by Mariner 10 in 1973. This time, however, Galileo was able to image the region in infrared, as well as visible, and provide new data on geologic processes. Exposed lava flows observed by Galileo indicate that volcanism during the Moon's early years was more frequent and widespread than had been thought. Galileo data also suggested that the Moon's far side had a thicker crust. The Galileo imaging team was surprised to find some hidden maria (lunar "seas" made up of frozen lava flows) that had been overlain by other features. These "cryptomaria" were only visible in certain spectral bands that Galileo's NIMS was fortunately able to detect.[50]

A Second Asteroid Encounter: Ida

Months after leaving Earth for the last time, Galileo once again entered the asteroid belt between the orbits of Mars and Jupiter. On 28 August 1993, Galileo flew to within 1,500 miles (2,400 kilometers) of a new asteroid: 243 Ida. This asteroid draws its name from a Greek myth in which the god Zeus was raised by Ida the nymph. The Ida asteroid is considerably larger than Gaspra but is also highly irregular, measuring 56 by 24 by 21 kilometers (35 by 15 by 13 miles). Like Gaspra, Ida is heavily cratered with small depressions (see figure 6.7). The cratering is more extensive on Ida, however, suggesting that it formed earlier than Gaspra.[51]

---

[49] " Farewell to Earth," pp. 3–4.

[50] "Observing the Moon in a Different Light," *Galileo Messenger* (February 1993): 2–3.

[51] "Revisiting the Asteroids: This Time It's Ida," *Galileo Messenger* (February 1994): 4; "Galileo Mission to Jupiter."

Figure 6.7. **The asteroids Ida (left) and Gaspra (right). Gaspra was imaged on 29 October 1991 at a range of 5,300 kilometers (3,300 miles). Ida was imaged on 28 August 1993 from a range of 3,000 to 3,800 kilometers (1,900 to 2,400 miles). The surfaces of Ida and Gaspra contain many small craters, evidence of numerous collisions. The fact that craters are more abundant on Ida suggests that it is older than Gaspra. (NASA image number PIA00332)**

Galileo detected changes in the interplanetary magnetic field in the vicinity of Ida; as with Gaspra, these changes implied that Ida may have a magnetic field. Prior to Galileo's journey, many space scientists did not believe that asteroids had magnetic fields. In planets such as Earth, magnetic fields are generated in dense metallic cores. But asteroids the size of Ida are too small and light to have gravitational fields strong enough to have pulled their dense metallic minerals into the center as the asteroids were forming. Nevertheless, other mechanisms could explain the formation of asteroidal magnetic fields. One theory is that the strong fields in the solar wind might have induced magnetism in the asteroids at some point in the past. Another theory is that asteroids may be pieces broken from larger parent bodies. These bodies could have had strong enough gravitational fields to form metal cores that generated magnetic fields. These fields then magnetized some of the minerals surrounding the cores. When chunks broke away from the parent bodies and formed asteroids, they carried magnetized elements with them.[52]

---

[52] "Revisiting the Asteroids," p. 4.

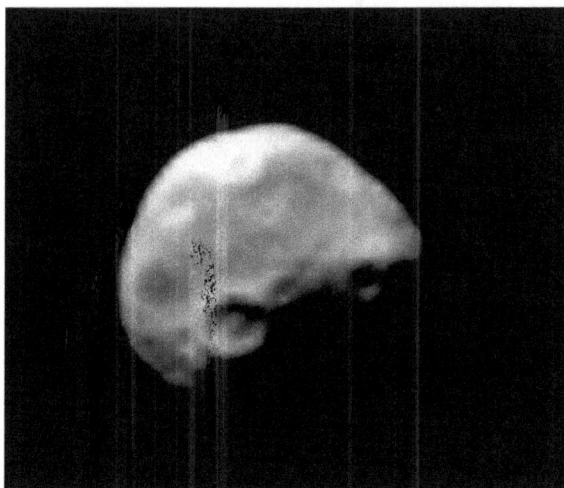

Figure 6.8. **Dactyl, satellite of the asteroid Ida. The little moon is egg-shaped and only about 1.6 kilometers (1 mile) across. The large crater is about 300 meters (1,000 feet) across. This image was taken on 28 August 1993 at a distance of 3,900 kilometers (2,400 miles) from the moon. (NASA image number PIA0029)**

Months after the Ida flyby, when Galileo's images were played back, project scientists got a big surprise. On 17 February 1994, a search of imaging data revealed a small moon orbiting Ida! Its existence was confirmed eight days later in a search of NIMS data. This was the first conclusive evidence that natural satellites of asteroids exist. The piece of rock, heavily cratered as are Ida and Gaspra, is only approximately 1.6 kilometers (1 mile) in diameter, and it orbits about 100 kilometers (60 miles) from Ida's center (see figure 6.8). The International Astronomical Union named it "Dactyl." The name is derived from the Dactyli, Greek mythological beings who lived on Mount Ida.[53]

Galileo's planetary gravity-assists and asteroid flybys were now completed, and the spacecraft continued on its way to Jupiter with a scheduled arrival date of 7 December 1995.

Summary of Galileo's Major Observations During Its Moon and Earth Flybys and Its Asteroid Encounters

Major accomplishments of Galileo's first Moon and Earth flybys included the detection of an impact basin almost 2,000 kilometers in diameter on the Moon's far side, along with a rare sighting of mesospheric clouds above Earth in December, possibly indicating changes in our planet's ozone layer.

---

[53] "Ida Data Return," *Galileo Home Page*, pp. 2–3, *http://www2.jpl.nasa.gov/galileo/iaf/iaf_sect3.html* (accessed 21 March 2000) (available in Galileo–Meltzer Sources, folder 18522, NASA Historical Reference Collection, Washington, DC); Donald J. Williams, "Jupiter—At Last!" *Johns Hopkins APL Technical Digest* 17, no. 4 (1996): 352–353; "High Resolution View of Dactyl," *NASA's Planetary Photojournal Catalog Page*, *http://photojournal.jpl.nasa.gov/cgi-bin/ PIAGenCatalogPage.pl?PIA00297* (available in Galileo–Meltzer Sources, folder 18522, NASA Historical Reference Collection, Washington, DC); "Galileo Mission to Jupiter."

Galileo made these significant observations during its encounter with the asteroid Gaspra:

- Image data revealed Gaspra to have an irregular shape, linear features related to fractures, and abundant small craters, but a dearth of large ones. These data suggest that Gaspra may be relatively young and that it was formed from a chunk of some larger body. The asteroid is about 20 kilometers (12 miles) long, 12 kilometers (7.4 miles) wide, and 11 kilometers (6.8 miles) thick, considerably smaller than the largest asteroids, which range up to 800 kilometers (500 miles) in diameter.

- Changes in direction of the solar wind a few hundred kilometers from the asteroid suggest that Gaspra may have its own magnetic field. If so, this field might provide data on the ancient solar system's magnetic environment and could give an indication of how rich Gaspra is in iron and iron-nickel alloys.

During its second Earth flyby, Galileo's accomplishments included the following:

- Detection of a huge geocorona bulge that extended nearly to the Moon's orbit.

- Another observation of mesospheric clouds, indicating that their existence during the first flyby was not an anomaly.

- Lunar lava-flow observations indicating more extensive volcanism during the Moon's early years than previously thought.

- Data collected that suggested a thicker farside lunar crust than expected.

Galileo's observations during its encounter with the asteroid Ida revealed the following information:

- The asteroid is 56 by 24 by 21 kilometers (35 by 15 by 13 miles), significantly larger than Gaspra.

- It is more heavily cratered than Gaspra, a condition that suggests greater age.

- The interplanetary magnetic field changes near Ida; this fact indicates that Ida may have its own magnetic field also.

- Ida has its own moon: a 1.6-kilometer-diameter (1-mile-diameter) rock 100 kilometers (60 miles) away, which has been named Dactyl.

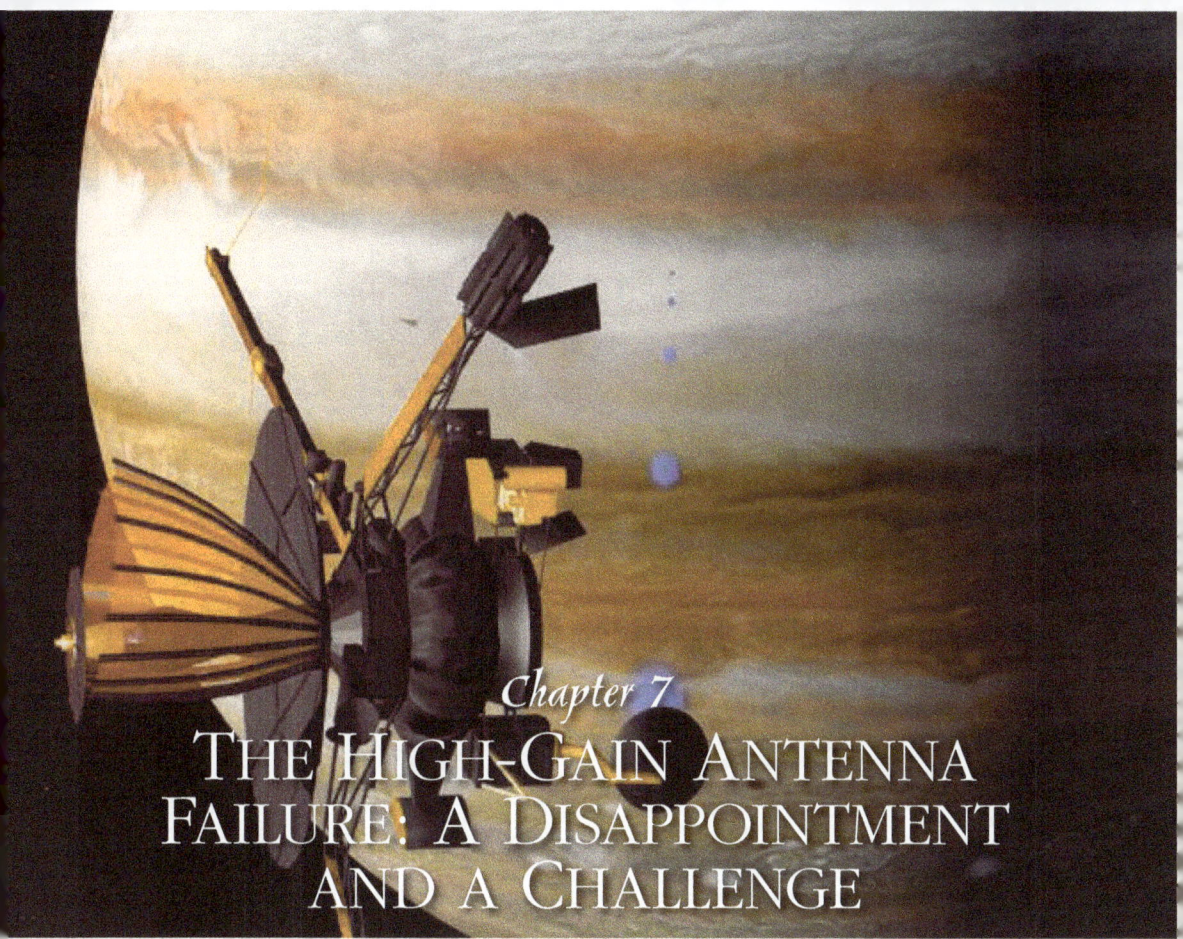

Chapter 7

THE HIGH-GAIN ANTENNA FAILURE: A DISAPPOINTMENT AND A CHALLENGE

IN APRIL 1991, GALILEO WAS ON ITS WAY FROM ITS FIRST EARTH FLYBY TO its encounter with the asteroid Gaspra. Since the October 1989 launch, the 4.8-meter-diameter (16-foot-diameter) parabolic dish of the spacecraft's high-gain antenna (HGA) had been folded up like an umbrella and kept stowed behind a sun shield. This configuration had helped protect it from the extreme forces of Galileo's launch and the intense solar radiation of the inner solar system tour. But Galileo was now far enough from the Sun that the HGA could be deployed, thus making Galileo capable of rapidly sending and receiving data. JPL's flight team prepared to send commands to unfurl the HGA. (Figure 7.1 depicts the open antenna.)

The flight team was nervous. Conducting a major mechanical operation many millions of kilometers from the nearest set of hands is risky. If anything goes wrong, it isn't possible to take out a toolbox and fix the problem. On 11 April 1991, the flight team initiated the deployment sequence. It radioed commands for Galileo to turn on two direct current (DC) electric motors in the mechanical drive system that was designed to open the HGA. The antenna deployment time, if both motors operated properly, was expected to be 165 seconds. Even if one of the motors or its gear train failed, the other motor should have been able to deploy the HGA in about 330 seconds.

The motors obeyed the commands and turned on, but they operated at higher-than-expected power levels for nearly 8 full minutes, as if they were laboring to open the

Figure 7.1. **The unfurled high-gain antenna. (Adapted from JPL image number 230-893B)**

antenna. The motors were supposed to turn a worm gear,[1] which in turn was supposed to push a ring up the axis of the antenna. This ring was connected to levers that were to open the antenna by spreading its ribs. The team expected a signal from Galileo confirming that the antenna had opened, but the signal never came.[2]

Analysis of the Problem

On the same day that the Galileo flight team tried to open the HGA and realized there was a problem, JPL formed an HGA deployment anomaly team from its staff of mechanical, electrical, thermal, materials design, reliability, and flight operations specialists. JPL also drew in personnel from Harris Corporation, which had built the HGA.

By 30 April, the anomaly team had reviewed all videos taken from the Shuttle of Galileo's deployment, trying to determine the status at that time of the HGA and its solar shade. The team studied all telemetry data from the spacecraft and evaluated the possible

[1] Worm gears resemble coiled worms and are used when large gear reductions are needed. A worm gear is typically employed to reduce speed. In the Galileo spacecraft's antenna assembly, each complete turn of the worm gear shaft was supposed to advance a ring a small amount up the shaft of the antenna. See Karim Nice, "How Gears Work," *HowStuffWorks, Inc.*, 2003, *http://computer.howstuffworks.com/gear5.htm*; "Worm Gear," *Flying Pig Co.*, United Kingdom, 2003, *http://www.flying-pig.co.uk/mechanisms/pages/worm.html*.

[2] "The High-Gain Antenna," *Galileo: The Tour Guide, http://www.jpl.nasa.gov/galileo/tour/2TOUR.pdf*, p. 6 (accessed 15 March 2000); "Unfurling the HGA's Enigma," *Galileo Messenger* (August 1991); Michael Dornheim, "Improper Antenna Deployment Threatens Galileo Jupiter Mission," *Aviation Week & Space Technology* (22 April 1991): 25; Paula Cleggett-Haleim and Frank O'Donnell, "Galileo Antenna Deployment Studied by NASA," *NASA News*, Washington, DC (19 April 1991); Michael R. Johnson, "The Galileo High Gain Antenna Deployment Anomaly," NASA Technical Report No. N94-33319 in the proceedings of the 28th Aerospace Mechanisms Symposium (held at NASA Lewis Research Center, May 1994), pp. 360, 364.

effects of the different thermal expansion characteristics of the antenna's various parts, as well as the effects of shock, vibration, and vacuum on the HGA.[3]

Mission staff received the following telemetry data from the spacecraft:

- Output from one of Galileo's sun sensors was reduced at certain "clock angles." The clock angle measured angular position on the spacecraft, with the origin at the craft's rotational axis.

- The spacecraft's spin-detector[4] output spiked 8 seconds after the HGA deployment attempt, indicating a sudden acceleration and deceleration of the spacecraft's spin rate.

- The spacecraft's spin rate further decreased over the remainder of the deployment attempt.

- Electric current data from the two motors designed to open the HGA indicated that they had stalled 56 seconds after the start of the deployment sequence.

- The wobble of the spacecraft had increased.[5]

The surface of the antenna dish was constructed of a gold-plated molybdenum wire mesh stretched across 18 graphite-epoxy support ribs. The anomaly team thought that the reduction in output from the sun sensor might be due to shadowing by one or more of the ribs. After examining the sensor's position relative to the HGA and the Sun, the team concluded that only one rib was capable of shadowing the sensor, and only at a partially deployed angle of 34 to 43 degrees. Analysis of the drop in output of the sun sensor narrowed this possible range of angles. The rib shadowing the sensor had deployed about 35 degrees from its fully stowed position.

Initially, the anomaly team thought that the sudden release of stuck ribs might have caused the spike in the spin-detector data, indicating a sudden change in spin rate. But the team noticed that the spin-detector spike occurred at the same time as an increased torque from the motors. Rib release was therefore discarded as an explanation for the spike because such a release should not have augmented the load on the motors.

The spin rate of the spacecraft's spun section was expected to decrease as the antenna dish opened, due to the increase in the antenna's moment of inertia. This effect is similar to the slowing of a spinning ice skater when she extends her arms, increasing her moment of inertia. A decrease was measured in Galileo's spin rate, but it was smaller than was expected from a fully deployed antenna. This was one more indication that the antenna had not completely opened.

As mentioned above, the electric motors stalled 56 seconds into the deployment sequence. The anomaly team used its knowledge of the antenna's mechanical drive

---

[3] "Unfurling the HGA's Enigma," *http://www.jpl.nasa.gov?galileo/messenger/oldmess/HGA.html* (accessed 15 March 2000).

[4] Galileo's spin detector was a sensitive accelerometer mounted on the spun section of the spacecraft. This detector measured spin rate changes, which were affected by whether the HGA was opened or closed.

[5] Johnson, "The Galileo High Gain Deployment Anomaly."

system, the amount of current that the motors drew during deployment, and several tests on an antenna at JPL to determine that the most likely configuration of Galileo's HGA dish was that three ribs on one side of the antenna had remained in their stowed position, stuck to the antenna's central mast (see figure 7.2 and the photograph at the beginning of this chapter). The stuck ribs would have prevented the antenna dish from fully opening but would have allowed a rib opposite them to deploy just enough to shade the sun sensor by the amount indicated in the telemetry data. This scenario of an asymmetric, partially opened antenna dish could also explain the increase in the spacecraft's wobble.[6]

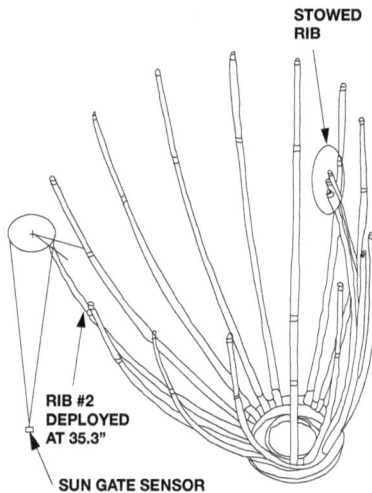

STOWED
RIB

RIB #2
DEPLOYED
AT 35.3"

SUN GATE SENSOR

Figure 7.2. **HGA's supporting ribs did not fully deploy. The HGA was not able to send usable data in its partially opened state.**

Once the anomaly team determined the probable shape of the semideployed antenna, they analyzed the HGA's design in great detail in order to identify what might have caused the problem. Four possible scenarios survived the team's rigorous analysis:

- The sunshade, mounted on the tip of the central mast of the HGA to protect the antenna from overheating during the early months of the mission (when Galileo flew within Earth's orbit), snagged in the wire mesh of the antenna dish.

- The mechanical drive system got restrained in some way.

- The tips of several of the antenna's ribs got stuck in their tuning-fork-like sockets.

- Restraint pins at the midpoints of several antenna ribs got stuck in their sockets.[7]

---

[6] Johnson, "The Galileo High Gain Antenna Deployment Anomaly"; "Unfurling the HGA's Enigma."

[7] Johnson, "The Galileo High Gain Antenna Deployment Anomaly."

Exhaustive tests that the team performed on an antenna at JPL failed to show any way in which the sunshade could snag the antenna dish's wire mesh such that ribs could be restrained in their stowed position. A restrained mechanical drive system was also eliminated as a likely cause of the antenna anomaly. The vicinity around the drive system had been closed off and inaccessible during successful prelaunch deployment tests of the flight antenna, as well as during installation of the antenna on the spacecraft, making damage to the drive system very unlikely. Based on the results of prelaunch testing, retention of the tips of the antenna dish's ribs in their tuning-fork sockets was also deemed improbable. The sockets would have to have been damaged after the final deployment test or in flight. The team considered that the chances of this happening had been quite small.

The anomaly team's analysis left a failure of restraint pins to retract from their sockets as the most likely anomaly scenario. Each graphite/epoxy rib had two titanium restraint pins. The titanium pins were finished with a ceramic anodized coating, using a process called Tiodizing. This helped provide a good bond surface for the next coating, a molybdenum disulfide dry lubricant called Tiolube 460. This material, which derived its lubricity from its weak sulfur bonds' sliding past each other, had been successfully used in the past on many different spacecraft.[8]

When the antenna was closed, the restraint pins remained seated in receptacles made of Inconel, an alloy composed of nickel, chromium, and iron. The function of the pins was to hold the antenna in a closed position while inside the Shuttle's cargo bay (if open, the antenna would not fit within the bay). But the pins were supposed to slide smoothly out of their receptacles when attempts were made to deploy the antenna. The deployment anomaly team analyzed the possibilities that a "cold weld," adhesion, or friction between pin and receptacle had prevented the antenna from unfurling.[9]

Cold welding can occur if metallic elements come into contact with each other under certain conditions, creating bonds across the interface. In order for any degree of cold welding to occur, surface oxide layers have to be removed. This could have happened on Galileo if the restraint pins and their receptacles rubbed against each other. Also necessary for cold welding is the plastic deformation[10] of at least one of the two contacting metal surfaces. Such deformation involves the sliding and rotation of individual grains of one of the metals, leading to better bonding across the interface.[11]

Unexpected friction between restraint pin and receptacle, possibly great enough to prevent the antenna from opening, might have occurred if the first attempt to deploy the HGA had bent the pins slightly (see figure 7.3). Another possibility was that the metal surfaces of pins and receptacles had been galled and pitted from rubbing against each other. Opportunities for pins and receptacles to rub together, which could have led either to cold welds or to increased friction between the two parts, occurred during the spacecraft's multiple truck rides across the country, under the stresses of launch, and during maneuvers in space such as Galileo's deployment from the Shuttle.[12]

---

[8] Ibid.

[9] Ibid.; Donald F. Lewis and Tim O'Donnell, *Materials Analysis in Support of the Galileo High Gain Antenna (HGA) Deployment Anomaly* (Pasadena, CA: JPL D-9814, November 1992), p. 10.

[10] Plastic deformation refers to the ability of a metal to change shape without breaking or fracturing, with the cohesion between the molecules of the metal sufficient to hold them together. See "Ductility," *Electric Library's Encyclopedia. Com, http://www.encyclopedia.com/printable/14228.html* (accessed 9 January 2002).

[11] Lewis and O'Donnell, p. 10; Johnson, "The Galileo High Gain Antenna Deployment Anomaly."

[12] Lewis and O'Donnell, p. 12; Johnson, "The Galileo High Gain Antenna Deployment Anomaly."

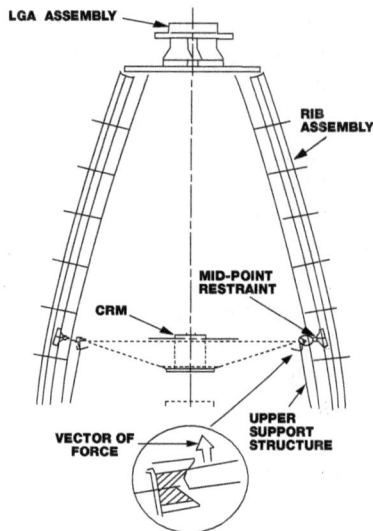

Figure 7.3. **A blowup view of the HGA in its closed, or "stowed," configuration illustrates the suspected misalignment of its restraint pins. Rather than fitting into the centers of their housings, the pins might have been bent so as to hang up on the edges of the housings, thus preventing deployment of the antenna. Note: "LGA Assembly" at the top of the figure refers to the low-gain antenna assembly that was eventually used in place of the HGA.**

Pin and Socket Analysis

The anomaly team used pin and socket pairs from an extra, non-flight-ready HGA at JPL to evaluate the various failure scenarios. Vibrational testing had previously been performed on the extra HGA, and so the pin and socket pairs already should have experienced considerable relative motion.

Visual inspection of pins from the spare HGA revealed some plastic deformation. X-ray diffraction scans of a number of the pins indicated that both the Tiodize ceramic coating and the molybdenum disulfide lubricant coating had been destroyed in areas contacting the socket. Scans of nondeformed pins did not show the same damage to the coatings, suggesting that deformation of the contact areas was involved in the destruction of the Tiodize and "drylube" layers.[13]

In order to better understand the friction behavior of the HGA's restraint pins as they rubbed against their sockets, JPL arranged with NASA Lewis Research Center to conduct a rigorous series of tests. JPL provided Lewis with Tiolubed and unlubricated titanium pins, as well as discs of the same Inconel material as the sockets. Lewis staff performed a series of friction tests, both under atmospheric conditions and in a vacuum, holding together the materials under load and displacing them relative to each other. The forces applied in these friction tests were adjusted so that the contact stresses between pins and discs were the same as those estimated for the HGA.[14]

---

[13] Johnson, "The Galileo High Gain Antenna Deployment Anomaly."

[14] Lewis and O'Donnell, p. 82; Johnson, "The Galileo High Gain Antenna Deployment Anomaly."

Lewis Research Center reported that the coefficient of friction[15] between the lubricated pins and Inconel in vacuum was 0.04, a very low value. In air, the sliding friction between lubricated pins and Inconel was nearly 10 times higher (a result that also had been obtained by testing at JPL). Lewis Research Center determined that the endurance life of the Tiolube coating on the pins was three orders of magnitude (roughly 1,000 times) greater in a vacuum than in air. This finding indicated that rubbing a Tiolubed pin against its socket within Earth's atmosphere wears away the Tiolube coating far faster than rubbing under vacuum conditions.[16]

Lewis Research Center also performed tests on bare titanium pins, simulating the behavior of HGA restraint pins whose Tiolube and Tiodized coatings had been worn away. As long as an atmosphere was present to react with the bare titanium and form oxide coatings, the coefficient of friction between pin and socket never exceeded 0.35, which was not high enough to trap the pins inside their sockets. By contrast, rubbing bare titanium pins against Inconel in a vacuum led to galling (a type of severe surface damage) and a coefficient of friction between the rubbing surfaces that was over three times greater than in an atmosphere.[17]

Why Galileo's HGA Rib Restraint Pins Stuck in Their Sockets

From the mass of data it collected, the anomaly team developed a credible scenario for the failure of Galileo's HGA to deploy. The first step in the failure mechanism was postulated to be deformation of the restraint pins. Whenever the antenna was folded into its stowed position, the restraint pins were pushed tightly into their sockets. Plastic deformation of the contact points on some of the pins may have resulted, damaging their Tiodized coatings, which were supposed to provide good bonding surfaces for the dry lubricant coating.

During the HGA's truck journeys across the country, the damage was compounded. The HGA made four long truck trips: first in 1982, when it was shipped from Harris Corporation's Florida manufacturing facility to JPL; then from JPL to Kennedy Space Center (KSC) in late 1985 for the planned launch; back to JPL in 1986 after the *Challenger* tragedy; and a return to KSC in 1989 for the actual launch. The antenna was kept in its stowed position during the journeys and was subjected to "enough of a vibration environment to cause relative motion between the pins and sockets."[18] The particular cantilevered mounting configuration of the HGA in its shipping container amplified this relative motion, the anomaly team believed. The folded-up antenna lay horizontal in its

---

[15] The force of friction between two objects is equal to a constant number times the force holding the objects together. The constant number is called the coefficient of friction. The higher it is, the greater the friction. The coefficient of friction for rubber sliding on concrete is 0.8 (relatively high), while the coefficient for Teflon sliding on steel is 0.04 (relatively low). Applying a layer of lubricant reduces the coefficient of friction by minimizing the contact between rough surfaces. A lubricant's particles slide easily against each other and cause far less friction than would occur between bare surfaces. See "Friction," *Microsoft Encarta Online Encyclopedia*, 2001, *http://encarta.msn.com*.

[16] Lewis and O'Donnell, p. 82.

[17] Johnson, "The Galileo High Gain Antenna Deployment Anomaly," p. 375; Kazuhisa Miyoshi and Stephen V. Pepper, "Properties Data for Opening the Galileo's Partially Unfurled Main Antenna," NASA Technical Memorandum 105355, February 1992, p. 1.

[18] Johnson, "The Galileo High Gain Antenna Deployment Anomaly."

container, and the restraint pins of ribs that were on the top and bottom of the prone antenna experienced the greatest relative motion with respect to their sockets. Because these truck journeys occurred in an atmosphere, the anomaly team believed that the Tiolube and Tiodized layers on the contact points of some of the pins were completely worn away by the time of Galileo's launch. Vibration testing of the flight antenna at JPL may have compounded the problem.

After the launch and deployment of Galileo from the Shuttle, the spacecraft was once again subjected to significant vibration when the Inertial Upper Stage fired, causing more relative motion of pins and sockets—but now the spacecraft was in a vacuum environment, with some titanium pins that had been rubbed bare of lubricant and Tiodizing. The anomaly team postulated that at this point, the rubbing of pin in socket galled the metal surfaces, resulting in a high coefficient of friction between the parts.

Possibly due to their orientations in the shipping container, some pins and sockets were less damaged than others. The antenna's mechanical drive system was able to eject most of the restraint pins from their sockets and deploy most of the HGA's ribs, but the drive system's motors stalled before the forces were large enough to eject the last three ribs. As a result, the antenna dish deployed only partially, assuming an asymmetric shape with the stuck ribs all on one side. This configuration created unexpected stresses on the antenna's mechanical drive system. The anomaly team conducted tests on a mock-up of the drive system in order to surmise "how it would respond to the odd loading condition created by the antenna." The team found that the drive system did not respond well to its unbalanced load. Parts in the mocked-up drive system jammed against each other, resulting in the loss of two-thirds of the torque that the system had been expected to generate. If Galileo's HGA drive system had experienced a comparable loss, it would have had very little available torque to free the ribs from their restraints. This reduced-torque condition and the higher-than-expected coefficient of friction between some pins and sockets were, the team believed, major factors in the HGA's malfunction.[19]

In summary, four sequential conditions had to be generated in the HGA to cause the deployment failure:

1. High enough contact stress to plastically deform the titanium restraint pins and break the Tiodized coating used to bond the dry lubricant to the pins.

2. Relative motion—under atmospheric conditions—between pins and sockets. This removed the dry lubricant and damaged Tiodized coating from the contact areas, exposing rough, bare metal surfaces on mating parts.

3. Relative motion—in a vacuum—leading to galling of the contact surfaces and a very high coefficient of friction.

4. Asymmetric rib deployment resulting in large loads on the antenna's mechanical drive system and loss of most of the system's available torque.[20]

---

[19] Ibid.

[20] Ibid.; Douglas Isbell, "Cooling Down Fails to Free Galileo Antenna," *Space News* (26 August–8 September 1991).

Attempts To Free the Antenna

In addition to modeling the root causes of the HGA failure, the anomaly team also developed possible strategies for overcoming the problem. Although the stuck ribs could have been easily freed had the spacecraft still been on Earth, it was a bit more complicated to set them loose from many millions of miles away. Mission staff could accelerate the spacecraft, change its orientation, and exert some control over the antenna's mechanical drive system, but they could do little else and had to build a rescue strategy based on these limited actions.

The first suggestion that the anomaly team made was to restow the antenna in its furled, "closed umbrella" configuration, then try the deployment sequence again. This strategy was quickly dropped because stowing the antenna required human intervention in order to carefully roll up the antenna dish's wire mesh so that it would not snag either on itself or on other parts of the HGA. Also, ground testing of the spare antenna revealed that each time the HGA was stowed and redeployed, a rotating steel ballscrew within the drive mechanism galled its aluminum housing, resulting in a reduction of the torque that the drive mechanism was able to generate.

The first actual attempt to break the stuck antenna ribs loose was to rotate the spacecraft toward and then away from the Sun. The anomaly team believed that the thermal expansion and contraction of much of the antenna's structure due to heating and cooling would be considerably greater than the expansion and contraction of the antenna dish's ribs, resulting in significant changes in the forces on the stuck restraint pins and sockets. Computer modeling indicated that after four to six thermal cycles of the antenna, the stuck pins might break loose from their sockets due to "infinitesimal sliding each time the forces changed."[21]

JPL staff had high hopes that the HGA would eventually be deployed, either through the heating and cooling turns or through other means. Franklin O'Donnell, a JPL spokesman, stated in April 1991, shortly after the HGA first failed to deploy, "There's a fair amount of optimism that [the problem] is correctable." And Bill O'Neil, the Galileo Project Manager, affirmed, "We're going to get on top of this. I'm pretty sure that we'll get the antenna open, because we've seen other cases of antennas temporarily hung up that responded to subsequent exercising."[22]

On 20 May 1991, the first of Galileo's recommended thermal turns was performed; the spacecraft was rotated 38 degrees to expose the HGA assembly to solar heating. The spacecraft performed with precision, attaining the desired attitude within 4 milliradians (about 0.23 degrees of arc). Unfortunately, the HGA ribs remained stuck, as confirmed by spacecraft wobble data. Before the heating turn, the wobble had an angular magnitude of 3.47 milliradians. Afterwards, it was nearly identical—3.52 milliradians. If the antenna had deployed symmetrically, the spacecraft's wobble should have shrunk to about 1 milliradian (0.057 degrees of arc).[23]

Galileo performed a cooling turn on 10 July 1991, rotating so that the HGA dish pointed almost directly away from the Sun. This orientation put the whole HGA assembly

---

[21] Johnson, "The Galileo High Gain Antenna Deployment Anomaly."

[22] John Noble Wilford, "Stuck Antenna Hinders Jupiter Mission," *New York Times* (17 April 1991): A16; "NASA Considers Mission to Aid Faulty Galileo Probe," *Washington Post* (30 April 1991): A4.

[23] "Unfurling the HGA's Enigma."

in the shade created by the rest of the spacecraft. There it sat for a 32-hour "cold soak." Galileo then rotated again to warm the antenna back up, but still it remained stuck. The spacecraft performed another cooling turn on 13 August 1991. This time, JPL staff reduced Galileo's power consumption in order to cool down the antenna even more than last time. The cold-soak duration was increased to 50 hours. Gary Coyle, the antenna task manager, determined that this more severe cold soak yielded a small fraction of an inch (0.004 inch) of greater contraction of the antenna tower, but it was not sufficient to free the trapped ribs. A check on 19 August 1991 of the spacecraft's wobble and of the partially obscured sun sensor revealed no change from the previous configuration.[24]

Flight controllers were still able to send commands and monitor spacecraft operations, utilizing Galileo's small low-gain antenna (LGA). But this antenna was only capable of transmitting and receiving data at a very slow rate. Project staff believed that the large HGA was essential for transmitting the voluminous data and numerous digitized photographs that Galileo would collect once it reached Jupiter.[25]

As the spacecraft approached the asteroid Gaspra, mission managers turned their attention to the upcoming flyby that was slated to occur in late October 1991. However, they made plans to try to free the stuck HGA with more thermal maneuvers in December. By that time, Galileo would have passed the asteroid and be starting its loop back toward Earth for its second flyby. The craft's December 1991 position would be the farthest it would get from the Sun in several years, and it would thus provide an opportunity for cooling the craft to temperatures even lower than during the last cooling turn in August. NASA managers believed that the December opportunity would provide the best chance for several years to free the crucial antenna.[26]

On 13 December 1991, the Galileo team commanded its spacecraft to turn so that the HGA faced 165 degrees away from the Sun and was in the shade of the craft's sun shield. The team kept Galileo in that position for 50 hours, hoping that lowering the antenna's temperature to -274°F would shrink its 78-inch central tower and free the hung-up pins. They calculated that it would be necessary for the tower to contract by only 0.073 inches.[27]

By 16 December 1991, preliminary data sent from Galileo indicated that the latest repair attempt had not worked. The information from the spacecraft "seems to have the same signature as before," said Bill O'Neil. He was referring to the piece of data from one of Galileo's sun sensors that said that the antenna was still casting the same shadow as before. This diagnosis was confirmed after further analysis of sun-sensor data and the spacecraft's wobble.[28]

NASA continued trying to deploy the HGA. Repeated thermal cycling attempts over the next several months were not successful, however, in freeing the HGA's stuck pins.[29] The anomaly team suggested another recovery technique. Galileo had two LGAs. LGA-2 was mounted on top of a mast that was about 2 meters long. This mast was able

[24] Ibid.

[25] "Stuck Antenna Hinders Jupiter Mission."

[26] Isbell, "Cooling Down Fails to Free Galileo Antenna."

[27] JPL Public Information Office, "Galileo Mission Status: December 11, 1991," 11 December 1991.

[28] Kathy Sawyer, "Galileo Antenna Apparently Still Stuck," *Washington Post* (17 December 1991): A14; Kathy Sawyer, "$1.4 Billion Galileo Mission Appears Crippled," *Washington Post* (18 December 1991): A3.

[29] "NASA Fails to Free Galileo's Antenna," *Washington Times* (11 February 1992): A2; "Spacecraft Antenna Stuck; NASA Will Keep Trying," *Philadelphia Inquirer* (17 April 1992); Douglas Isbell, "NASA Plans to Get Most Out of Disabled Galileo," *Space News* (20–26 April 1992): 9.

to swing 145 degrees, at the end of which it hit a hard stop and imparted a significant impulse to the spacecraft structure. The Galileo team tried swinging LGA-2 six times, but the maneuver was not able to free the stuck HGA ribs.[30]

Leonard Fisk, NASA's Associate Administrator for Space Science, said that more radical solutions, such as using Galileo's thrusters to shake the spacecraft, might free the antenna. Some NASA staff were reluctant to try such drastic measures, worrying that Galileo's long appendages such as its 11-meter (36-foot) boom might be damaged. Such an occurrence could conceivably end the mission.[31]

In spite of some staff members' misgivings, the anomaly team developed an HGA recovery strategy that involved pulsing the mechanical drive system's dual electric motors at 1.25 and 1.875 hertz (cycles per second). During testing of the HGA mechanical drive system, the team had found that the system had a mode of oscillation that resulted from a coupling of the electric motor armature inertia with the gearbox. As a result, the drive system was able to produce pulsing torques that were 40 percent greater than its normal maximum torques. When the spare HGA had been subjected to the pulsing torque, the team had observed that it oscillated at the same frequencies. The team hoped that if this pulsing was tried on Galileo's HGA, the augmented force on the stuck restraint pins might be enough to free them.[32]

The anomaly team made aggressive plans to free the antenna. They set their sights on what might have been one of the best opportunities to deploy the antenna—Galileo's flyby of Earth in December 1992. This would be the closest it would get to the Sun for the remainder of the mission, and it would be a good time to try a combination of thermal turns and pulsed torques. As 1992 drew to a close, project staff sent commands to Galileo to rotate so that the HGA faced 45 degrees from the Sun (turning directly toward the Sun might overheat and damage the antenna further). In addition, staff initiated a "hammering" process, rapidly pulsing the drive motors over 13,000 times during a three-week period between December 1992 and January 1993. Disappointingly, even these radical efforts did not deploy the antenna. The staff also tried other approaches, such as spinning the spacecraft up to its maximum rotation rate of 10 rpm and then pulsing the antenna drive motors once again, but that strategy did not work either.[33]

After numerous unsuccessful efforts to deploy Galileo's HGA, it appeared likely that scientists on Earth would not be able to free the remote antenna. The massive size of Galileo—nearly 2.5 tons—made such a task especially difficult. There did not seem to be any way to shake the spacecraft hard enough to slide the stuck pins out of their niche. Project Manager Bill O'Neil described the situation as "trying to dislodge something caught in the smokestack of the Queen Mary by using a rowboat pushing from the stern.[34]

---

[30] Johnson, "The Galileo High Gain Antenna Deployment Anomaly."

[31] Isbell, "Cooling Down Fails to Free Galileo Antenna"; Michael A. Dornheim, "Galileo Engineers Plan Aggressive Actions to Free Stuck Antenna," *Aviation Week & Space Technology* (29 June 1992): 48.

[32] Johnson, "The Galileo High Gain Antenna Deployment Anomaly."

[33] Sean McNamara, "Space Mission: Mend Galileo Glitch," *USA Today* (29 December 1992): 3A; "Galileo Spacecraft Antenna Fails to Open," *Washington Post* (21 January 1993): A10; "New Telecommunications Strategy," *Galileo Home Page*, July 1995, http://www.jpl.nasa.gov/galileo/hga_fact.html.

[34] Kathy Sawyer, "$1.4 Billion Galileo Mission Appears Crippled," *Washington Post* (18 December 1991): A3.

Should the HGA Anomaly Have Been Prevented?

An analysis of the HGA anomaly by John F. Kross in the periodical *Ad Astra* concluded that human error led to the antenna's failure to open. "The problem should have been anticipated," Kross wrote. "After all, the [molybdenum disulfide] lubricant was applied just once, nearly a decade before launch. Travel is notoriously fraught with risk . . . . When asked if this excess travel was ever addressed as a potential problem, project manager William O'Neil claims, 'I believe that (the travel induced lubricant problem) was not raised as a concern; it was not recognized that . . . these cross country trips of the antenna in the stowed position . . . would be deleterious to the lubrication.'"[35]

The *Ad Astra* article pointed fingers for the HGA anomaly not only at NASA technical and managerial staff, but also at Washington, DC, legislators. "Funding uncertainties and political machinations by multiple congresses and administrations left their scars," said the article. "A perceived lack of funds sent Galileo down its lubricant-wearing journey, symbolic perhaps of the rocky fiscal road the project always faced . . . shrinking slices of the budgetary pie meant reliance on success-oriented missions with little room for error."[36]

One of the results of limited budgets was that no "proof-test model" Galileo spacecraft could be constructed. Proof-test spacecraft served as valuable sources of spare parts for the flight spacecraft. Although the immediate causes of the HGA's difficulties were damaged restraint pin coatings, a *Space News* analysis suggested that the root of the problem may well have been NASA's failure to provide a backup flight-ready HGA. According to Robert Murray, Galileo program manager in 1991, the flight-ready antenna "was a one-of-a-kind unit and we had no spare." JPL did have an extra HGA that it used for testing, but it had no unit that could be substituted for the HGA aboard Galileo. As a result, the HGA that was installed on Galileo "was protected more than" usual from rigorous testing, Murray said. The flight-ready HGA never received the full thermal evaluation that, according to the *Space News* analysis, might have revealed the deployment problem. In 1983 and 1984, the HGA developed prelaunch problems in the form of snags in the antenna dish's fragile metal mesh. To guard the flight-approved HGA from further damage, project managers decided to perform the majority of additional prelaunch testing on a damaged backup unit. According to antenna task manager Gary Coyle, the actual flight-approved HGA was only unfurled approximately a half-dozen times before launch.[37]

Former Galileo Project Manager John Casani had a different view of the reasons leading to the HGA anomaly. He had wanted more flight-ready spare parts to be available for Galileo. In fact, he had wanted a complete proof-test spacecraft to be fabricated, as had been done on previous missions. Such a backup craft would have served as a "spares bed" for the HGA, as well as for other Galileo systems. A proof-test craft would have given the Galileo mission "some robustness and assurance that you wouldn't lose time in the schedule if you have a system or subsystem that has to be refurbished or fails. If you have spares, you can keep going with the test program." Nevertheless, Casani did not believe that having a proof-test model would have prevented the HGA failure. Nor did he think that, in the case of the HGA, funding restrictions were to blame. "The problem [with

---

[35] John F. Kross, "Silent in the Heavens," *Ad Astra* (December 1991): 14–15.

[36] Ibid.

[37] Douglas Isbell, "Galileo Distress Apparent in '80s Tests by NASA," *Space News* (27 May–2 June 1991): 4.

the HGA] was that we didn't recognize the physics of what was going on at the time," he said. "It was a problem that wouldn't show up in tests. Because it was a combination of wearing that lubricant off and then launching (the spacecraft) and it being in a hard vacuum for a period of time." Casani noted that the Galileo team was never able to get the HGA failure to recur in tests on Earth. "The combination of losing the lubricant during transportation, the vibration under vacuum during the launch phase, and the period of time under vacuum in space . . . [all] contributed to it. So, I don't think we would have found the problem. I guess you can say we should have discovered it through analysis or inspection, but I don't think there was anything that happened in the way of budget restrictions or anything else that would have revealed that."[38]

Using air instead of truck transport for the spacecraft's trips between JPL and Florida might have prevented the HGA anomaly. Antennas similar to Galileo's HGA had been successfully deployed on Earth-orbiting Tracking and Data Relay Satellites (TDRSs). But these spacecraft were typically flown to their launch site, rather than driven. Tom Williams, NASA's Deputy Project Manager in 1991 for Advanced TDRSs, commented that "air shipping generally places less loads, less vibrations on a satellite." Air transport was expensive, however. Williams estimated that continental flights of a TDRS cost about $65,000 per plane ride. Such a flight was considered for Galileo, but it appeared more cost-effective to use ground transportation.[39]

There were also other reasons for choosing truck transport. JPL had a tradition of shipping by land that went back a long time. Not only is air transportation very expensive, but the plane also must be scheduled far in advance, eliminating much-needed flexibility. The Galileo team might have found it quite difficult either to delay or to move up the shipment of the spacecraft. In addition, there is no place to land a plane at JPL, so mission staff would have had to load the spacecraft onto a truck anyway in order to transport it to the airport. There, the spacecraft would have been taken off the truck and put onto the airplane. The plane would have had to fly to Patrick Air Force Base because there was no suitable landing strip at Kennedy Space Center. The antenna would then have been taken off the airplane and put onto another truck. John Casani commented, "We always felt that not only was . . . [air transport] more expensive, but it was actually more hazard-ous, because once you get . . . [the spacecraft] onto a truck, you can move it down to Florida safely in a matter of three or four days . . . you were exposing the spacecraft to less damage potential from the handling than by putting it on an airplane, which required a lot more handling."[40]

The antenna deployment failure was a huge setback for the mission. But thanks to the ingenuity and creativity of the Galileo team, the event proved not to be a mission-ending failure. In fact, one of the most notable achievements of the mission was the innovation of systems to work around the antenna failure.

---

[38] John Casani interview, tape-recorded telephone conversation, 29 May 2001.

[39] Kross, p. 14. The $65,000 for cross-country plane transportation was a 1991 estimate. Air costs for Galileo's 1985 and 1989 cross-country trips might have been a little less.

[40] Casani interview.

Even with a Stuck Antenna, the Mission Could Go On

As attempts to free the stuck HGA failed, Galileo planners attacked the difficult question of what to do if the antenna could not ultimately be opened. An operational HGA would have been able to transmit data from Jupiter at 134,000 bits per second, versus the LGAs' more limited capability of only 10 to 40 bits per second. The difference between the two types of antennas was comparable to the difference between a powerful searchlight and a dim incandescent bulb. The HGA's reflecting dish was supposed to unfurl into a parabolic shape, the same shape as the reflector of a searchlight or an automobile head-light. This parabolic reflector would have focused the HGA's energy into a tight beam directed toward Earth. The LGAs, on the other hand, emitted unfocused broadcasts in the same way that a bare lightbulb gives off light in all directions. Only a tiny fraction of the LGA's broadcast would be directed toward Earth. Because the received signal would be extremely faint, data would have to be sent at a much lower rate in order to ensure that the contents were decipherable. The minuscule data transmission rate would especially impact the number of digitized photos that could be sent. Project managers had hoped to receive 50,000 images of the Jovian system—enough to make movies of the planet's swirling weather patterns and document their rapidly changing dynamics.[41]

In spite of the LGA's slow transmission rate, project staff devised ways to complete 70 percent of the mission's scientific goals. "The exciting news we have today," said Bill O'Neil, "is that we have found a way to return a majority of the science."[42] Many of the scientific studies were able to do just fine using the LGA to transmit their data. Some experiments required only a few images. The magnetic fields investigation required none at all.

Project scientists worked hard on developing approaches for compressing the data that Galileo would collect so that maximum use of the LGA's limited data-transmission capacity could be realized. Through squeezing more data into fewer computer bits, the scientists hoped to increase the transmission capacity of the LGAs at least a hundred times. Lead scientist Torrence Johnson estimated that the LGAs would then be able to transmit 2,000 to 4,000 images during the two years that were originally planned for Galileo to explore the Jovian system. These would be the highest priority pictures, with fine resolution of detail. What would mostly be lost were lower resolution images, motion pictures of Jupiter's atmosphere, and details of the magnetic field and associated particles.[43]

Project engineers developed two sets of new flight software that would help compensate for the HGA difficulties. The Phase 1 software was radioed to Galileo first, and it began operating in March 1995 as the craft was making its approach to Jupiter. NASA designed Phase 1 to ensure that the Atmospheric Probe's most important data made it safely to Earth. Because the data could not be sent as quickly to Earth with the LGAs

[41] "NASA Is Optimistic on Salvaging Jupiter Missions," *NY Times* (12 June 1992): D17; Kathy Sawyer, "Galileo's Mission to Jupiter Can Be Salvaged, Scientists Say," *Washington Post* (12 June 1992): A2; "Effort Begins to Free Space Probe's Balky Antenna," from UPI, 29 December 1992; Michael A. Dornheim, "Prospects Dim for Stuck Antenna," *Aviation Week & Space Technology* (11 January 1993); "New Telecommunications Strategy," *Galileo Home Page*, July 1995, http://www.jpl.nasa.gov/galileo/hga_fact.html.

[42] "Galileo's Mission to Jupiter Can Be Salvaged, Scientists Say," p. A2.

[43] Ibid., p. A2; Robert C. Cowen, "Galileo on Course, But One Antenna Still Won't Unfurl," *Christian Science Monitor* (13 January 1993): 6.

as with the HGA, dependable data storage aboard Galileo took on new importance. The Phase 1 software provided a partial backup for Probe data.[44]

After the Probe's valuable data were received by Earth, NASA radioed Galileo the Phase 2 software in March 1996. Its job was data compression—shrinking the enormous files of information that Galileo was collecting while retaining the most scientifically important material. Only then would the modified data be transmitted to Earth. Phase 2's first task was to process data collected during the Orbiter's final approach to the Jovian system in late 1995, including data from its close encounters with the satellites Europa and Io.[45]

Phase 2 used two different approaches to data compression. One was termed "lossless compression," in which the data package could be expanded to its original state once it reached Earth and all information in the original file could be recovered. This approach is one that personal computers commonly employ (in "zipping" and "unzipping" files) in order to augment their data-transmission capabilities. The other approach used in the Galileo mission was called "lossy" compression, in which mathematical approximations were employed to abbreviate the total amount of data sent to Earth. One way the software did this was to minimize or eliminate the least valuable part of data sets, such as the dark background of space in many of the images. The approximations had to be carefully chosen so that the information lost was not critical.[46]

Gerry Snyder was a systems engineer who worked on the Galileo project in the late 1970s but then left. He was brought back to the project in 1992, when the trouble with the HGA occurred and data flow through the LGA had to be maximized. The spacecraft had been handling data using an inefficient system called "time division multiplex," or TDM. If, for instance, one of the spacecraft's instruments was generating the same data time after time, TDM would send the repetitive information again and again to Earth, wasting valuable transmitting time. This method wouldn't have been a problem if the HGA had been operating, for it would have sent data at a high enough rate (134,000 bits per second) that Earth would have received all the data it wanted, even if some of it was repetitive. But the LGA could only send a trickle of information (about 10 to 40 bits per second), and it all had to count. What Snyder did when he rejoined the Galileo project was to implement a "packetized data" system in which the spacecraft computer kept a watch on the data flow and only sent a packet to Earth when there was something new. The packetized system selected just the data that mission personnel on Earth needed to reconstruct what the spacecraft was doing and observing.[47]

When the spacecraft computer used the packetized-data approach, an image taken by Galileo's SSI camera could be reconstructed on Earth, employing as little as one-eightieth of the data that TDM would have sent, with very good visual quality. The Galileo spacecraft only had two or so months after an encounter to send what it had collected to Earth before the tape recorder had to store information from the next encounter. Because the spacecraft had to transmit data at a very slow rate, it could not send all the information it collected. Galileo scientists had to make tough decisions

---

[44] "New Telecommunications Strategy."

[45] Ibid.

[46] Ibid.

[47] Galileo mission support crew interview, JPL, 5 August 2001.

about what to transmit and what to erase. Packetized data were important because they allowed significantly more of the observational data to be sent to Earth.[48]

The Galileo team also augmented data flow by improving the ground antenna systems of Earth's Deep Space Network. This provided "a larger, more sensitive ear"[49] that could extract additional information out of the faint signals that Galileo's LGAs transmitted. Of the DSN's three stations around the world, the one in Canberra, Australia, received particular attention during the upgrades. Australia is in the Southern Hemisphere, and Jupiter was in the southern sky during much of the mission. Enhancements to the DSN included the linking of Canberra's 70-meter and two 34-meter antennas so that all could receive Galileo's signals concurrently, thereby allowing the electronic signals generated by each antenna to be combined. Another antenna in Australia, the Parkes Radio Telescope, was also arrayed with the Canberra antennas, as was the DSN's 70-meter antenna in Goldstone, California, when its view of Galileo overlapped with Australia's. The intercontinental array began operations just in time for Galileo's flyby of the Jovian moon Callisto on 4 November 1996. The event occurred when Galileo was at one of its most distant locations from Earth, which added to the difficulty of picking up the spacecraft's whisper of a signal. With the array, more of Galileo's signal was captured, allowing a greater rate of data transmission.[50]

These creative maneuvers for doing more with less were geared toward achieving 80 percent of the originally planned atmospheric research, 60 percent of the magnetic field studies, and 70 percent of the work on observing Jupiter's satellites. Overall, the improvements enabled Galileo to attain 70 percent of its original science goals. Bill O'Neil summed up the Galileo recovery effort by saying, "All in all, it's still an excellent mission."[51] In addition, Galileo's hardware, software, and antenna-arraying modifications significantly enhanced deep space communications for other missions. This benefit was foreseen by Paul Westmoreland, Director of Telecommunications and Mission Operations for JPL at the time, who stressed that Galileo's "data compression and encoding techniques will be especially useful for the new era of our faster, better, cheaper interplanetary missions . . . . This opens the way for mission developers to reduce future spacecraft and operations costs by using smaller spacecraft antennas and transmitters."[52]

[48] Gerry Snyder, Galileo mission support systems engineer, interview, JPL, 5 August 2001.

[49] "Galileo's Mission to Jupiter Can Be Salvaged," p. A2.

[50] "New Telecommunications Strategy"; Mary Beth Murrill, "NASA Links Antennas To Maximize Galileo Data Return," JPL Public Information Office news release, 7 November 1996, accessed at *http://www.jpl.nasa.gov/ galileo/status961107.html*.

[51] "Galileo Mission to Jupiter Can Be Salvaged, Scientists Say"; Cowen, "Galileo on Course, but One Antenna Still Won't Unfurl"; Murrill, "NASA Links Antennas To Maximize Galileo Data Return."

[52] Murrill, "NASA Links Antennas To Maximize Galileo Data Return."

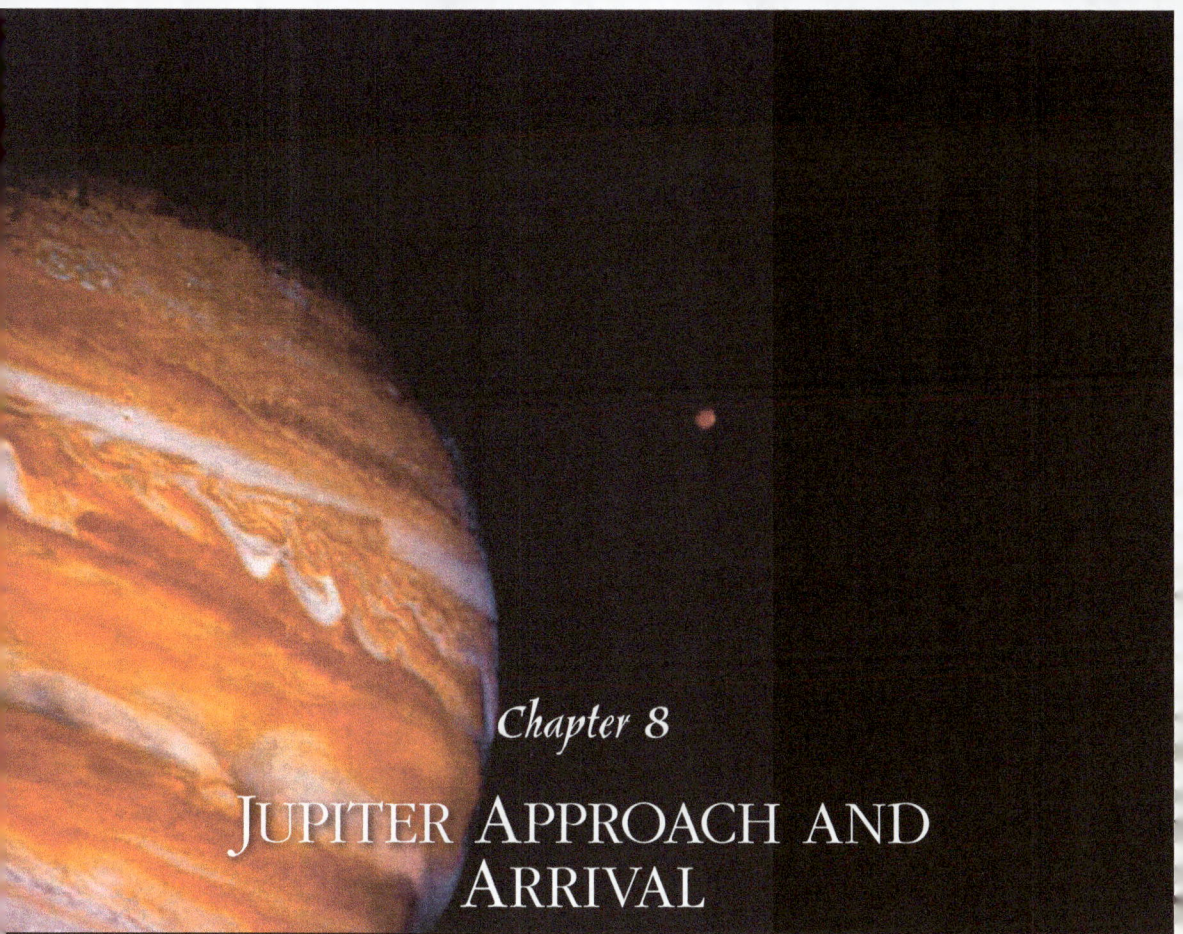

Chapter 8

JUPITER APPROACH AND ARRIVAL

GALILEO FLEW BY EARTH FOR THE LAST TIME on 8 December 1992. Finished with its Venus and Earth gravity-assists, Galileo headed toward the outer solar system on a leg of its voyage that lasted three years and ultimately brought it to Jupiter (see figure 8.1). Two experiences of this long journey were described in previous chapters: Galileo's encounter with the asteroid Ida and its little moon (chapter 6) and the continued efforts to deploy the spacecraft's high-gain antenna (chapter 7). Other major mission events that occurred during the long space voyage and immediately after arrival at Jupiter included the following:

- A serendipitous encounter with a comet.

- A trip through the most intense interplanetary dust storm that a spacecraft had ever confronted.

- A serious malfunction of Galileo's tape recorder, the spacecraft's main data-storage device, which threatened the success of the entire mission.

- The descent of Galileo's Probe into the atmosphere of Jupiter.

- The insertion of the Galileo Orbiter into a stable trajectory around Jupiter.

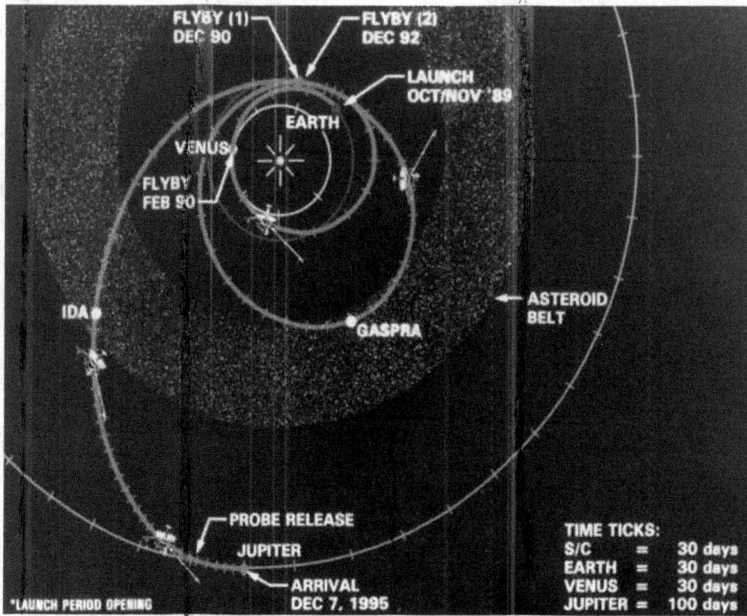

Figure 8.1. **The Galileo trajectory from launch to Jupiter arrival. This drawing charts Galileo's flightpath to Jupiter through its three planetary gravity-assists and two asteroid encounters. Gravity-assists included those from Venus (February 1990) and from Earth (December 1990 and December 1992). Encounters with the asteroids Gaspra and Ida are also depicted. Galileo arrived at Jupiter on 7 December 1995. (NASA photo number S89-44173)**

A Midmission Opportunity: The Comet Spectacle

In July 1992, five months before Galileo made its last pass by Earth, a comet was being torn to pieces as it skimmed by Jupiter, passing less than one-third of a Jovian radius (about 21,000 kilometers or 13,000 miles) above the planet's surface. The comet was passing closer to the planet than its Roche limit. Within the Roche limit, the unequal gravitational attractions of a planet on a body's near and far sides will rip that body apart.

Astronomer Fred L. Whipple has compared comets to "dirty snowballs"[1] in that their nuclei, which contain almost all of their mass, are typically conglomerates of ice and dust, held together by mutual gravitational attraction. In particular, Whipple suggested that within comet nuclei, icy cores lie inside thin, insulating layers of dust. It was this ice-and-dust nucleus that Jupiter's gravitational force was fragmenting. Afterwards, the shattered comet continued on its path away from Jupiter and toward interplanetary space. It could

---

[1] Sonoma State University Department of Physics and Astronomy, "The Bruce Medalists—Fred Lawrence Whipple," 10 September 2003, *http://www.phys-astro.sonoma.edu/BruceMedalists/* (available in Galileo—Meltzer Sources, folder 18522, NASA Historical Reference Collection, Washington, DC); "At Age 92 Astronomer Fred Whipple Is Joining NASA Space Team to Explore Three Comets in 2002," *Cornell News* (26 July 1999), *http://www.news.cornell.edu/releases/July99/Whipple.bpf.html* (available in Galileo—Meltzer Sources, folder 18522, NASA Historical Reference Collection, Washington, DC).

go only so far, however. Although its pieces traveled many millions of kilometers, they could not escape Jupiter's massive gravity and would eventually be pulled back.[2]

On 24 March 1993, eight months after the comet's fragmentation, world-class comet-seekers Eugene and Carolyn Shoemaker and David Levy were taking a photo of the night sky with the Palomar Observatory's 0.4-meter Schmidt telescope. Two days after the observation, Carolyn Shoemaker examined the photographic plates and saw a strange linear object with a fan of light on one side. Subsequent analyses by observatories in Hawaii, Arizona, and Texas confirmed that the object was a comet orbiting Jupiter that had made a very close approach on 7 July 1992. Imaging conducted by the Spacewatch Telescope on Arizona's Kitt Peak and the Mauna Kea Observatory in Hawaii exposed the comet's unusual form—a line of numerous pieces resembling a "string of pearls."[3]

In the past, the Shoemakers and David Levy had jointly discovered eight other comets, so they named this one Shoemaker-Levy 9 (S/L-9). Further analyses showed that the comet's fragments would return to Jupiter approximately two years after their encounter with the planet, during the period from 16 through 22 July 1994. This time, the comet would not fly by the planet; it would smash into it. When this happened, the Galileo spacecraft, though still far away, would have an excellent view of the collision sites. This was fortuitous, because the comet's fragments would hit Jupiter's night side, away from the Sun. Earth-based telescopes would be able to see evidence of the impact, such as ejecta blown into space, but would not be able to observe the impact sites themselves.[4]

More energy was released during the Shoemaker-Levy spectacle than from the largest nuclear warhead explosions ever conducted on Earth. The event represented the first time in human history when people had discovered a body in the sky and been able to predict its collision with a planet more than seconds in advance. The Shoemaker-Levy crash into Jupiter also gave credence to astronomers' predictions that similar events may occur on our home planet. This was one of the reasons why thousands of astronomers on Earth, both professional and amateur, were so interested in studying the details of Shoemaker-Levy's demise. The comet's crash on Jupiter demonstrated one of the solar system's basic physical processes. Better knowledge of the phenomenon could help predict how serious the damage to Earth might be if various-sized bodies struck it.

A similar event to Shoemaker-Levy might in fact have occurred on Earth 360 million years ago, leaving evidence that can still be studied today. Radar images of Earth have revealed a chain of craters in the central African country of Chad that may have been created by a fragmented comet or asteroid comparable in size to Shoemaker-Levy. Such an impact may also have contributed to the mass biological extinctions that were occurring

---

[2] "Comet Shoemaker-Levy Background," *Comet Shoemaker-Levy Collision with Jupiter, http://www2.jpl.nasa. gov/sl9/background.html* (accessed 13 March 2000) (available in Galileo—Meltzer Sources, folder 18522, NASA Historical Reference Collection, Washington, DC); David M. Harland, *Jupiter Odyssey: The Story of NASA's Galileo Mission* (United Kingdom: Springer-Praxis Books, 2000); "Comet," *Microsoft Encarta Online Encyclopedia, http:// encarta.msn.com*, 2001 (available in Galileo—Meltzer Sources, folder 18522, NASA Historical Reference Collection, Washington, DC).

[3] JPL, "Comet Shoemaker-Levy Background"; Harland, *Jupiter Odyssey*; "Comet Freight Train to Collide with Jupiter," *Comet Shoemaker-Levy Collision with Jupiter, http://www2.jpl.nasa.gov/sl9/back1.html* (accessed 13 March 2001) (available in Galileo—Meltzer Sources, folder 18522, NASA Historical Reference Collection, Washington, DC).

[4] JPL, "Comet Shoemaker-Levy Background"; Harland, *Jupiter Odyssey*.

at the time. In addition, there is growing evidence that the energy from a large comet or asteroid crash caused the extinction of dinosaurs 65 million years ago.[5]

The Collision

In the days prior to impact, fragments of the broken-up comet stretched in a line across 5 million kilometers (3 million miles) of space, more than 12 times the distance from Earth to our Moon. They sped like a runaway freight train toward their rendezvous with Jupiter. This "train" expanded in length as it flew through space, due to the different velocities between fragments. Some astronomers, believing it was unlikely that the comet crash would be seen through small amateur telescopes, discouraged the public from expecting too much of the impact. An article in *Nature* entitled "The Big Fizzle Is Coming"[6] hypothesized that each comet fragment would ablate and burn up like a meteoroid in Jupiter's upper atmosphere. The article said that because the fragments did not have sufficient momentum or structural integrity, they would probably not penetrate deeply into the atmosphere, where they might have exploded with thousands of megatons of energy. The result was thus predicted to be a spectacular meteor shower, but not the giant fireball explosions that other researchers thought would occur.[7]

How wrong this prediction turned out to be. From 16 through 22 July 1994, 21 discernible pieces of Comet Shoemaker-Levy 9 collided with Jupiter in an event that was called "simply spectacular and beyond expectations."[8] When the first comet fragment hit Jupiter at over 320,000 kilometers per hour (200,000 mph), it generated an intensely bright fireball at least 3,000 kilometers (1,900 miles) high. Although the impact site was hidden from Earth because it was on Jupiter's night side, the fireball was so large and bright that even 2-inch amateur refractor telescopes on Earth could see it. The Shoemaker-Levy event was the first collision of two solar system bodies ever observed. Impacts of the cometary fragments, which ranged in size up to 2 kilometers, resulted in the generation of plumes spewing material many thousands of kilometers up from Jupiter; hot bubbles of gas in the atmosphere; and large, dark, scarlike spots on the atmosphere that persisted for weeks. Some of these spots were two to three times the size of Earth.[9]

---

[5] "Comet Freight Train to Collide with Jupiter"; Douglas Isbell and Mary Hardin, "Chain of Impact Craters Suggested by Spaceborne Radar Images," *http://www2.jpl.nasa.gov/sl9/news80.html* (accessed 17 March 2001) (available in Galileo–Meltzer Sources, folder 18522, NASA Historical Reference Collection, Washington, DC); Ray L. Newburn, Jr., "Background Material for Science Teachers: July 1994–Periodic Comet Shoemaker-Levy 9 Collides with Jupiter," *http://www.seds.org/sl9/Educator/intro.html* (accessed 27 December 2001) (available in Galileo–Meltzer Sources, folder 18522, NASA Historical Reference Collection, Washington, DC).

[6] Paul Weissman, "The Big Fizzle is Coming," *Nature* (14 July 1994), *http://www.jpl.nasa.gov/sl9/news9.html* (available in Galileo–Meltzer Sources, folder 18522, NASA Historical Reference Collection, Washington, DC).

[7] Amara Graps, *Comet Shoemaker Levy 9 Impact: Analysis and Interpretation (6 Months After Event)* (presentation at Macworld Exposition, San Francisco, CA, 4 January 1995).

[8] JPL, *Comet Shoemaker-Levy Collision with Jupiter*, *http://www2.jpl.nasa.gov/sl9/sl9.html* (accessed December 2004) (available in Galileo–Meltzer Sources, folder 18522, NASA Historical Reference Collection, Washington, DC).

[9] Graps, presentation at Macworld; JPL, *Comet Shoemaker-Levy Collision with Jupiter, http://www2.jpl.nasa.gov/sl9/sl9.html* (accessed 17 March 2001) (available in Galileo–Meltzer Sources, folder 18522, NASA Historical Reference Collection, Washington, DC); "Comet Shoemaker-Levy 9 Collision with Jupiter," *National Space Science Data Center (NSSDC)* Web site, *http://nssdc.gsfc.nasa.gov/planetary/comet.html* (accessed 13 March 2001) (available in Galileo–Meltzer Sources, folder 18522, NASA Historical Reference Collection, Washington, DC); "Results of SL9's Impacts Thrill Scientists," *JPL Universe, 29 July 1994, http://www2.jpl.nasa.gov/sl9/uni1.html* (available in Galileo–Meltzer Sources, folder 18522, NASA Historical Reference Collection, Washington, DC).

Effects of the comet-fragment collisions were discerned by most Earth-based observatories, as well as by spacecraft that included Galileo, the Hubble Space Telescope, Ulysses, and Voyager 2. Each fragment impact began with a flash that gradually declined in brightness (see figure 8.2). When Galileo observed a fragment impact event in ultraviolet light, the total lifetime of the event appeared to last about 10 seconds, while in infrared light, events appeared to continue for 90 seconds or more.

Figure 8.2. **Comet Shoemaker-Levy 9 fragment W impact with Jupiter. These four images of Jupiter and the luminous night-side impact of fragment W of comet Shoemaker-Levy 9 were taken by the Galileo spacecraft on 22 July 1994, when it was 238 million kilometers (148 million miles) from Jupiter. The images were taken at intervals of 2⅓ seconds using a green filter (visible light). The first image, taken at an equivalent time to 8:06:10 Greenwich mean time (1:06 a.m. Pacific daylight time, or PDT), shows no impact. In the next three images, a point of light appears, brightens so much as to saturate its picture element, and then fades again, 7 seconds after the first picture. (NASA image number PIA00139)**

The fragments released enormous energy when they crashed into Jupiter. Ultraviolet data taken by Galileo showed a 20-percent increase in the total brightness of the entire planet during the impact of one large comet fragment (the fragment labeled "G"). During the collisions of other fragments, the brightness of Jupiter typically increased 10 percent. Fragment G's impact was followed by the eruption of spectacular plumes of hot gases. Fragment G created a fireball 7 kilometers (5 miles) in diameter that burned with a temperature of at least 8,000 K (14,000°F), hotter than the Sun's surface. Galileo's near infrared mapping spectrometer (NIMS) observed the fireball's evolution. In only 1.5 minutes, it expanded to a size hundreds of miles across and cooled swiftly to 400 K (260°F).[10]

---

[10] *Comet Shoemaker-Levy 9*; C. W. Hord, W. R. Pryor, A. I. F. Stewart, K. E. Simmons, J. J. Gebben, C. A. Barth, W. E. McClintock, L. W. Esposito, W. K. Tobiska, R. A. West, S. J. Edberg, J. M. Ajello, and K. L. Naviaux, "Direct Observations of the Comet Shoemaker-Levy Fragment G Impact by Galileo UVS," *Geophysical Research Letters* 22, no. 12 (15 June 1995): 1565; James Wilson, "Galileo Comet SL9 Observations," JPL Public Information Office news release, 31 October 1994, *http://www2.jpl.nasa.gov/sl9/gll22.html* (available in Galileo—Meltzer Sources, folder 18522, NASA Historical Reference Collection, Washington, DC).

Scientists divided Shoemaker-Levy's collision events into three phases:

- Meteor phase: The incoming comet fragments hit the Jovian atmosphere and heated.

- Fireball phase: The fragments exploded into a fireball of extremely hot gas and debris that expanded out of the atmosphere, then cooled to form plumes that many observing stations recorded.

- Splash phase: The plume material fell back into the Jovian atmosphere and heated it, which led to intense thermal emissions.[11]

Preliminary analysis of Galileo NIMS data indicated that the splash-phase material might have contained the hot remnants of water molecules that originated either from the comet's vaporized ice or from Jupiter's presumed water clouds. The splash material also contained carbon and hydrogen, derived either from Jovian atmospheric methane or from the comet's hydrocarbons. Either Jupiter's atmosphere or the comet may have contributed micrometer-sized carbon particles (a micrometer is 1 millionth of a meter, or about 40 millionths on an inch) that formed the immense black patches that persisted in Jupiter's high atmosphere for weeks after the comet impact. The patches may also have been caused by sulfur that erupted during the comet impact from a lower Jovian cloud layer of condensed ammonium hydrosulfide.[12]

Interpretation of the Data

In the months that followed Shoemaker-Levy's crash into Jupiter, data collected by Galileo, other spacecraft, the Hubble Space Telescope, and Earth observatories were painstakingly analyzed. Scientists believe that the comet was a body that had orbited the Sun for 4.5 billion years and had only recently been captured by Jupiter, possibly just a few decades earlier. It was the first comet ever observed orbiting Jupiter (comets generally orbit the Sun).

The composition of Shoemaker-Levy's impact plumes was investigated in many different wavelength bands using spectroscopic techniques. Surprisingly, the data indicated that the plumes were devoid of water. Since 1949, when the astronomer Fred L. Whipple proposed that comets were "dirty snowball" conglomerates of dust and ices of various compounds, the space science community has believed that comets typically contain large percentages of water ice. Halley's Comet, for instance, was found in 1986 to contain about 80 percent water ice. Did Shoemaker-Levy's hydrogen and oxygen molecules combine with other elements to form new molecules? Or, alternatively, could the "comet" have actually been an asteroid? Comets and asteroids have much in common. They are primordial in origin, in that they were both formed 4.6 billion years ago, along with the solar system's planets and their satellites. Also, both comets and asteroids are found near Jupiter. But the main difference between the two types of bodies is that

---

[11] A Comet's Fiery Dance at Jupiter," *Galileo Messenger* (May 1995), *http://www2.jpl.nasa.gov/sl9/gll29.html* (available in Galileo—Meltzer Sources, folder 18522, NASA Historical Reference Collection, Washington, DC).

[12] Ibid.

comets are composed largely of ices, while asteroids are virtually devoid of ices because they formed too close to the Sun.[13]

In spite of the lack of water detected, some scientists believe that Shoemaker-Levy very likely had the composition of a comet. Sensitivity limitations of receiving stations may have been responsible for the lack of typical comet emissions being observed spectroscopically. The strongest evidence that Shoemaker-Levy was indeed a comet was the existence of a persistent, circularly symmetric coma (an envelope of gas and dust) around each of its fragments. A comet's coma is formed when the Sun sublimates some of its ices into gases, which also results in a release of a portion of the comet's dust.[14]

Ultraviolet image data suggested that the comet initiated enhanced auroral phenomena in Jupiter's atmosphere. Auroras are luminous atmospheric activities caused by charged particles entering an atmosphere and colliding with and exciting its gas molecules, which then emit electromagnetic radiation in the visible as well as other parts of the spectrum. Jovian auroral displays were first observed near the planet's northern pole. Scientists believe that the comet fragments' collisions with Jupiter formed many charged particles, which then moved rapidly along magnetic field lines, plunged into the Jovian atmosphere near the poles, and created auroral displays.[15]

Images taken in the hours after the larger fragments of Shoemaker-Levy hit Jupiter show transient dark rings, reminiscent of the circular ripples that propagate outward when a stone is thrown into a pond, surrounding the collision sites. The dark rings are likely some sort of wave phenomenon. It is not clear, however, why we can see them. The rings occur high in the Jovian atmosphere; the gases there are normally transparent. The wave phenomenon may cause partial condensation of the gases into visible droplets or particles. Either an upward displacement or a rarefaction could initiate condensation by lowering the temperature. Alternatively, the comet fragment impacts might render normally transparent gases visible due to the chemical changes and condensation occurring during the explosive collision.[16]

---

[13] Graps; Donald Savage, Jim Elliott, and Ray Villard, "Hubble Observations Shed New Light on Jupiter Collision," NASA news release number 94-161, 29 September 1994, Comet Shoemaker-Levy Home Page (JPL), *http://www.jpl.nasa.gov/sl9/hst15.html* (available in Galileo—Meltzer Sources, folder 18522, NASA Historical Reference Collection, Washington, DC); R. M. West, "Comet Shoemaker-Levy 9 Collides with Jupiter: The Continuation of a Unique Experience," *ESO Messenger* (European Southern Observatory, Germany, September 1994; "Comet," *Microsoft Encarta Online Encyclopedia*, 2001, *http://encarta.msn.com* (available in Galileo—Meltzer Sources, folder 18522, NASA Historical Reference Collection, Washington, DC).

[14] H. A. Weaver, M. F. A'Hearn, C. Arpigny, D. C. Boice, P. D. Feldman, S. M. Larson, P. Lamy, D. H. Levy, B. G. Marsden, K. J. Meech, K. S. Noll, J. V. Scotti, Z. Sekanina, C. S. Shoemaker, E. M. Shoemaker, T. E. Smith, S. A. Stern, A. D. Storrs, J. T. Trauger, D. K. Yeomans, and B. Zellner, "The Hubble Space Telescope (HST) Observing Campaign on Comet Shoemaker-Levy 9," *Science* 267 (3 March 1995): 1287; "Astronomy: A.2 Comets and Asteroids," *Microsoft Encarta Online Encyclopedia*, 2001, *http://encarta.msn.com/find/print.asp?&pg=8&ti=1741502444&sc=23&pt=1&pn =5* (available in Galileo—Meltzer Sources, folder 18522, NASA Historical Reference Collection, Washington, DC).

[15] "Aurora (phenomenon)," *Microsoft Encarta Online Encyclopedia*, 2001, *http://encarta.msn.com* (available in Galileo—Meltzer Sources, folder 18522, NASA Historical Reference Collection, Washington, DC); R. M. West, "Comet Shoemaker-Levy 9 Collides with Jupiter."

[16] H. B. Hammel, R. F. Beebe, A. P. Ingersoll, G. S. Orton, J. R. Mills, A. A. Simon, P. Chodas, J. T. Clarke, E. De Jong, T. E. Dowling, J. Harrington, L. F. Huber, E. Karkoschka, C. M. Santori, A. Toigo, D. Yeomans, and R. A. West, "HST Imaging of Atmospheric Phenomena Created by the Impact of Comet Shoemaker-Levy 9," *Science* 267 (3 March 1995): 1293.

The Orbiter and Probe Separate

On 12 July 1995 (PDT), Galileo released its 750-pound Jupiter Atmospheric Probe (see figure 8.3). The spacecraft was 83 million kilometers (52 million miles) and five months away from Jupiter at the time.[17] In December 1995, the Probe would become the first "Earth emissary" to plunge into the atmosphere of one of the solar system's gas-giant planets. Alvin Seiff, principal investigator for the Probe's atmospheric-structure experiments, had proposed back in 1969 that a Jupiter entry vehicle be designed and constructed. He commented that the temperatures to which the Probe would be subjected, which would exceed those at the surface of the Sun, would be "unprecedented, enormous, inconceivable," and he went on to say that conducting a successful Probe mission into Jupiter's atmosphere would involve "the most challenging entry problem in the solar system, except for [entering] the Sun itself."[18]

Figure 8.3. **An artist's concept of the Galileo Probe inside its aeroshell being released from the mother ship, the Galileo Orbiter, prior to arrival at Jupiter. (NASA photo number ACD95-0229)**

News of the Probe's release took 37 minutes to travel at the speed of light across the 664 million kilometers (412 million miles) of space from Galileo to its control crew in California. At 11:07 p.m. PDT on Wednesday, 12 July 1995, points on a Mission Support Area monitor displaying Doppler-shift data took a sudden dip. This tiny frequency shift in the signal resulted from a small change in the Orbiter's velocity along the Earth-line of some tens of millimeters per second, and it indicated that the Probe had been sent on its freefall path toward Jupiter. The news was greeted with jubilation. William J. O'Neil, Galileo Project Manager, commented at the time, "We're just elated. It's been a long haul for us."[19]

[17] JPL, "Probe Release Status Updates," JPL news release, 13 July 1995, *http://www2.jpl.nasa.gov/galileo/release.html* (available in Galileo—Meltzer Sources, folder 18522, NASA Historical Reference Collection, Washington, DC).

[18] Kathy Sawyer, "Galileo Sends Probe Toward Jupiter Study," *Washington Post* (14 July 1995): A3.

[19] JPL, "Probe Release Status Updates."

Critical Course Corrections

On 27 July 1995, two weeks after the Probe's release, the Galileo Orbiter fired its 400-newton main engine for 5 minutes and 8 seconds, performing an important maneuver that would deflect it away from a collision course with Jupiter. NASA maintained a real-time telemetry link with the spacecraft throughout the burn, tracking changes in the Orbiter's velocity. These changes could be calculated from the Orbiter's accelerometer data, as well as from Doppler shifts in the craft's radio signals. From this information, the Orbiter's engine performance was deduced.[20]

This was the first major burn by Galileo's German-manufactured engine, and it demonstrated the engine's operability. Firing the engine had to wait until the release of the Atmospheric Probe because the Probe had been mounted in front of the engine nozzle. The burn altered the Orbiter's speed by 61 meters per second (about 140 mph), within a meter per second of the planned change. The trajectory modification resulting from this maneuver was enough for the Orbiter to miss Jupiter by 230,000 kilometers (140,000 miles) rather than colliding with the planet. The course change targeted the spacecraft to fly by Jupiter's volcanic moon Io on 7 December 1995, the same day that the Probe would plunge into Jupiter's atmosphere.[21]

The Orbiter had a critical task to perform in conjunction with the Probe. The Orbiter had to fly high overhead and receive data from the Probe as it fell through the Jovian atmosphere, then relay the information back to Earth. This information would have to be transmitted at a far slower rate than originally intended because of the failure to deploy the Orbiter's high-gain antenna (see chapter 7 on the high-gain antenna failure for more details). However, the Probe's Project Manager, Marcie Smith of Ames Research Center, was optimistic that the Orbiter's low-gain antennas would be sufficient to eventually transmit all of the data collected by the spacecraft.[22]

Interplanetary Dust Storm

In August 1995, when the Galileo Orbiter was 63 million kilometers (39 million miles) from Jupiter, it entered the most intense interplanetary dust storm ever observed. The storm was so large that it took several months for the spacecraft to journey through it. During normal interplanetary travel, a spacecraft comes across about one dust particle every three days. In plowing through the most severe of the several storms that Galileo encountered, the spacecraft counted up to 20,000 particles per day. The particles' impact trajectories on Galileo's dust detector clearly indicated that they originated from somewhere in the Jovian system.[23]

---

[20] "Orbiter Deflection Maneuver Status July 27," *JPL Galileo Home Page, http://www2.jpl.nasa.gov/galileo/odm.html* (accessed 22 March 2001) (available in Galileo—Meltzer Sources, folder 18522, NASA Historical Reference Collection, Washington, DC).

[21] JPL, "Critical Engine Firing Successful for Jupiter-Bound Galileo," JPL Public Information Office news release, 27 July 1995, *http://www2.jpl.nasa.gov/galileo/oldpress/odm.html* (available in Galileo—Meltzer Sources, folder 18522, NASA Historical Reference Collection, Washington, DC); "Orbiter Deflection Maneuver Status July 27.

[22] Sawyer, "Galileo Sends Probe Toward Jupiter Study."

[23] JPL, "Galileo Flying Through Intense Dust Storm on Way to Jupiter," JPL Public Information Office news release, 29 August 1995, *http://www2.jpl.nasa.gov/galileo/duststorm.html* (available in Galileo—Meltzer Sources, folder 18522, NASA Historical Reference Collection, Washington, DC); A. L. Graps, E. Grun, H. Svedhem, H. Kruger, M. Horanyi, A. Heck, and S. Lammers, "Io as a Source of the Jovian Dust Streams," *Nature* 405 (4 May 2000): 48–50; M. Horanyi, G. Morfill, and E. Grun, "Mechanism for the Acceleration and Ejection of Dust Grains from Jupiter's Magnetosphere," *Nature* 363 (13 May 1993): 144–146.

These interplanetary streams of dust were first discovered by the Ulysses spacecraft in 1992 and were greeted with much surprise by the space science community. Interplanetary space had been known to contain dust, but the presence of discrete streams, at the densities observed, had been totally unexpected. How could large populations of dust escape Jupiter's immense gravitational field? The answer appears to be that the dust first encountered extensive plasmas, electrically charged ionized gases, in Jupiter's magnetosphere, and these induced charges on the dust particles. The dust's motion could then be controlled not only by Jupiter's gravitational field, but also by its electromagnetic fields, which exert strong forces on charged particles. Models of these field-particle interactions developed at the University of Colorado and Germany's Max Planck Institute indicated that only particles in a certain size range could gain sufficient energy from electromagnetic forces to escape Jupiter's gravity. This constraint proved useful in identifying the sources of the particles.[24]

Scientists originally suspected that the Shoemaker-Levy comet could be the source of the dust, emitting particulates either when it flew close by Jupiter in 1992 and fragmented or when it smashed into the planet in 1994. But no correlation could be found between the dust storms and comet events. Some of the dust streams occurred before known comet events, while others occurred long after these events.[25]

The scientific community also thought that Jupiter's gossamer ring, one of the faint particle rings surrounding the planet, could be the dust source. The gossamer ring is made up mainly of micrometer-sized particles that are located in a disc between 1.8 and 3.0 Jupiter radii. These particles orbit Jupiter like a multitude of tiny moons. Collisions commonly occur between ring particles, and the sub-micrometer-sized particles that result from these collisions are accelerated away from Jupiter by the planet's powerful electromagnetic fields. The gossamer ring generates a fairly constant flux of escaping particles, but the dust streams observed by Ulysses and Galileo varied dramatically with time. Magnetic field effects could modulate the emissions of dust, but Galileo measurements of the interplanetary field showed little variation between when a dust storm was observed and when the spacecraft saw no dust streaming. This suggested that Jupiter's gossamer ring was not the source of the dust storms.[26]

Particulates from the satellite Io were another potential source of the dust. To investigate a possible Io origin of the particles, Amara Graps of the Max Planck Institute led a team that analyzed several years of Galileo dust-detector data. The team found that during a dust storm, oscillations in the rate of particle impacts on Galileo's detector were related to the rotational periods of Jupiter and Io. The fact that Io's rotation period appeared to influence the dust particles' behaviors pointed to the moon as a likely source of the dust. Also, calculations showed that Io had usually been located somewhat near (within 50° longitude) the origin of dust particles' trajectories.[27]

Scientists on Graps's team considered several mechanisms that might have generated the particulates. They could have been created as a result of impacts on Io's surface

---

[24] Horanyi, Morfill, and Grun, "Mechanism for the Acceleration and Ejection of Dust Grains."

[25] E. Grun, M. Baguhl, D. P. Hamilton, R. Riemann, H. A. Zook, S. Dermott, H. Fechtig, B. A. Gustafson, M. S. Hanner, M. Horanyi, K. K. Khurana, J. Kissel, M. Kivelson, B. A. Lindblad, D. Linkert, G. Linkert, I. Mann, J. A. M. McDonnell, G. E. Morfill, C. Polanskey, G. Schwehm, and R. Srama, "Constraints from Galileo Observations on the Origin of Jovian Dust Streams," *Nature* 381 (30 May 1996).

[26] Ibid.

[27] "Galileo Flying Through Intense Dust Storm"; Horanyi, Morfill, and Grun, "Mechanism for the Acceleration and Ejection of Dust Grains"; Graps et al., "Io as a Source."

that pulverized and ejected fragments of minerals out into space. An impact-generated dust cloud existed around another Jovian moon, Ganymede, and scientists studied it in order to understand better whether a similar mechanism on Io was a likely source of the dust storm's particles. The study concluded that there was little chance that impact-generated particulates from Io were the dominant source of the intense dust streams that Galileo encountered.[28]

The Max Planck Institute team also revisited the possibility that the dust originated from Jupiter's ring system. If dust particles then passed close by Io, could the satellite have influenced their trajectories and in some way ejected the particles from the Jovian system? The team concluded that this scenario, too, was very unlikely.[29]

Io's volcanoes appeared to be a much more likely source of the dust. The sub-micrometer sizes of the particles observed by Galileo in the interplanetary dust storms corresponded well to particle sizes in Io's volcanic plumes. Io's volcanic dust production rate, as measured by Voyager, was large enough that the escape of only a small fraction of it would be sufficient to generate the dust streams observed by Galileo. Furthermore, plausible models existed for the ejection of Io's sub-micrometer-sized volcanic dust into interplanetary space by Jovian electromagnetic forces.

The characteristics of Jovian-created interplanetary dust storms may provide glimpses into Jupiter's environment that cannot be easily obtained in other ways. In the future, dust-stream measurements might be used to monitor Io's volcanic plume activity and composition; they might also serve as a source of data on the dynamics of Jupiter's magnetosphere.[30]

Galileo's dust detector was roughly the size of a large kitchen colander. In addition to counting particle impacts, it measured their energy and direction of flight, from which sizes and speeds were calculated. Dr. Eberhard Grun of Germany's Max Planck Institute of Nuclear Physics was the principal investigator on Galileo's dust-detector experiment. He also placed dust detectors aboard the Ulysses spacecraft that flew by Jupiter in 1992 on its way to the Sun and the Cassini spacecraft that passed by the Jovian system in 2000 and 2001 en route to Saturn.[31]

The Tape-Recorder Crisis

Galileo's reel-to-reel tape recorder was originally intended to serve only as a backup to the high-gain antenna (HGA), which was supposed to send images to Earth in near-real time. But the failure of the HGA to unfurl left the Galileo team dependent on data storage and slow data transmission through the spacecraft's low-gain antennas (LGAs). The tape recorder became a vital link in the strategy developed to work around the HGA's problems. Critical data remained stored on the recorder's 560 meters (1,850 feet) of 6-millimeter-wide (or a quarter of an inch) Mylar tape until they got painstakingly sent to Earth at an average rate of two to three images per day. After the images were received on Earth, their section of tape could be overwritten with new data. It was only with a

---

[28] Graps et al., "Io as a Source."

[29] Ibid.

[30] Ibid.

[31] "Galileo Flying Through Intense Dust Storm."

working tape recorder that the mission could still attain approximately 70 percent of its original science objectives.[32]

As Galileo began its final approach to Jupiter, the Orbiter took a global image of the planet on 11 October 1995 that included the Probe atmospheric-entry site. As usual, the Orbiter digitally stored this image on its tape recorder. Shortly after NASA sent the command for the tape recorder to rewind so that the image could be transmitted to Earth, project staff got an unwelcome surprise: Signals from the Orbiter indicated that the reel-to-reel recorder had rewound for playback as intended but then failed to stop spinning. What did that indicate? Had the tape broken so that a reel was turning endlessly? Or were the capstans, the recorder's speed-regulating spindles that were supposed to pull the tape along, slipping instead of pulling? Perhaps the recorder actually had stopped and the signals from the Orbiter were incorrect.[33]

Within hours of the tape recorder failure, in what Project Manager William J. O'Neil called "an amazing coincidence," an Earth-based duplicate of Galileo's recorder also malfunctioned during a routine simulation of operations to be performed at Jupiter. Due to faulty circuitry, the recorder failed to sense that it had reached the end of its Mylar tape. According to O'Neil, "The thing just tore the tape off the reel and left it flapping."[34] The Galileo team feared that the same thing had happened aboard the spacecraft, which was now 870 million kilometers (540 million miles) from Earth.[35]

For 15 hours, Galileo's tape recorder remained stuck in its rewind setting. NASA staff, including Project Manager William J. O'Neil, "sank into despair." After all that Galileo had been through and survived, it now appeared that a broken tape recorder would irreparably damage the mission. The recorder was the spacecraft's main data-storage system; if it could not be fixed, much of the imaging that was so vital to the mission might not be transmittable to Earth.[36]

Commands were sent for the recorder to stop, and project engineers began an intricate analysis of the problem. Meanwhile, other staff began developing a strategy to keep the mission going even if the tape recorder proved to be permanently broken. Once again, Galileo's redundant systems and the project staff's ingenuity provided a way out of the crisis. The staff found that memory storage sites in Galileo's main on-board computer could partially replace the tape recorder's function. O'Neil said of this effort, "Some of us did calculations and became convinced that, remarkably, there was a way. We were really buoyed up by the realization that we had a really good mission even without the tape recorder. That's when we came out of our depression . . . ." Using the computer's memory sites instead of the tape recorder, the Galileo team estimated that it would still achieve over half of the mission's original research goals. All of the Atmospheric Probe's objectives could still be achieved, as well as the Orbiter's survey of the Jovian magnetic and charged-particle environment. The principal loss of data would be "the number of images and

---

[32] "Rewind Almost Unhinges Galileo Mission," *Washington Post* (29 October 1995): A6.

[33] "The Journey to Jupiter," section 2 in *Galileo Tour Guide, JPL Galileo Home Page*, p. 10, *http://www2.jpl.nasa.gov/galileo/tour/tourtoc.html* (accessed 22 March 2001) (available in Galileo—Meltzer Sources, folder 18522, NASA Historical Reference Collection, Washington, DC).

[34] "Rewind Almost Unhinges Galileo."

[35] Ibid.

[36] Ibid.

other high-speed spectral data that could be returned by the spacecraft,"[37] according to project scientist Torrence Johnson. The spacecraft would be able to send to Earth perhaps only hundreds instead of thousands of images of Jupiter and its satellites.[38]

On 20 October, NASA performed the first test of Galileo's tape recorder since its breakdown by sending a play command. The recorder played back a few seconds of data, and this sparked tremendous elation among the staff. The tape had not broken and the recorder could still operate!

O'Neil speculated that an "uphill/downhill problem" had caused the tape recorder malfunction. On 10 October, just before the problem was noticed, engineers had rewound the tape back to its very beginning. When the tape was near either of its ends, it took more force from the recorder's capstans to move the tape along. O'Neil compared the situation to "a car trying to make it up the hill when the road is iced over." He was fairly confident that the recorder's three capstans had just started spinning like a car spinning its wheels and had not moved the tape along at all.[39]

Why had the capstans lost traction? Galileo's tape recorder was the same type as those installed in other spacecraft, where they had good records of reliability. But on other missions, the recorder had been constantly in use. On Galileo's six-year-long cruise to Jupiter, it was rarely used. O'Neil believed that the grease in the recorder's bearings might have gotten stiff, thus making the reels harder to turn and leading to the capstan slippage. He thought that a period of running the recorder might help redistribute the grease and make the recorder less prone to another malfunction.[40]

NASA staff worried that the section of tape involved in the recorder failure might have been weakened by "being polished by the capstan for 15 hours"[41] and be ready to break. To guard against this, they sent commands for the recorder to wind tape onto one reel, wrapping 25 times around the potentially damaged section. This secured that area of tape against more stresses that could tear it. Throughout the rest of the mission, that section of tape remained "off-limits" for further recording. One consequence of this decision was that some of the approach images of Jupiter became permanently unavailable.[42]

Project engineers still had questions about what the recorder's safe modes of operation were and did not want to unknowingly damage the recorder further. They did not know how fast the tape could safely move or whether it could reliably move in both directions. Until these questions were answered, Galileo's management decided to use the recorder as little as possible. In the immediate future, it would be devoted to the mission's highest priorities, which were to help the Orbiter attain a delicate insertion into a Jupiter orbit and to store data sent from the Probe as it plunged down through Jupiter's atmosphere. In order to have the maximum probability of attaining these objectives, Galileo's management sacrificed certain of the tape recorder's less essential operations.

---

[37] Franklin O'Donnell, "Galileo Spacecraft Tape Recorder To Be Tested," JPL Public Information Office news release, 20 October 1995, *http://www2.jpl.nasa.gov/galileo/dmsoct20.html* (available in Galileo—Meltzer Sources, folder 18522, NASA Historical Reference Collection, Washington, DC).

[38] "Rewind Almost Unhinges Galileo"; O'Donnell, "Galileo Spacecraft Tape Recorder to be Tested"; Donald J. Williams, "Jupiter—At Last!" *Johns Hopkins APL Technical Digest* 17, no. 4 (1996): 346.

[39] "Rewind Almost Unhinges Galileo."

[40] Ibid.

[41] Ibid.

[42] "The Journey to Jupiter."

These included the storage of the first-ever closeup images of the satellites Europa and Io when the Orbiter flew by them during its 7 December 1995 Jupiter arrival.

Bill O'Neil, Galileo Project Manager from 1992 to 1997, commented that the most unpleasant task he ever had to perform on the mission was to tell his staff that there was no way the imaging of Europa and Io, which had been planned and anticipated for years, could be carried out on Jupiter Arrival Day, given the uncertain condition of the spacecraft's tape recorder. The 7 December 1995 arrival day had, in fact, been picked because Galileo would have the opportunity to fly far closer than Voyager had to two of Jupiter's most interesting moons—Europa and Io—on the same day. Canceling the imaging operations was a hurtful decision that O'Neil regretted to make, but it was clear to him what had to be done. He could not risk losing the Atmospheric Probe data and possibly crippling the spacecraft for the remainder of the mission.[43]

At the time that O'Neil made this decision, it was a huge disappointment to many scientists on his staff because the Orbiter was not scheduled to return to Io during Galileo's primary mission. Mission scientists feared that their only chance to obtain closeup pictures of the moons might have been lost. NASA was considering an extended Galileo mission that could include another encounter with Io, but this mission could only be carried out if the spacecraft continued to operate. With all that had gone wrong with it already, it was highly questionable that the spacecraft would survive long enough to return to Io.[44]

Galileo's Final Approach

In the weeks before Arrival Day, telescopes around the world prepared to make as many observations as possible of Jupiter in order to best determine what the Atmospheric Probe was heading into. The international effort included the Swedish Telescope at La Palma in the Canary Islands, Mt. Wilson in California, the French observatory Pic-du-Midi in the Pyrenees, Mauna Kea in Hawaii, Yerkes Observatory near Chicago, the McMath Solar Telescope at Kitt Peak in Arizona, and others.[45]

According to Galileo Project Manager William J. O'Neil, the moment of truth for the project would occur at about 2 p.m. PST on Friday, 7 December 1995, when the spacecraft would begin "the most hazardous arrival at a planet ever attempted."[46] The 750-pound Probe would plunge into the Jovian atmosphere at a speed about 50 times greater than that of a rifle-shot bullet, then survive an ordeal that scientists compared to a trip through a nuclear fireball. Meanwhile, NASA would await the crucial signal, relayed by the Orbiter, that the Probe had survived and was still operating. This would

[43] Bill O'Neil interview, telephone conversation, 15 May 2001.

[44] "Arrival Day—December, 1995," *Frequently Asked Questions—Jupiter, JPL Galileo Home Page, http://galileo.jpl.nasa. gov/jupiter/arrival_day.html* (accessed 25 March 2001) (available in Galileo—Meltzer Sources, folder 18522, NASA Historical Reference Collection, Washington, DC); "Rewind Almost Unhinges Galileo."

[45] Glenn Orton, "Online from Jupiter—Field Journal from Glenn Orton," *NASA Quest*, 30 November 1995, *http://quest. arc.nasa.gov/galileo/bios/fjournals/orton-ofj13.html* (available in Galileo—Meltzer Sources, folder 18522, NASA Historical Reference Collection, Washington, DC).

[46] Kathy Sawyer, "Jupiter Probe Set for Big Plunge," *Washington Post* (4 December 1995): A3.

be the first spacecraft signal ever received from inside the atmosphere of one of the giant outer planets.[47]

The Probe's science instruments would be focused on analyzing Jupiter's complex chemical makeup, measuring the temperatures and pressures of its atmosphere, determining the amounts of water contained in its cloud banks, observing its lightning, and experiencing its intense winds, which were believed to race around the planet at speeds of up to 480 kilometers per hour (300 mph). Within hours of its arrival, after its descent of more than 650 kilometers (400 miles) below Jupiter's cloud tops, the Probe, which would have long since discarded its heatshields, would then be destroyed by temperatures high enough to vaporize metal. After the Orbiter, which would be flying 214,000 kilometers (133,000 miles) directly overhead, ceased to receive signals from the descending Probe, it would then fire its main engine and move on to its next task—inserting itself into Jupiter orbit. From there, it would begin its years-long study of the planet and its satellites.[48]

Because of the distance from Earth to Jupiter, over 800 million kilometers (half a billion miles), mission control could not quickly repair any problems that arose. Radio transmissions, which travel at the speed of light, take almost an hour each way at that distance. In most cases, when signs of trouble arise, a robotic spacecraft is designed to go into a "safing mode" and await further instructions from Earth. But during the arrival sequence, things would be happening too quickly for such a strategy. William O'Neil said that if problems arose during arrival, "we can't have . . . [the spacecraft] stop. We can't be in that loop. It would take an hour to see that something's wrong, and another hour to transmit a fix back to the spacecraft. For anything that's likely to happen, the spacecraft will sense it and keep going."[49] NASA's task had been to make the arrival sequence fail-safe, designing the spacecraft to accomplish its objectives as independently as possible from mission staff back on Earth.[50]

On 16 November 1995, the Orbiter's magnetometers indicated that the spacecraft was crossing Jupiter's bow shock, the shock wave preceding its magnetosphere, which is the region where Jupiter's magnetic field, rather than the interplanetary field, predominates. The bow shock in front of a magnetosphere is very similar to the wave generated in front of a boat as it travels through the water. A planet's bow shock helps to guide the charged particles of the solar wind around the planet's magnetosphere and, by so doing, aids in forming a boundary between the interplanetary magnetic field environment and that of the planet.[51]

The bow shock detected by the Orbiter was not stationary; it moved back and forth in response to gusts of the solar wind. According to Dr. Margaret Kivelson of UCLA, principal investigator of Galileo's magnetometer experiment, "As the solar wind velocity increased, the bow shock moved inside the position of the spacecraft, leaving Galileo again in the solar wind." This crossing and recrossing of the bow shock occurred several

[47] Ibid.; Ames Research Center, "Galileo Probe Mission Events," *http://ccf.arc.nasa.gov/galileo_probe/htmls/probe_events.html* (accessed 11 March 2000) (available in Galileo–Meltzer Sources, folder 18522, NASA Historical Reference Collection, Washington, DC); "Galileo Less Than Three Weeks From Jupiter," *JPL Universe*, 17 November 1995, *http://www2.jpl.nasa.gov/galileo/status951117.html* (available in Galileo–Meltzer Sources, folder 18522, NASA Historical Reference Collection, Washington, DC).

[48] Sawyer, "Jupiter Probe Set for Big Plunge"; "Galileo Less Than Three Weeks From Jupiter."

[49] Sawyer, "Jupiter Probe Set for Big Plunge."

[50] Ibid.

[51] "Bow Shock," *Windows to the Universe*, University Corporation for Atmospheric Research (UCAR), Regents of the University of Michigan, 2001, *http://www.windows.ucar.edu/cgi-bin/tour_def/glossary/bow_shock.html* (available in Galileo–Meltzer Sources, folder 18522, NASA Historical Reference Collection, Washington, DC).

times between the first encounter on 16 November 1995, when Orbiter was about 15 million kilometers (9 million miles) from Jupiter, and 26 November, when the spacecraft had approached to a distance of 9 million kilometers (6 million miles). At that point, the Orbiter passed for the time being into "Jupiter space," close enough to the planet that its bow shock could no longer be blown back far enough by the solar wind to cross the spacecraft's position again. William J. O'Neil said of the event, "With the spacecraft now in the magnetosphere, we begin our first direct measurements of the Jupiter system."[52]

Arrival Day: 7 December 1995

The date 7 December was a day of anticipation and tension as the men and women who ran the Galileo Project watched monitors and readouts and waited for news that the mission was going as planned. The 24 hours of Arrival Day were the most eventful, and possibly the busiest, of the whole mission. The Orbiter flew close by two Galilean moons; the Probe discovered a new radiation belt and dived into Jupiter's thick atmosphere at a velocity far faster than that of a speeding bullet; and then the Orbiter performed a long, critical burn to enter a Jovian orbit.[53]

Two of the first events of Arrival Day were the Orbiter's flybys of the moons Europa (at 3:09 a.m. PST) and Io (at 7:46 a.m. PST). The Orbiter passed 32,500 kilometers (20,200 miles) from Europa, but only 890 kilometers (550 miles) from Io, closer than originally planned (see figure 8.4). The Orbiter's trajectory had been altered to make maximum use of Io's gravitational field for decelerating the spacecraft, thereby conserving propellant for later in the mission. Approximately 4 hours after its Io encounter, at 11:54 a.m. PST, the Orbiter made the closest pass it would perform of Jupiter's cloud tops, traveling through a region of intense radiation. The Orbiter was heavily shielded to protect its electronics, but nevertheless, the radiation levels during this time shot up more abruptly than expected and almost reached the craft's design limit. One of the Orbiter's navigation systems temporarily lost its bearings, but a backup navigation system took over. Once again, the value of NASA's policy of building redundant systems into its spacecrafts was demonstrated.[54]

At about 8 a.m. PST, 6 hours before atmospheric entry, an internal alarm "woke up" the Probe and turned on its instruments to begin warming them up. When the Probe passed between Jupiter's ring and the upper atmosphere at about 11 a.m. PST (3 hours before entry), its energetic particle instrument (EPI) detected unexpected high-energy helium atoms and a belt of radiation 10 times stronger than Earth's Van Allen belts. This new phenomenon provided data related to Jupiter's high-frequency radio emissions, as well as to emissions from other celestial bodies with magnetospheres and trapped radiation.[55]

---

[52] JPL, "Galileo Crosses into Jupiter's Magnetosphere," JPL Public Information Office news release, 1 December 1995, *http://www2.jpl.nasa.gov/galileo/status951201.html* (available in Galileo—Meltzer Sources, folder 18522, NASA Historical Reference Collection, Washington, DC).

[53] JPL, "Arrival at Jupiter," section 6 in *Galileo: The Tour Guide*, June 1996, p. 47, *http://www2.jpl.nasa.gov/galileo/tour/6TOUR. pdf* (available in Galileo—Meltzer Sources, folder 18522, NASA Historical Reference Collection, Washington, DC).

[54] Kathy Sawyer, "Galileo Starts Tour of Giant Planet," *Washington Post* (8 December 1995): A1; "Arrival at Jupiter."

[55] Sawyer, "Galileo Starts Tour of Giant Planet"; JPL, "Probe Science Results," section 7 in *Galileo: The Tour Guide*, June 1996, p. 57, *http://www2.jpl.nasa.gov/galileo/tour/7TOUR.pdf* (available in Galileo—Meltzer Sources, folder 18522, NASA Historical Reference Collection, Washington, DC).

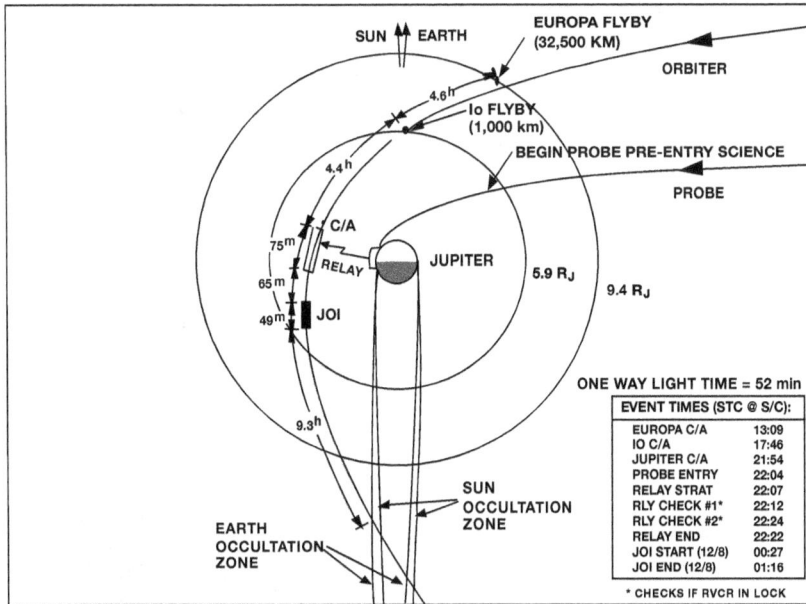

Figure 8.4. **Arrival Day events. Notice the two different trajectories of the Probe and Orbiter. (Adapted from image number P-45516A)**

Descent of the Probe

Minutes after 2:00 p.m. PST, the Probe sped by Jupiter's cloud tops on its way down into the planet's atmosphere. If it survived its incandescent deceleration, it was expected to transmit a message within a few minutes. The first signal from the Probe was to be a single data bit, relayed from the Orbiter to Earth. The trip across interplanetary space would take 52 minutes, and the signal should have arrived by just after 3 p.m. PST—but it did not. Faces in JPL's control room "drained of color as the minutes stretched on and nothing appeared on their screens."[56] At 3:13 p.m. PST, when a voice finally announced, "We have confirmation of telemetry lock,"[57] it meant that the Probe had survived its hellish atmospheric entry and was still operating. Galileo staff "broke into whoops of joy and wild applause. There were even a few tears."[58]

"It's wonderful," said David Morrison, chief of the space sciences division at Ames Research Center in Mountain View, California. "We have a spacecraft inside the atmosphere of a giant planet for the first time. We have been waiting twenty years for this moment."[59]

The one-bit signal had actually arrived at 3:04 p.m. PST, right on schedule, but the receiving equipment on Earth had taken 10 minutes longer than expected to process the signal and send it to Galileo's staff. Due to Galileo's slow transmission rate

---

[56] Sawyer, "Galileo Starts Tour of Giant Planet."

[57] Ibid.

[58] Bob Holmes, "Tears of Joy as Galileo Meets Jupiter," *New Scientist* (16 December 1995): 5.

[59] Ibid.

through its low-gain antennas, it would take days to get even a preliminary look at the Probe's findings and months before all the data were received. Project staff did know that the Orbiter had received the Probe's long-awaited transmissions for over an hour, so extensive data on the Jovian atmospheric environment were expected to be transmitted eventually to Earth receiving stations.[60]

Figure 8.5 and table 8.1 portray the times, atmospheric pressures, altitudes, and temperatures at which key events occurred during the Probe's approach, entry, and parachute descent into Jupiter's atmosphere. Since Jupiter does not have a solid surface, altitudes were measured relative to the level at which Jovian atmospheric pressure equals Earth sea-level atmospheric pressure (1.0 bar pressure). The listed times are relative to the moment when the Probe crossed the 450-kilometer altitude level ("atmospheric entry time"), which occurred at 22:04:44 universal time (UT). Note the shallow angle (only 8.4 degrees below horizontal) at which the Probe entered the Jovian atmosphere. The radio signal from the Probe ended 61.4 minutes into the entry, when the high atmospheric temperatures caused the radio transmitter to fail.

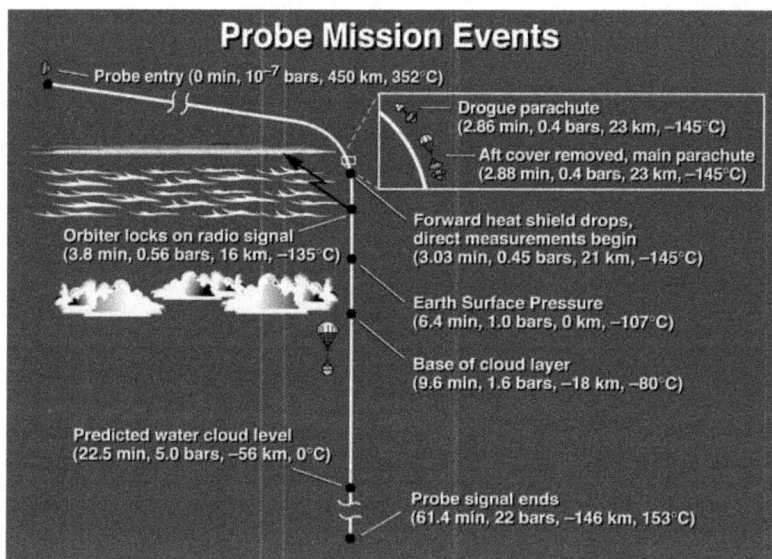

Figure 8.5. **Probe entry and descent into the Jovian atmosphere. (NASA photo number ACD96-0313-4)**

Deployment of the Probe parachutes (see figure 8.6) occurred only after the Probe had slowed sufficiently (to about 1,600 kilometers per hour or 1,000 mph) to prevent air friction from immediately ripping the parachutes apart. First, the pilot or drogue chute was deployed from inside the aft heatshield (or aft cover) with the aid of a mortar. The aft heatshield was released from the Probe, and the drogue parachute pulled off the aft cover and extracted the main chute. These events all occurred within

[60] Sawyer, "Galileo Starts Tour of Giant Planet"; Ames Research Center, "Galileo Probe Mission Events," http://ccf.arc. nasa.gov/galileo_probe/htmls/probe_events.html (accessed 11 March 2000) (available in Galileo—Meltzer Sources, folder 18522, NASA Historical Reference Collection, Washington, DC).

a period of 1.26 seconds. The forward heatshield was dropped 9 seconds later from the rapidly slowing descent module. When the forward shield had fallen to a distance of 30 meters below the descent module, the module's science experiments were begun. The main parachute eventually decreased descent module velocity to 430 kilometers per hour (265 mph).[61]

Figure 8.6. **Atmospheric Probe parachute deployment sequence and separation of deceleration module components from descent module. After the high-speed entry phase during which the atmosphere slowed down the Probe's speed, a mortar deployed a small drogue parachute, the drogue chute opened, the aft cover of the deceleration module was released, the drogue parachute pulled the aft cover off and deployed the main parachute, and the forward heatshield was dropped. (NASA photo number AC89-0146-2)**

[61] Ames Research Center, "Probe Mission Events," *http://spaceprojects.arc.nasa.gov/Space_Projects/galileo_probe/htmls/Probe_Mission.html*, and "Artwork of Parachute Deployment," *http://spaceprojects.arc.nasa.gov/Space_Projects/galileo_probe/htmls/Parachute_deployment.html* (accessed 17 October 2004) (hard copies available in Galileo—Meltzer Sources, folder 18522, NASA Historical Reference Collection, Washington, DC).

Table 8.1 includes a timeline of the life of the Probe, from launch through cessation of the Probe signal.

Table 8.1. **Timeline of Atmospheric Probe events.**[62]

TIME BEFORE OR AFTER ATMOSPHERIC ENTRY (E)	DATE	ACTION	DISTANCE FROM JUPITER IN KILOMETERS (MILES)	SPEED RELATIVE TO JUPITER IN KILOMETERS (MILES) PER HOUR
E - 6 years, 50 days	18 October 1989	Galileo launch.		
E - 147 days	12 July 1995 (PDT)	Probe separation: Orbiter/Probe spacecraft spins at up to 10 rpm for stability. Spin axis is lined up with Probe trajectory. Explosive nuts accomplish separation of Probe from Orbiter. Three springs push Probe away. Orbiter rocket engine fires to alter craft trajectory so that it will not impact Jupiter but will orbit the planet in position to pick up the Probe's signals.	81,520,000 (50,660,000)	20,448 (12,706)
	7 December 1995	Probe Atmospheric Entry Day		
E - 6 hours		Probe timer initiates craft's operation. Batteries are activated, and EPI is biased for operation.	600,000 (373,000)	76,700 (47,600)
E - 3 hours		Measurement of Jupiter's inner radiation belts begins.	360,000 (224,000)	97,200 (60,400)

[62] Julio Magalhães (ARC), "Structure of Jupiter's Atmosphere," 19 September 1997, *http://spaceprojects.arc.nasa.gov/ Space_Projects/galileo_probe/htmls/ASI_results.html* (accessed 13 December 2004) (available in Galileo—Meltzer Sources, folder 18522, NASA Historical Reference Collection, Washington, DC); Julio Magalhães, "Probe Mission Events," 19 September 1997, *http://spaceprojects.arc.nasa.gov/Space_Projects/galileo_probe/htmls/Probe_Mission. html* (accessed 13 December 2004) (available in Galileo—Meltzer Sources, folder 18522, NASA Historical Reference Collection, Washington, DC); Julio Magalhães, "Galileo Probe Mission Events Timeline," *Expresní Astronomické Informace*, 30 November 1995, *http://astro.sci.muni.cz/pub/galileo/probe_mission_events.html* (accessed 13 December 2004) (available in Galileo—Meltzer Sources, folder 18522, NASA Historical Reference Collection, Washington, DC); JPL, "Probe Mission Time Line," *Galileo Probe Background*, NASA Ames Research Center Public Affairs news release, *http://ccf.arc.nasa.gov/dx/basket/storiesetc/GalBkgd.html* (accessed 11 March 2000) (available in Galileo—Meltzer Sources, folder 18522, NASA Historical Reference Collection, Washington, DC); "Galileo Probe Mission Science Summary," 30 December 1996, *http://spaceprojects.arc.nasa.gov/Space_Projects/galileo_probe/htmls/Science_ summary.html* (accessed 13 December 2004) (available in Galileo—Meltzer Sources, folder 18522, NASA Historical Reference Collection, Washington, DC).

Table 8.1. **Timeline of Atmospheric Probe events.** (continued)

TIME BEFORE OR AFTER ATMOSPHERIC ENTRY (E)	DATE	ACTION	ALTITUDE* IN KILOMETERS (MILES)	SPEED IN KILOMETERS (MILES) PER HOUR	PRESSURE IN BARS	TEMPERATURE IN DEGREES CELSIUS (FAHRENHEIT)
E + 0		Probe atmospheric entry. Rapid deceleration and intense heating of atmospheric gases near the Probe begin. Atmospheric structure instrument begins recording; atmospheric pressure, density, and temperature inferred from data.	450 (280)	171,000 (106,000)	0.0000001	352 (666)
E + 35 seconds		Period of maximum heat generation and ablation begins. Heatshield burnoff sensors begin measurements of heatshield mass loss.	220 (1400		0.0001	Near Probe: 14,000 (25,000); away from Probe: -100 (-150)
E + 55 seconds		Peak aerodynamic stresses and maximum forces of up to 250 g's experienced.			0.006	Away from Probe: - 120 (-184)
E + 80 seconds		Main heat pulse and heatshield ablation end.				
E + 172 seconds		Pilot parachute deployed	23 (14)		0.4	-145 (-229)
E + 173 seconds		Aft heatshield separates and pulls out main parachute	23 (14)		0.4	-145 (-229)
E + 3.0 minutes		Main parachute slows descent module. Three explosive separation bolts fire, and forward heatshield falls away.	21 (13)	650 (400)	0.45	-145 (-229)
E + 3.8 minutes		Orbiter locks on Probe radio signal.	16 (10)		0.56	-135 (-211)
E + 6.4 minutes		Zero altitude attained (defined as the point where pressure = 1 bar, or Earth surface pressure).	0 (0)		1	-107 (-161)

Table 8.1. **Timeline of Atmospheric Probe events.** (continued)

TIME BEFORE OR AFTER ATMOSPHERIC ENTRY (E)	DATE	ACTION	ALTITUDE* IN KILOMETERS (MILES)	SPEED IN KILOMETERS (MILES) PER HOUR	PRESSURE IN BARS	TEMPERATURE IN DEGREES CELSIUS (FAHRENHEIT)
			-20 (-12)		2	-66 (-87)
			-96 (-60)		12	84 (183)
E + 61.4 minutes		Probe signal no longer detected. Pressure is approximately equal to that at an Earth ocean depth of 700 feet.	-146 (-91)		22	153 (307)

Altitude is measured relative to a point in the atmosphere with pressure of 1 bar, the pressure on Earth's surface.

As the Probe continued to fall into the planet after completing its mission, it was slowly destroyed by the increasing heat. This was not the short-lived heat of atmospheric friction and compression against which the Probe's now-discarded heatshields protected it during the first minutes of entry. This was a steady-state thermal barrage that would claim one part of the craft after another. Soon after the Probe stopped transmitting, mission scientists hypothesized that its parachute melted, then its aluminum fittings. A couple of hours later, the aluminum vaporized. Four hours after that, even the ultratough titanium shell of the Probe had melted, and, after an additional 2 hours, the titanium evaporated as the temperature climbed to 1,700°C (3,100°F) and the pressure rose to 5,000 times that of Earth's surface. The Probe had been reduced to its atomic components and become "one with the atmosphere of Jupiter."[63]

Jupiter Orbit Insertion (JOI)

After the Orbiter finished receiving data from the Atmospheric Probe, it had to adjust its trajectory quickly to avoid heading off into deep space, away from the planet and satellites it was supposed to study. The Orbiter had to fire its 400-newton engine in such a way as to slow itself down relative to Jupiter and allow the giant planet's gravitational field to capture it. NASA scientists scheduled this burn when the Orbiter was close to Jupiter, only four Jupiter radii (RJ) distant, for the further the JOI maneuver was performed from the planet, the less the gravitational force and the greater the spacecraft's propellant requirements would have been. The Orbiter's burn sequence began at 5:20 p.m. PST and reduced the craft's velocity by over 600 meters per second until the Orbiter was finally captured. JOI, the actual capture of the Orbiter by Jupiter, occurred approximately 45 minutes into the 49-minute burn.[64]

[63] "Arrival at Jupiter," p. 51; "The Probe Story: Secrets and Surprises from Jupiter," *Galileo Messenger* (April 1996).

[64] "Jupiter Orbit Insertion Burn Status," *Galileo Jupiter Orbit Insertion December 7, 1995, http://www2.jpl.nasa.gov/galileo/countdown/plot.html* (accessed 29 March 2000) (available in Galileo—Meltzer Sources, folder 18522, NASA Historical Reference Collection, Washington, DC); JPL, "Galileo Mission Status," JPL Public Information Office news release, 7 December 1995, *http://www2.jpl.nasa.gov/galileo/status9512071.html* (available in Galileo—Meltzer Sources, folder 18522, NASA Historical Reference Collection, Washington, DC).

Once the Orbiter was captured, it became the first artificial satellite to join Jupiter's family of moons and particles. Figure 8.7 depicts the elongated elliptical shape of the Orbiter's 7-month-long initial orbit, as well as the shapes of the spacecraft's next 10 orbits. During most of the initial orbit, the spacecraft slowly transmitted data to Earth that had been received from the Probe during its brief plunge into the Jovian atmosphere and stored on the craft's tape recorder. Mission engineers tested the tape recorder to determine its present capabilities and loaded and tested new software sent from Earth that would make data transmission more efficient. Data taken during the first Io flyby (on Arrival Day, 7 December 1995) were also sent to Earth. Because of uncertainty with the tape recorder's capabilities, images of Io, which would have required large storage volumes, had not been stored on the tape recorder, but other Io-related data that required less storage volume had been collected. Scientific findings from Io and other Jovian moons are discussed in detail in chapter 9.

When the spacecraft approached the apojove (furthest point from Jupiter) of its orbit on 14 March 1996, it fired its main engine once again. This maneuver changed the craft's trajectory so that its perijove (closest point to Jupiter of its orbit) was raised from 4 RJ to about 10 RJ, above the altitude of Jupiter's intense radiation belts. Nearly 50 percent of the radiation dosage allowed for Galileo's primary mission had been received during its initial low-altitude Jupiter flyby. To ensure that the spacecraft's instruments lasted as long as possible, it was important to limit future radiation exposure.[65]

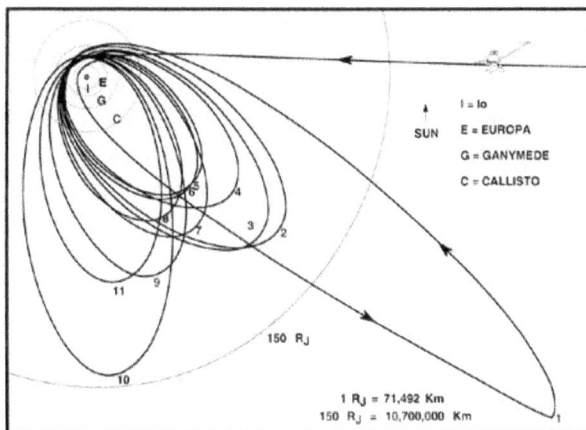

Figure 8.7. **Orbiter's first 11 orbits around Jupiter. Notice that the apojove of the first orbit is over 250 Jupiter radii (RJ) away from the planet. (Adapted from JPL image number P45516B)**

Scientific Results from the Probe

Data collected during the Probe's descent into Jupiter's atmosphere made it necessary for scientists to revisit many of their beliefs about the formation and evolution of our solar system's giant gaseous planets. Measurements of atmospheric composition, wind

[65] Donald J. Williams, "Jupiter—At Last!" *Johns Hopkins APL Technical Digest* 17, no. 4 (1996): 347; William J. O'Neil, "Project Galileo," paper number AIAA–90-3854 in *AIAA Space Programs and Technologies Conference Proceedings* (held in Huntsville, AL, 25–28 September 1990).

velocities, temperatures, cloud characteristics, electrical storms, and elemental and molecular abundances painted a very different picture of Jupiter from what was expected.[66]

Where Is the Water?

One of the most puzzling mysteries that the Probe's findings raised was, where was the
water that everyone expected to see? Data from the Voyager spacecraft that had flown
by Jupiter in 1979 suggested a water abundance such that the amount of oxygen present
in the atmosphere (most of which would be in the water molecules) would amount to
twice the oxygen abundance of the Sun. Analysis of the atmospheric waves propagating
across Jupiter's cloud tops as a result of the Shoemaker-Levy comet's impact indicated
that Jupiter might have a water abundance such that the oxygen level would be 10 times
the solar level. Preliminary data released six weeks after the Probe's 7 December 1995
descent, however, revealed a planetary atmosphere far dryer than expected, with a water
content corresponding to an oxygen abundance no greater than that of the Sun, and possibly far less.[67]

   A virtual absence of water and, in particular, water clouds was eventually
confirmed by several of the Probe's science instruments. Its nephelometer, which was
designed to analyze cloud formations near the Probe, saw none of the dense water
clouds that were expected based on flyby spacecraft and Earth-based telescope observations. The net flux radiometer, which could make measurements of distant clouds, did
not observe a thick water cloud deck either. The atmospheric structure instrument (ASI),
which determined the temperature, pressure, and density structure of Jupiter's atmosphere
from its uppermost regions down to a region where the pressure was about 22 times that
on Earth (or 22 bars), found evidence that the deep Jovian atmosphere (100 to 150 kilometers, or 60 to 90 miles, below visible, nonwater clouds) was dryer than expected, at
least in the region that the Probe visited. The ASI measured a vertical temperature gradient
indicative of a dry atmosphere without condensation. The lightning and radio emission
detector (LRD), which searched for optical lightning flashes in the Probe's vicinity and
found none, also supported an absence of extensive water clouds. The neutral mass
spectrometer (NMS), whose job was to determine the chemical composition of the
atmosphere, measured far less oxygen than in the Sun's atmosphere. Since most of Jupiter's
oxygen would be found in water molecules in the form of water vapor, the NMS data
implied a surprisingly dry atmosphere.[68]

   The water situation launched a spirited debate on the details of Jupiter's formation and evolution. Some theories of planet formation postulated that collisions with
small, water-bearing bodies such as comets and asteroids had played an important role
not only in Jupiter's formation, but also in that of other planets. But if Jupiter had indeed

----

[66] Williams, "Jupiter—At Last!" p. 353.

[67] Kathy Sawyer, "Jupiter Retains Atmosphere of Mystery," *Washington Post* (23 January 1996): A3; Douglas Isbell and
David Morse, "Galileo Probe Data Spurs New Concepts for Jupiter's Circulation and Formation," NASA news release
96-103, 21 May 1996, *http://www2.jpl.nasa.gov/galileo/status960521.html* (available in Galileo—Meltzer Sources,
folder 18522, NASA Historical Reference Collection, Washington, DC); M. Roos-Serote, P. Drossart, Th. Encrenaz, R.
W. Carlson, and L. Kamp, "New NIM 5 Micron Observations: Where Is the Water on Jupiter?" (paper presented at the
31st Annual Meeting of the Division for Planetary Sciences (DPS), session 51: "Outer Planet Atmospheres Posters,"
14 October 1999).

[68] "Galileo Probe Mission Science Summary"; Isbell and Morse, "Galileo Probe Data Spurs New Concepts for Jupiter's
Circulation and Formation."

captured significant amounts of water from these collisions, then where was it? Jupiter's powerful gravitational force should have prevented the water from escaping. If the lack of water observed by the Probe was typical of Jupiter as a whole, not just of the area the Probe examined, then collisions with water-bearing bodies may not have contributed much at all to the planet's composition. If such collisions were insignificant for Jupiter, they also may have been unimportant for the solar system's other planets. Comets, for instance, might not have played as large a part in the delivery of water for Earth's oceans as some studies had suggested.[69]

Theories of how the planets formed were also reevaluated. According to nucleosynthesis theories, our solar system's Sun and planets were created from the same primitive interstellar cloud of matter. This cloud consisted of a primordial component (one that had existed since the beginning of time)[70] generated in the Big Bang that began the universe, as well as younger material manufactured in stars and then injected into interstellar space through supernova explosions and stellar winds (streams of particles similar to the solar wind). But the Probe had measured only a fraction of the Sun's oxygen content in Jupiter's atmosphere. Where was the water (and its inherent oxygen) that should have remained from Jupiter's creation in the same nebula from which the Sun and the other planets emerged?[71]

One theory suggested that Jupiter actually did contain water levels at or above those of the Sun, with the bulk of the water trapped deep in the planet's interior. According to this theory, Jupiter began as a solid protoplanet of ice and rock that grew to 8 to 10 times the mass of Earth through accretion of ice grains and dust from the original solar nebula. As the solid proto-Jupiter grew, its increasing gravity attracted surrounding lighter gases such as helium and hydrogen from the nebula, producing a dry atmosphere similar to that detected by the Probe. Such a process could have permanently locked water ice in the solid core of Jupiter while keeping atmospheric regions nearly devoid of water.

This theory had to account for the enhanced atmospheric carbon, sulfur, and nitrogen levels measured by the Probe, which were notably enriched compared to solar abundances. In order to explain such abundances, scientists postulated that methane, ammonia, hydrogen sulfide, and other volatiles that were originally locked in the core escaped as it heated up. The theory had serious problems, however. Dr. Tobias Owen of the University of Hawaii questioned how water ice remained trapped in a planetary core believed to be very hot when many other compounds were supposed to have volatized, escaped, and entered the Jovian atmosphere.[72]

Another theory, one that has gained considerable acceptance in the space science community, proposed that Jupiter's atmosphere was not globally dry and that the Probe had just stumbled into an atypical entry site comparable to Earth's desert regions. This theory was strongly supported by Earth-based observations of Jupiter, including

[69] Sawyer, "Jupiter Retains Atmosphere of Mystery"; Isbell and Morse, "Galileo Probe Data Spurs New Concepts for Jupiter's Circulation and Formation"; Christopher F. Chyba, "Impact Delivery and Erosion of Planetary Oceans in the Early Solar System," *Nature* 343 (11 January 1990): 129–133.

[70] *Encarta Online Encyclopedia, http://encarta.msn.com* (available in Galileo—Meltzer Sources, folder 18522, NASA Historical Reference Collection, Washington, DC).

[71] B. J. Conrath, D. Gautier, R. A. Hanel, and J. S. Hornstein, "The Helium Abundance of Saturn from Voyager Measurements," *Astrophysical Journal* 282 (15 July 1984): 807–815.

[72] "Isbell and Morse, "Galileo Probe Data Spurs New Concepts for Jupiter's Circulation and Formation"; "Astronomy," *Microsoft Encarta Online Encyclopedia, 2001, http://encarta.msn.com* (available in Galileo—Meltzer Sources, folder 18522, NASA Historical Reference Collection, Washington, DC).

infrared telescope measurements taken by JPL's Glenn Orton, which showed the Probe's entry site to be situated on the edge of a prominent "hotspot," a patch of atmosphere that was clearer and drier than that of most of the rest of the planet, with a marked absence of clouds. Hotspot areas cover about 1 percent of the planet and about 15 percent of the surface in equatorial areas.[73]

Clouds

Most space science researchers predicted that the Probe would encounter three substantial Jovian cloud layers: an upper ammonia ice-particle cloud layer at an altitude at which atmospheric pressure was 0.5 to 0.6 bar, a middle ammonium hydrosulfide (NH_4SH) ice cloud with its base at 1.5 to 2 bars, and a lower water cloud deck, expected to be the densest of all, with its base at 4 to 5 bars (see figure 8.8). Scientists envisioned this scenario based on the known thermodynamic properties of these compounds, Jovian atmospheric models, and past Earth-based and spacecraft-based observations of large regions of Jupiter. Once again, the Probe sent back data that surprised the scientific community. Cloud decks were far more tenuous than expected or completely absent.[74]

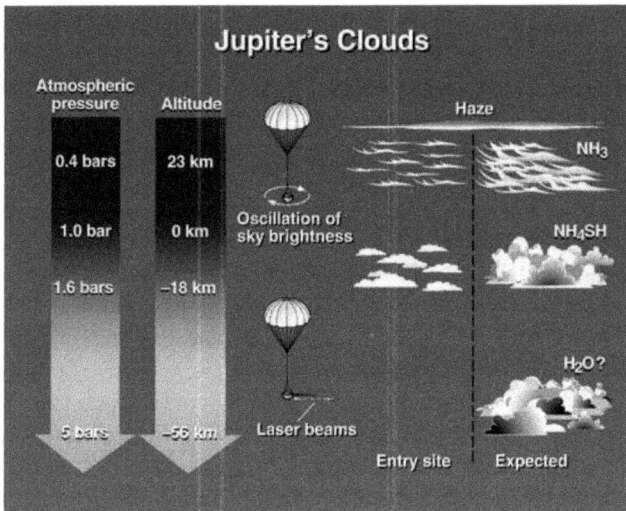

Figure 8.8. **Jupiter's expected cloud layers, along with the clouds actually observed at the Probe entry site. (NASA image number ACD96-0313-7)**

Two experiments on the Probe used different approaches to detect possible clouds. The net flux radiometer (NFR), designed to measure energy balance between

---

[73] "The Probe Story: Secrets and Surprises from Jupiter," *Galileo Messenger* (April 1996); "Probe Mystery Solved: Jupiter as Wet (and Dry) as Earth," *Galileo Messenger* (July 1997); Richard E. Young, "The Galileo Probe Mission to Jupiter: Science Overview," *Journal of Geophysical Research* 103, no. E10 (25 September 1998): 22,775–22,776.

[74] Sawyer, "Jupiter Atmosphere of Mystery"; Young, "Galileo Probe Mission to Jupiter: Science Overview," p. 22,780. Caption information from NASA image number ACD96-0313-7.

solar flux from above and thermal energy from below, recorded variations in sky brightness in the 0.4- to 0.6-bar level of the atmosphere. This phenomenon corresponded to altitudes from 23 down to 14 kilometers (14 to 9 miles) above the level at which pressure equaled 1.0 bar,[75] the pressure found at sea level on Earth. The sky brightness variations and altitude suggested scattered clouds that were likely composed of ammonia ice particles.[76]

The Probe's nephelometer (NEP) detected a tenuous layer between 0.46 and 0.55 bar—nearly the same altitude range as above—that could have been the predicted ammonia cloud. NEP data, however, had an uncertainty associated with them. The NEP operated by sending out a laser beam and observing scattered laser light from cloud particles passing between the body of the Probe and a deployed 0.1-meter arm. The NEP measured scattering only from matter in the near vicinity of the Probe, so if the Probe happened to be passing through even a small inhomogeneous region, the measurements could lead to a fallacious picture of the surrounding atmosphere.

The NEP recorded denser particle concentrations at pressures ranging from 0.76 to 1.34 bars, or altitudes from roughly 9 to -10 kilometers (i.e., from 9 kilometers above to 10 kilometers below the 1-bar level in the atmosphere). Though far more tenuous than predicted, these particle populations were possibly the expected ammonium hydrosulfide clouds. The NFR detected a similar structure at 1.35 bars. The NEP also found one well-defined tenuous cloud structure with a base at the 1.6-bar pressure level.

Scientists had predicted the occurrence of the thickest clouds of all, composed of water, in the 1.9- to 4.5-bar range (roughly in the region of -20 to -50 kilometers). The NEP, however, detected only a weak structure in that region which might have been a thin water cloud, although the data were inconclusive. The NFR did not detect this structure at all. In summary, the Probe's search for Jovian clouds did not identify any thick, dense cloud layers, nor did it conclusively identify any water clouds.[77]

Helium Abundance and the Evolution of the Solar System

One of the Probe's primary objectives was to analyze the composition of the Jovian atmosphere. Jupiter's strong gravitational field (due to its large mass), its low atmospheric temperatures, and the low strength of the solar wind at the planet's distance from the Sun all have contributed to the retention of even the lightest elements, such as hydrogen and helium. The mean escape time for light elements in Jupiter's atmosphere is far longer than the age of the solar system. Thus, Jupiter should contain close to the original abundance of elements that it had when it was formed from the primitive solar nebula, the cloud of

---

[75] Altitudes are arbitrarily measured from the level at which the pressure is 1 bar. Negative altitudes denote levels beneath the 1-bar level.

[76] "The Probe Story: Secrets and Surprises from Jupiter."

[77] Boris Ragent, David S. Colburn, Kathy A. Rages, Tony C. D. Knight, Philip Avrin, Glenn S. Orton, Padmavati A. Yanamandra-Fisher, and Gerald W. Grams, "The Clouds of Jupiter: Results of the Galileo Jupiter Mission Probe Nephelometer Experiment," *Journal of Geophysical Research* 103, no. E10 (25 September 1998): 22,891–22,909; Alvin Seiff, Donn B. Kirk, Tony C. D. Knight, Richard E. Young, John D. Mihalov, Leslie A. Young, Frank S. Milos, Gerald Schubert, Robert C. Blanchard, and David Atkinson, "Thermal Structure of Jupiter's Atmosphere Near the Edge of a 5-μm Hot Spot in the North Equatorial Belt," *Journal of Geophysical Research* 103, no. E10 (25 September 1998): 22,857–22,889; L. A. Sromovsky, A. D. Collard, P. M. Fry, G. S. Orton, M. T. Lemmon, M. G. Tomasko, and R. S. Freedman, "Galileo Probe Measurements of Thermal and Solar Radiation Fluxes in the Jovian Atmosphere," *Journal of Geophysical Research* 103, no. E10 (25 September 1998): 22,929–22,977; "Lightning," *Microsoft Encarta Online Encyclopedia, http://encarta.msn.com* (available in Galileo—Meltzer Sources, folder 18522, NASA Historical Reference Collection, Washington, DC); Young, "Galileo Probe Mission to Jupiter: Science Overview," p. 22,781.

matter out of which the solar system was created. If Jupiter has indeed retained most of its original elemental abundances, then the study of the planet's composition should provide not only a view of its current conditions, but also a glimpse back in time to the early solar system environment (see figure 8.9).[78]

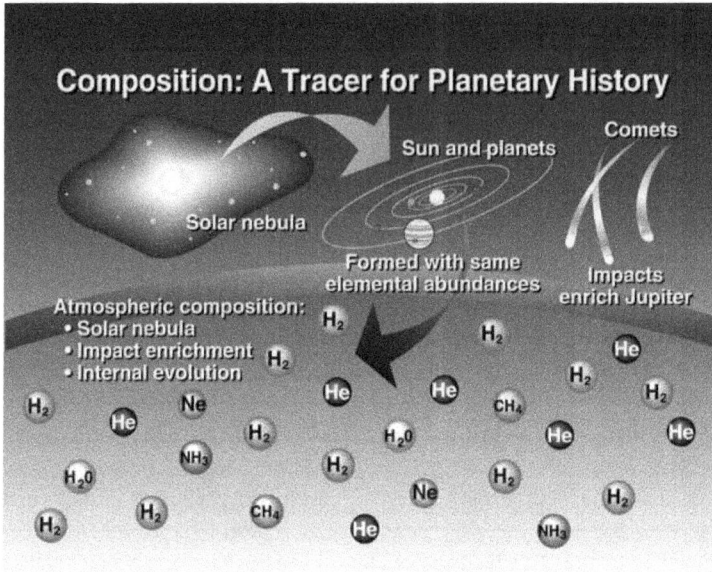

Figure 8.9. **The Jovian atmospheric composition provides a "tracer" for planetary history. (NASA photo number ACD96-0313-9)**

Space scientists believe that conditions in the primitive solar nebula were such that helium and hydrogen did not separate from each other during the process of planetary formation, but remained homogeneously mixed. Jupiter's atmosphere still consists mainly of hydrogen and helium. What the space science community did not know, however, was whether Jovian atmospheric helium abundance had remained the same as in the primitive solar nebula. A significantly lower helium percentage in the present Jovian atmosphere would indicate that segregation had occurred, perhaps through gravitational settling of the helium, which is heavier than hydrogen.

Theories of planetary evolution predict a layer of hydrogen, forming at high pressures in Jupiter's and Saturn's interiors, that has the characteristics of a metal. As Jupiter grows older, it continuously cools. If its interior cools sufficiently, helium would become immiscible in the metallic hydrogen layer. Droplets of helium would form and fall toward the center of the planet, separating from hydrogen and eventually leading to a depletion of helium in the planet's atmosphere. Such a process may be quite appreciable on Saturn, where Voyager data have indicated that the atmospheric helium abundance is less than one-quarter that of the primitive solar nebula environment.[79]

---

[78] Conrath et al., "Helium Abundance of Saturn from Voyager Measurements," pp. 807–815; Richard E. Young, "The Galileo Probe Mission to Jupiter: Science Overview," pp. 22,776–22,778.

[79] Conrath et al., "Helium Abundance of Saturn from Voyager Measurements," pp. 807–815; Young, "Galileo Probe Mission to Jupiter: Science Overview."

The helium mass abundance of the primitive solar nebula had been estimated, based on stellar evolution models, at 27.5 percent, with the remaining percentage mainly hydrogen. The best pre-Galileo estimates of Jovian atmospheric helium abundance were derived from Voyager data and calculated to be about 18 percent. This estimate suggested that significant gravitational settling of helium out of the atmosphere may have occurred, although not as much as on Saturn.[80]

Two independent Galileo Probe measurements taken by the helium abundance detector and the neutral mass spectrometer indicated a helium abundance in Jupiter's atmosphere of 24 percent, higher than the 18 percent expected from Voyager data and close to the primitive solar nebula value of 27.5 percent. Jupiter is so large that a 24-percent helium abundance implies that there are on the order of 10 Earth masses more helium in Jupiter's atmosphere than had been predicted from Voyager data.

The difference between Jupiter's 24-percent atmospheric helium abundance and Saturn's 6-percent abundance is dramatic, but perhaps not surprising. Jupiter, with its greater mass and dimensions, has retained more primordial thermal energy in its interior than Saturn has. Theoretical calculations of the solubility curve of helium in metallic hydrogen suggest that Jupiter is still too warm for helium droplets to form within its metallic hydrogen region. If this is indeed the case, differentiation of the planet's helium from its hydrogen through gravitational settling of the helium droplets will not occur.[81]

Other Substances in the Jovian Atmosphere

In the Jovian atmosphere, the heavier elements carbon, nitrogen, oxygen, and sulfur occur principally in the compounds methane, ammonia, water, and hydrogen sulfide, respectively. These may have been deposited in part from encounters with meteorites and comets. Some hydrocarbons with two or three carbon atoms appear to be present. Only upper limits could be estimated for heavier hydrocarbons, and these were extremely small. The dearth of complex hydrocarbons makes it very unlikely that life similar to that of Earth will be found in the Jovian atmosphere.

Also detected in the atmosphere were noble gases, including argon, krypton, and xenon, in amounts up to three times the levels found in the Sun. This was surprising because noble gases are difficult for a planet to trap. Detection of relatively large concentrations raised questions in the scientific community regarding what in the planet's formation process allowed Jupiter to acquire such amounts of the gases. One explanation suggested that if Jupiter trapped the gases through cooling processes that caused them to condense, this would have required extremely cold temperatures of about -240°C (-400°F), colder than the surface of Pluto. For such low temperatures to exist, Jupiter might have been formed out in the area around the Kuiper Belt and later dragged inward to its present location closer to the Sun.

Another possibility is that the solar nebula, the huge cloud of gas and dust that gave birth to our solar system, was much colder than scientists currently believe. A third proposal was that the solid, icy materials that transported noble gases to Jupiter began forming in the original interstellar cloud of gas and dust long before it collapsed to form

---

[80] "Galileo Probe Mission to Jupiter: Science Overview," pp. 22,776–22,777.

[81] Young, "Galileo Probe Mission to Jupiter: Science Overview," p. 22,777; Williams, "Jupiter—At Last!" p. 353; Conrath et al., "Helium Abundance of Saturn from Voyager Measurements," p. 807.

the solar nebula. That would make those icy materials far older and more primitive than had been thought. The last two hypotheses suggest that giant planets such as Jupiter might form closer to their stars than current theories predict. This would help explain the new observations of planetary systems around other stars, in which giant planets are commonly located close to their stars.[82]

Galileo's neutral mass spectrometer determined that the carbon-to-hydrogen ratio in Jupiter's atmosphere was almost three times the solar value. In other words, the Jovian atmosphere has been enriched in carbon relative to the Sun. Two possible explanations for this are an internal-to-Jupiter mechanism and an external influence on the planet. The internal theory proposes that as solar nebula material accreted over many years to form Jupiter, the pressure in the planet's core increased and caused carbon in the form of methane to outgas from the core. The external theory postulates that Jupiter's overabundance of carbon relative to the Sun came from an influx of late-accreting planetesimals impacting the planet.[83]

One study of Galileo mission atmospheric science predicted that ammonia density would vary considerably with altitude in the Jovian atmosphere. Solar energy–initiated photochemical reactions in the upper atmosphere, condensation of ammonia gas into cloud droplets when temperatures dropped below about -130°C, dissolution of ammonia in water clouds, and chemical reactions with hydrogen sulfide to form ammonium hydrosulfide clouds were all expected to affect ammonia concentrations at different altitudes.[84]

Observations of radio signals broadcast at 1,387 megahertz from Probe to Orbiter showed attenuation, or lessening of signal strength, that was presumed to result from constituents of the Jovian atmosphere. Clouds had been expected to attenuate radio signals, but the Probe descended through a relatively cloud-free region, so clouds could have had little effect. High concentrations of atmospheric water and hydrogen sulfide might have had a marked attenuation effect, but at the concentrations of these chemicals that were measured by the Probe's neutral mass spectrometer, their effect would be negligible. JPL scientists inferred that the observed radio attenuation must be mainly due to atmospheric ammonia concentrations. From the data, they were able to derive an ammonia concentration profile as a function of depth in the atmosphere. This profile showed that for atmospheric pressures greater than 6 bars, corresponding to depths below -60 kilometers, ammonia abundance rose to four times the predicted value, based on the solar abundance of nitrogen, one of the components of ammonia.

The Probe's net flux radiometer, although not able to quantify ammonia abundance explicitly, collected data consistent with an ammonia abundance significantly higher than solar. The neutral mass spectrometer was not able to determine ammonia abundance, possibly due to a contamination problem.

The high ammonia concentration inferred from radio attenuation data does not agree with data from Voyager measurements or Earth-based microwave spectroscopy,

---

[82] Young, "Galileo Probe Mission to Jupiter: Science Overview," p. 22,777; Ames Research Center, "Jupiter's Atmospheric Composition," 19 September 1997, *http://spaceprojects.arc.nasa.gov/Space_Projects/galileo_probe/ htmls/Composition.html* (accessed 31 October 2004) (available in Galileo—Meltzer Sources, folder 18522, NASA Historical Reference Collection, Washington, DC); Robert Roy Britt, "Jupiter's Composition Throws Planet-formation Theories into Disarray," *Space.com*, 17 November 1999, *http://www.space.com/scienceastronomy/solarsystem/ jupiter_elements_991117.html* (available in Galileo—Meltzer Sources, folder 18522, NASA Historical Reference Collection, Washington, DC); Jane Platt and Kathleen Burton, "Galileo Probe Results Suggest Jupiter Had an Ancient, Chilly Past," JPL news release, 17 November 1999.

[83] Young, "Galileo Probe Mission to Jupiter: Science Overview," p. 22,777.

[84] Ibid.

which predicted abundances closer to the solar value. Voyager studies, however, probably measured average values in the Jovian atmosphere rather than values in an anomalous region such as the one into which the Probe descended.[85]

Dynamics of Jupiter's "Hotspots" and Their Relation to Atmospheric Composition Profiles

Concentrations of hydrogen sulfide and water, as well as ammonia, rose noticeably with depth in the atmosphere. One of the most puzzling features of the Probe's atmospheric data was the lack of correlation between cloud decks and abundances of the water, hydrogen sulfide, and ammonia thought to form or be involved in the formation of clouds. Abundance increases of hydrogen sulfide, for instance, did not occur as expected near the altitude of the ammonium hydrosulfide cloud deck, but in a region that scientists had thought to be well mixed, and thus where more constant abundances were expected.[86]

Scientists postulated that downdrafts at the Probe entry site would suppress abundances of water, ammonia, and hydrogen sulfide in the upper atmosphere while concentrating them at lower depths. The occurrence of downdrafts in the relatively dry hotspot into which the Probe plunged would be analogous to phenomena in Earth's desert regions, where dry air descends after drying out during ascent. A downdraft phenomenon at the Probe entry site is difficult to explain for several reasons, however. Dry air generally has a lower density than moist air, which means that if masses of moist and dry air are next to each other, the heavier, moist air will tend to sink to the lower altitude and displace the dry air upward. Downflowing air in the dry hotspot that the Probe entered would be more plausible if the air surrounding the hotspot were as dry or nearly as dry. Researchers have asked, however, what would make the surrounding regions dry.

Another proposed mechanism for the downdrafting was that the less dense, dry air of the hotspot was entrained in large-scale convection currents (flows initiated by differences in density due to temperature variations) and carried down into the atmosphere. This process would rely on descending air currents' merging and trapping less dense, dry air between them. This model was derived from fluid simulations of intense convection. A similar entrainment process may occur in Earth's atmosphere, with convection currents responsible for dragging stable stratospheric air down into the troposphere. Convection currents, which typically form separate regions, or cells, of upflows and downflows, can possibly explain the regular spacing of hotspots observed on the Jovian surface—if the hotspots are associated with downdrafting cells.[87]

Later modeling efforts by groups at NASA Ames Research Center and the University of Louisville suggested that the downwellings in hotspots were part of large-scale, long-lived Jovian circulation patterns, which helped to explain why hotspots such as the one at the Probe entry site persisted for many years. Ames's Adam Showman and the University of Louisville's Timothy Dowling, director of the school's Comparative Planetology Laboratory, generated models indicating that if large-scale pressure differences

---

[85] W. M. Folkner, R. Woo, and S. Nandi, "Ammonia Abundance in Jupiter's Atmosphere Derived from the Attenuation of the Galileo Probe's Radio Signal," *Journal of Geophysical Research* 103, no. E10 (25 September 1998): 22,847, 22,854; Young, "Galileo Probe Mission to Jupiter: Science Overview," p. 22,777.

[86] Young, "Galileo Probe Mission to Jupiter: Science Overview," pp. 22,778–22,779.

[87] Young, "Galileo Probe Mission to Jupiter: Science Overview," p. 22,779; Seiff et al., "Thermal Structure of Jupiter's Atmosphere," p. 22,885.

are present, the air that regularly flows west to east just north of Jupiter's equator also moves "dramatically up and down every few days."[88] As the air moves up and drops in temperature, the water, ammonia, and other vapors it contains condense into liquid droplets that form clouds, and they are left behind. The air that then flows downward is very dry, devoid of material that can form more clouds, and thus produces a clear area such as the one the Probe entered (see figure 8.10). According to JPL's Dr. Robert Carlson, principal investigator for the NIMS experiment, the air in these dry, clear areas contains only 1 percent of the water that the air around the dry spots has.

Jupiter's hotspots actually are not particularly hot; the temperature at their visible depth is typically about 0°C (32°F), though that is far warmer than the -130°C (-200°F) temperatures found at surrounding cloud tops. The hotspots might more accurately be called "clear spots" or "bright spots" because their transparency allows increased infrared radiation emitted from the planet below to pass through the atmosphere and be received by both spacecraft and Earth-based instruments.

These spots are typically the size of North America. Researchers' computer simulations recreated some observed features of the spots and of Jupiter's equatorial cloud plumes. The simulations showed that winds on the rim of a hotspot can get stronger and stronger with depth, a characteristic that the Probe noticed as well.[89]

Jupiter's equatorial region is depicted in figure 8.10. The dark region near the center of the mosaic is an equatorial hotspot similar to the Galileo Probe's entry site.

Jupiter's Winds and the Energy Sources That Drive Them

Pre-Galileo models of Jovian meteorology predicted that wind speeds at the cloud tops would be 360 to 540 kilometers per hour (220 to 330 mph). If these winds were generated as winds on Earth are—by solar energy and the release of latent heat as water vapor condenses—the wind speeds would be expected to drop with atmospheric depth and reach zero at some point. This is not what was found at Jupiter. The Probe's Doppler wind experiment, which correlated Doppler shifts in the frequency of the Probe's radio signal with wind speed, produced totally unexpected data. The winds below Jupiter's cloud tops were fierce, blowing at speeds of up to 720 kilometers per hour (450 mph), and these speeds remained fairly constant to depths far below the clouds. Scientists sought to determine an energy source that would plausibly drive this wind pattern and be compatible with other observations, such as the Probe's atmospheric entry point's being on the edge of a desertlike Jovian hotspot. Energy sources considered included the following:

- Conversion of ortho/para states of hydrogen to attain equilibrium.

- Latent heat release by condensation of water and other compounds.

[88] JPL, "Computer Simulation Reveals Ups and Downs of Jupiter's Winds," JPL Media Relations Office news release, 8 September 2000, *http://www.jpl.nasa.gov/releases/2000/glhotspot.html* (available in Galileo—Meltzer Sources, folder 18522, NASA Historical Reference Collection, Washington, DC).

[89] JPL, "Computer Simulation Reveals Ups and Downs of Jupiter's Winds"; Douglas Isbell and Jane Platt, "Galileo Finds Wet Spots, Dry Spots and New View of Jupiter's Light Show," NASA news release 97-123, 5 June 1997, *http://nssdc. gsfc.nasa.gov/planetary/text/gal_wet_dry.txt* (available in Galileo—Meltzer Sources, folder 18522, NASA Historical Reference Collection, Washington, DC).

- Direct heating of the atmosphere through solar-energy absorption.

- Heat fluxes from inside the planet.[90]

Figure 8.10. **A mosaic of Jupiter's equatorial region, photographed in the near-infrared and showing Jupiter's main visible cloud deck. The smallest resolved features are tens of kilometers in size. The dark hotspot near the center is a hole in the equatorial cloud layer and resembles the Probe's entry site. Circulation patterns and composition measurements suggest that dry air is converging and sinking over these holes, maintaining their cloud-free appearance. The bright oval in the upper right of the mosaic and the other, smaller, bright features are examples of moist air upwellings resulting in condensation. (NASA photo number PIA01198)**

Ortho/Para Conversion. Quantum mechanical constraints produce a coupling between the rotational states of molecular hydrogen and the spin states of its protons. This coupling results in two distinct spin states of molecular hydrogen. Molecules with antiparallel nuclear spins are termed para-hydrogen, while molecules with parallel nuclear spins are called ortho-hydrogen. In Jupiter's atmosphere, ortho and para hydrogen are not in equilibrium with each other, and there are significant energy differences between the two states. Converting from one hydrogen state to another can release significant energy. One study of this energy source has indicated, however, that the ortho/para conversion rate likely to occur in Jupiter's atmosphere is too slow to be a major factor in driving the observed Jovian winds.[91]

---

[90] Williams, "Jupiter—At Last!" p. 353; "The Probe Story: Secrets and Surprises from Jupiter."

[91] Michael D. Smith and Peter J. Gierasch, "Convection in the Outer Planet Atmospheres Including *Ortho-Para* Hydrogen Conversion," *Icarus* 116 (July 1995): 159; Young, "Galileo Probe Mission to Jupiter: Science Overview," p. 22,783.

Latent Heat Release from Water Vapor Condensation. Latent heat release through condensation of water vapor was investigated as a possible driving force for large-scale convection of the Jovian atmosphere. Various researchers developed a range of models approximating such a system, but their results did not exhibit good agreement with observations of the Jovian atmosphere. A review of these models found that they tended to oversimplify the vertical structure, boundary conditions, and dynamics of Jupiter's atmosphere so much that the models couldn't reproduce vital properties of the atmosphere.[92]

Direct Solar Heating of the Atmosphere. The Probe observed high wind speeds that extended significantly below the region of clouds as well as the depth to which solar energy penetrates. The downward solar flux measured by the Probe's net flux radiometer was effectively zero by the time the Probe had descended to the 10-bar pressure level, while the winds that the Probe observed at cloud level extended deep into the atmosphere to at least the 22-bar pressure level. No dynamic model has convincingly explained how solar energy absorbed in the upper reaches of the atmosphere would be likely to transport momentum to the atmosphere's deep interior in order to maintain the wind flows there. Thus, direct solar heating does not appear to be a likely driving mechanism for the observed wind patterns.

Heat Fluxes from Inside the Planet. An internal heat source could provide the thermal energy to drive convective motions in the atmosphere from below, leading to the observed zonal wind structure. Such a mechanism would explain the atmospheric flows that the Probe observed, in which the winds extend deep into the Jovian atmosphere. A heat source internal to the planet may well prove to be the dominant energy source for observed winds.[93]

Lightning and Radio-Frequency Signals

The Probe's lightning and radio emissions detector (LRD) sought evidence of both nearby and distant electrical discharge events from cloud to cloud or between cloud and planet. The LRD was the only one of the Probe's instruments able to perform in situ measurements of Jovian atmospheric characteristics, as well as remote sensing observations. The LRD searched for optical flashes from nearby electrical discharges and radio waves emitted from more distant events.

The thick Jovian clouds that scientists thought the Probe would encounter suggested that the LRD would see many cloud-to-cloud lightning flashes, but it did not. Considering that the ammonia and ammonium hydrosulfide cloud decks were far more tenuous than predicted and that the anticipated dense water clouds were largely absent, it was not surprising that the LRD did not record any optical flashes. The LRD did, however, receive radio signals that may have been signatures of as many as 50,000 distant lightning strikes that took place up to an Earth diameter away. The signals indicated the existence of lightning strikes 100 times stronger than a typical lightning discharge on Earth. But based on the data that the LRD was able to collect, the occurrence rate of Jovian lightning events (measured in number of flashes per square kilometer of atmosphere per year) was

---

[92] Peter J. Gierasch, "Jovian Meteorology: Large-Scale Moist Convection," *Icarus* 29 (December 1976): 445–454.

[93] Young, "Galileo Probe Mission to Jupiter: Science Overview," pp. 22,783–22,785.

only about one-hundredth the frequency of discharges on Earth. This dearth of lightning events may have been due to the atypical location in the Jovian atmosphere through which the Probe descended.[94]

Thermal Structure of the Jovian Atmosphere at the Probe Entry Site

The Galileo Probe atmospheric structure instrument (ASI) conducted the first in situ study of the Jovian atmosphere and measured its temperatures, pressures, and densities from 1,029 kilometers (638 miles) above the 1-bar level to 133 kilometers (82 miles) below it. ASI data for these thermal characteristics versus height in the Jovian atmosphere served as an altitude scale for other experiments. For instance, if a certain atmospheric pressure was observed in another experiment, this could be correlated with an altitude by using the ASI data.

The primary goal of the ASI experiment was to measure thermal characteristics of the atmosphere below cloud level, a region that was inaccessible to remote sensing by flybys such as Voyager or by Earth-based instruments. A second goal of the ASI was to define upper-atmosphere thermal structure starting in the exosphere, a tenuous outer layer of the atmosphere, and continuing down to cloud-top level. The ASI found temperatures in the upper Jovian atmosphere hotter than could be explained by a heating source of sunlight alone. Scientists postulated the precipitation of energetic particles and dissipation of waves traveling through the atmosphere as two possible mechanisms for additional upper-atmosphere heating. Temperature oscillations observed from the tropopause (the altitude where the troposphere ends and the stratosphere begins) up to the highest altitude measured by the ASI were consistent with wave phenomena. The viscosity of the atmosphere would, over time, damp the amplitudes of the waves, converting the energy they carried into thermal energy. Scientists' calculations show that the energy dissipated by the postulated waves would be comparable to the amount needed to heat the Jovian atmosphere to its observed levels.[95]

Upper-atmosphere pressure measurements showed a region that was more dense than expected. A surprisingly thick layer of a constant temperature of roughly -113°C (171°F) was encountered in Jupiter's stratosphere, from an altitude of 290 kilometers (180 miles) down to 90 kilometers (56 miles). A second isothermal (constant-temperature) region was found as well, just above the tropopause. This region was 25 kilometers (16 miles) thick, with a temperature of about -161°C (-258°F). In the lower regions of the atmosphere measured by the Probe, temperature increased with pressure, although at a slightly slower rate than expected.[96]

---

[94] K. Rinnert, L. J. Lanzerotti, M. A. Uman, G. Dehmel, F. O. Gliem, E. P. Krider, and J. Bach, "Measurements of Radio Frequency Signals from Lightning in Jupiter's Atmosphere," *Journal of Geophysical Research* 103, no. E10 (25 September 1998): 22,979–22,992; Klaus Rinnert and L. J. Lanzerotti, "Radio Wave Propagation Below the Jovian Ionosphere," *Journal of Geophysical Research* 103, no. E10 (25 September 1998): 22,993; "The Probe Story: Secrets and Surprises from Jupiter."

[95] Seiff et al., "Thermal Structure of Jupiter's Atmosphere," p. 22,857; JPL, "Probe Science Results," section 7 in *Galileo: The Tour Guide*, June 1996, p. 54, *http://www.jpl.nasa.gov/galileo/tour/6TOUR.pdf* (available in Galileo—Meltzer Sources, folder 18522, NASA Historical Reference Collection, Washington, DC).

[96] Seiff et al., "Thermal Structure of Jupiter's Atmosphere"; JPL, "Probe Science Results."

We're able to observe lava flowing on a surface that is similar to what we had here on Earth two billion years ago. We can almost provide a window into our own past.

—Dr. Eilene Theilig, Galileo Project Manager, 8 May 2001

Chapter 9
THE ORBITER TOUR

THE JUPITER ORBITER TOUR WAS MORE THAN SIMPLY A series of flybys of the planet's moons, where photographs were taken and fields measured. The tour was the culmination of an exploration plan developed over 20 years earlier to study the Jovian system thoroughly. The multifunction Galileo spacecraft was intended to explore the various sectors of the Jupiter system in an integrated manner, examining interactions among all sectors so as to better understand the whole. The first part of the exploration plan, involving in situ measurements of the parent planet, had been accomplished by Galileo's Atmospheric Probe. The second part of the plan depended on the Orbiter to analyze comprehensively the planet's satellites, fields and particles, and magnetosphere[1]. The Orbiter concentrated its satellite observations on Jupiter's four major moons, the ones that Galileo Galilei discovered nearly four centuries ago, although some of the Orbiter's time was also spent observing several of Jupiter's smaller moons.

Discovery of the Galilean Satellites

Exploration of the Jovian system began long before the 20th century. During the summer of 1609, a mathematics lecturer at the University of Padua named Galileo Galilei heard of a new optical instrument, a telescope, that had been displayed in Holland. He constructed one for himself and verified its ability to discern "sails and shipping that were so far off

---

[1] JPL, "The Tour," section 8 in *Galileo: The Tour Guide*, June 1996, p. 59, *http://www.jpl.nasa.gov/galileo/tour/8TOUR. pdf*; Torrence V. Johnson interview, tape-recorded telephone conversation, 31 July 2001.

that it was two hours before they were seen with the naked eye." Galileo was not satisfied with the capabilities of his first model telescope, however, and he quickly developed improved versions that could observe new features on the Moon and find many more stars in the night sky than had been seen before. It was on one of the evenings when he was searching the sky that he saw something that startled him:[2]

> On the seventh day of January in this present year 1610, at the first hour of night, when I was viewing the heavenly bodies with a telescope, Jupiter presented itself to me; and because I had prepared a very excellent instrument for myself, I perceived (as I had not before, on account of the weakness of my previous instrument) that beside the planet there were three starlets, small indeed, but very bright.[3]

Galileo continued to watch these starlets over the next several days. At first, he could not understand how, overnight, Jupiter could move from a position west of two of them to a new position east of all three of them. The movements of Jupiter that had been recorded by other astronomers predicted that the planet should have moved in the opposite direction. Galileo's initial thoughts were that the other astronomers had somehow miscalculated Jupiter's movement.[4]

On 11 January 1610, a fourth starlet appeared. After a week of observation, Galileo discovered that these four bright bodies never left the vicinity of Jupiter; instead, they appeared to be carried along with the planet as it journeyed through the sky. The starlets did change their positions with respect to each other and to Jupiter, though. Finally, Galileo deduced that he was not observing stars, which do not change their positions relative to each other, but planetary bodies that were in orbit around Jupiter. By showing that not all bodies revolved around Earth, Galileo's revelation helped to support the Copernican view of the solar system, which places the Sun rather than Earth at the center.[5]

The planetary bodies that Galileo discovered turned out to be Jupiter's four largest moons, now known as the Galilean satellites. They were called Io, Europa, Ganymede, and Callisto, which are the Greek names of four lovers of the god Zeus in Greek mythology (or Jupiter in Roman mythology).[6,7]

Although Galileo Galilei discovered Jupiter's four major moons, his telescope was capable of seeing them only as points of light. Subsequent telescopes and space vehicles garnered much valuable data regarding these moons, but it was not until the Galileo Orbiter tour that the different characteristics of these bodies and their interactions

---

[2] J. R. Casani, "A Name for Project Galileo," JPL interoffice memo GLL-JRC-78-53, 6 February 1978, John Casani Collection, Galileo Correspondence 12/77–2/78, folder 43, box 5 of 6, JPL 14, JPL Archives. Biographical sketch of Galileo Galilei included in this memo was written by Mary Jo Smith, Galileo program engineer for NASA Headquarters.

[3] Galileo Galilei, *The Starry Messenger (Sidereus Nuncius)*, 1610. One of many places this book can be accessed is at *The History Guide, http://www.historyguide.org/earlymod/starry.html*, 2002.

[4] *Galileo to Jupiter: Probing the Planet and Mapping Its Moons*, JPL Document No. 400-15 7/79 (Washington, DC, GPO No. 1979-691-547), p. 1.

[5] The Discovery of the Galilean Satellites," *JPL Galileo Project Home Page*, p. 1, *http://www.jpl.nasa.gov/galileo/ ganymede/discovery.html* (accessed 12 March 2000).

[6] Casani, "A Name for Project Galileo."

[7] NASA, "Welcome to the Planets—Glossary," *Planetary Data System*, last updated 15 November 2004, *http://pds.jpl. nasa.gov/planets/special/glossary.htm* (accessed 12 June 2005).

with the parent planet could be repeatedly evaluated at close range. The results of these evaluations have sparked the imaginations of scientists around the world and have radically altered our previous picture of the Jovian system.

Attaining Scientific Goals During the Tour

Each Galileo orbit of Jupiter usually included a close encounter with one of Jupiter's satellites[8] and a cruise period that typically lasted a couple of months.[9] During each encounter period, the Orbiter collected a range of data at high rates. The Orbiter returned some real-time data to Earth on fields-and-particles characteristics, but it had to store most of the other data on its tape recorder due to limitations in the transmission rate of the low-gain antenna. The stored data included satellite images, ultraviolet and infrared spectra, and additional fields-and-particles measurements, especially those taken at the closest approach to the satellite. Subsequent cruise periods gave the spacecraft the chance to play back and send the recorded data to Earth, as well to take additional fields-and-particles measurements.

Project management established three working groups for collecting three types of scientific data: magnetospheric, atmospheric, and satellite. In situ measurements of Jupiter's magnetosphere were acquired nearly continuously during the Orbiter tour, but with higher resolution during satellite encounters as well as during the spacecraft's journey down Jupiter's magnetotail (which occurred between the close flybys in orbits C9 and C10). The spacecraft performed remote sensing of Jupiter's atmosphere and of the planet's satellites primarily within a few days of each satellite flyby.[10]

Each of the scientific working groups had its own aims and needs for acquiring and storing data. Unfortunately, these needs could not always be met simultaneously by the spacecraft. This shortcoming was due in part to the limited data-storage capacity of the tape recorder and the slow transmission rate of the low-gain antenna. These factors restricted the amount of data that could be gathered during each satellite encounter and returned to Earth before the next encounter. Another restriction was the limited amount of memory available to retain computer code sequences. Such sequences directed the spacecraft to perform the necessary operations for attaining scientific objectives (such as orienting the scan platform in a specific direction). Due to the Orbiter's operational limitations, negotiations and bargaining among the scientific working groups for access to the spacecraft became an important part of how mission business was conducted.[11]

Mission Support Activities During Satellite Encounters

Key to the success of the Galileo project were the vital activities performed by the engineers, scientists, and technicians of the mission support staff, whose job it was to ensure

---

[8] During part of Galileo's fifth orbit after Jupiter Orbit Insertion (JOI), the spacecraft had to be out of communication with Earth due to a solar conjunction; no satellite encounter was planned for that orbit.

[9] An exception to this occurred on the first orbit after JOI, in which Galileo's orbit was very elongated and its cruise period was nearly half a year in duration.

[10] JPL, "The Tour," pp. 60–61.

[11] Ibid., p. 61.

the "health and safety of the vehicle."[12] Their activities intensified just before and during a close encounter. Mission support activities surrounding an Io encounter in 2001 serve as examples of typical actions taken during a flyby. They give a picture of the many data streams that teams had to watch and the various actions they needed to take.

Tape-Recorder Preparation

In the week before the 5 August 2001 encounter with Io (the abbreviation for this encounter was "I31"), mission support guided the spacecraft through several preparatory housekeeping tasks. Routine maintenance of the on-board tape recorder was a critical task. Playback and transmission of the recorder's data from the previous flyby had to be stopped and the tape wound halfway down its length and back again. This action helped prevent the tape from sticking to the recorder's head, as happened in October 1995 when Galileo was first approaching Jupiter. A tape jam would have eliminated most of the spacecraft's data-storage potential.[13]

Targeting Maneuvers

Mission support closely oversaw the spacecraft's final targeting maneuvers on its approach to a satellite. This oversight involved precisely controlled firings of Galileo's 12 thrusters to ensure that the spacecraft reached its scheduled satellite rendezvous at the correct time and place. Galileo's thrusters delivered only a modest amount of propulsion (about 2.2 pounds of thrust each) and had to fire for several hours to nudge the 1,300-kilogram (2,900-pound) spacecraft onto a slightly different trajectory. With some thrusters pointing forward, some backward, and some to the sides, the choice of which to fire and when determined in what direction the spacecraft moved. Mission support personnel also used the thrusters to rotate the spacecraft, with no change to its orbital path, in order to point its antenna in new directions.

Typically, the final targeting maneuvers changed Galileo's speed by only a few tenths of a meter per second. This was minuscule compared to the craft's speed of 7.1 kilometers per second (16,000 mph) as it flew by Io.[14]

Coping with Radiation Effects

Galileo was a robotic spacecraft that needed little attention when working correctly. Its control sequences had all been programmed into its computer, and if it followed them without incident, the crew mainly sat back and tracked the craft's progress. But when Galileo malfunctioned, the crew got busy and sometimes had to work marathon hours through the night.

---

[12] Nagin Cox interview, tape-recorded conversation, 15 May 2001.

[13] "This Week on Galileo: Galileo Continues Preparations for Next Week's Io Flyby," JPL DOY 2001/211-215, 30 July 2001, *http://www2.jpl.nasa.gov/galileo/news/thiswk/today010730.html*; "This Week on Galileo July 30–August 3, 2001," SpaceRef.Ca, 2 August 2001, *http://www.spaceref.com/news/viewsr.html?pid=3369* (available in folder 18522, NASA Historical Reference Collection, Washington, DC); "Today on Galileo—Sunday, August 5, 2001—Day 2 of the Io 31 Encounter," SpaceRef.Ca, 5 August 2001, *http://www.spaceref.ca/news/viewsr.html?pid=3385* (available in folder 18522, NASA Historical Reference Collection, Washington, DC).

[14] "This Week on Galileo—July 30–August 3, 2001."

At 6:00 on Sunday evening, 5 August 2001, the spacecraft seemed to be performing as hoped as it neared its closest approach to Io. But the moon exists within an intense radiation environment whose high-energy particles had degraded instrument and spacecraft performance repeatedly during past encounters. Mission support was concerned about additional radiation incidents. Each near-Jupiter orbit increased the odds of more serious damage from exposure to the planet's radiation belts. Galileo's solid state imaging (SSI) camera was one component that had not been operating at full performance. The camera's passage through radiation belts had resulted in completely saturated, overexposed images. Project engineers had developed procedures for minimizing the loss of images that had plagued the camera since summer 2000. Cycling the power—turning it off, then on again—sometimes helped, at least temporarily. So did reloading the camera's software. Mission support personnel implemented these procedures eight times during encounter I31, just before key observations.[15]

Another Galileo component affected by radiation was the star scanner, whose measurements determined spacecraft orientation. According to Gerry Snyder, a systems engineer and member of the mission support team, Galileo's attitude-control system normally used light collected by the star scanner from three or four stars to accurately determine the direction in which the spacecraft was pointing. But in the intense radiation environment close to Jupiter, high-energy electrons flooded Galileo's star scanner detector and masked the light from the fainter stars. This noise could cause serious problems. Spacecraft and sequence team chief Duane Bindschadler explained that when the background counts from radiation went up, the star scanner saw intermittent flashes. The attitude-control system software could mistake the flashes for dim stars and try to act on incorrect knowledge. Because the attitude-control system would not know which way the spacecraft was pointing, the system would crash, restart, and crash again. Meanwhile, the scan platform would freeze up. The spacecraft computer would hold it stationary to prevent any instruments from getting damaged. During these times, the remote sensing instruments mounted on the scan platform would not be able to perform observations that Galileo's scientists were counting on, and much of the encounter would be wasted.

To avoid such problems during passes through high-radiation regions, mission personnel commanded the star scanner to temporarily lock in on only one bright star—a star with a signal well above the noise produced by the radiation. Then the attitude-control software would not be confused and could safely maintain Galileo's orientation. The star used during I31 was Alpha Eridani, the brightest star in the constellation of Eridanus, the River.[16]

Jupiter Occultation and Deep Space Network Limitations

On I31, the relative positions of spacecraft, Jupiter, and Earth allowed less time than other encounters for mission support staff to deal with problems. Gerry Snyder explained that

---

[15] Guy Webster, "Spacecraft to Fly Over Source of Recent Polar Eruption on Io," JPL news release 2001-161, July 2001; "Today on Galileo—Saturday, August 4, 2001: The Encounter Begins," *SpaceRef.com* Web site, 4 August 2001, *http://www.spaceref.com/news/viewsr.html?pid=3380* (accessed 13 June 2005) (available in Galileo—Meltzer Sources, folder 18522, NASA Historical Reference Collection, Washington, DC).

[16] Gerry Snyder (mission support systems engineer) and Duane Bindschadler (spacecraft and sequence team chief) interviews, JPL, 5 August 2001; "This Week on Galileo: The Io 31 Encounter Begins," JPL DOY 2001/216, 4 August 2001, *http://www2.jpl.nasa.gov/galileo/news/thiswk/today010804.html* (available in folder 18522, NASA Historical Reference Collection, Washington, DC).

Galileo had been occulted behind Jupiter for 3 hours (a fairly long occultation) and out of radio contact with Earth until about 6 p.m. PDT on 5 August—just hours before the craft's closest approach to Io. During this occultation, mission support personnel instructed the spacecraft to stay in "fill data" mode, which meant that the craft ceased sending valuable information to Earth until after occultation was over and data could actually be received.[17]

At about 8 p.m. PDT, 2 hours before the closest approach, the craft again traveled out of radio contact with Earth, but for a different reason from before. It flew into a region of space that was not in sight of two of NASA's three Deep Space Network (DSN) stations (the ones in California and Australia). The large antennas of NASA's DSN typically provide the communication link with the Agency's interplanetary spacecraft, which included Voyagers 1 and 2, Ulysses, the Mars Global Surveyor, Galileo, and others. Normally, at least one of the DSN's three stations stayed in contact with Galileo. During the close encounter, only the station in Spain had a line-of-sight view of Galileo and could potentially communicate with it, but the Spanish station's large antenna that typically tracked Galileo had been put temporarily out of service. It was being upgraded to handle the increased number of missions anticipated for 2003 and 2004. Data concerning Galileo's status during the I31 flyby had to be received several hours afterward by the DSN's other stations, and any rescue attempts on the part of mission support had to wait until then.

The Deep Space Operations Center

During flybys, mission support crews carefully monitored many types of data to determine the health of the spacecraft. Information received by the three DSN stations was sent to JPL's Deep Space Operations Center. This facility is called the "darkroom" because of its subdued lighting, which allows better viewing of its banks of monitors. One bank of monitors reports on the state of communications between JPL and the DSN. Another bank enables operators to check that the appropriate data get routed to support personnel for the right project—the darkroom monitors track many interplanetary missions, and it would not do for Martian data to get sent to Galileo personnel or vice versa.[18]

Galileo data arrived as packets of information, signified on the screens by various number and letter codes. The darkroom's computers are programmed to display certain signs when data change beyond prearranged limits. The data codes might suddenly turn bold, or the surrounding screen might change to red or some other color. If the out-of-spec data happened to be, for instance, the temperature of the spacecraft, mission support needed to determine whether the craft was heating up or cooling down, then take appropriate actions before sensitive instruments were damaged or a flyby was ruined.[19]

Galileo sent information down to Earth in compressed packets of data that allowed more information than originally planned to be transmitted each second. Gerry Snyder, who had started with the Galileo project in the late 1970s but then left, was brought back on the project in 1992 when the trouble with the high-gain antenna occurred and data flow through the low-gain antenna had to be maximized. The spacecraft had been handling data using an inefficient system called "time division multiplex" (TDM). If,

---

[17] Gerry Snyder interview.

[18] Galileo mission support crew interview, JPL, 5 August 2001; JPL, "Galileo Millennium Mission Status," JPL Media Relations Office news release, 6 August 2001.

[19] Galileo mission support crew interview.

for instance, one of the spacecraft's instruments was generating the same data time after time, TDM would send the repetitive information again and again to Earth, wasting valuable transmitting time. This procedure would not have been a problem if the high-gain antenna had been operating, for it would have sent data at a high enough rate (134,000 bits per second) for Earth to receive all the data it wanted, even if some of the information was repetitive. But the low-gain antenna could send only a trickle of information (about 10 to 40 bits per second), and it all had to count. What Snyder did when he rejoined the Galileo project was to implement a "packetized data" system in which the spacecraft computer kept a watch on the data flow and sent a packet to Earth only when there was something new. The packetized system selected just the data that mission personnel on Earth needed to reconstruct what the spacecraft was doing and observing.

Using the packetized-data approach, an image taken by Galileo's SSI camera could be reconstructed on Earth with very good visual quality, employing as little as one-eightieth of the data that TDM would have sent. The Galileo spacecraft had only two or so months after an encounter to send what it had collected to Earth before the tape recorder had to store information from the next encounter. Because the spacecraft had to transmit data at a very slow rate, not all the information collected could actually be sent. Tough decisions had to be made as to what to transmit and what to erase. Packetizing data was important because it allowed significantly more of the information collected to be sent on to Earth.[20]

Galileo's Data Transmissions

The data Galileo transmitted to Earth were of two basic types: 1) engineering data that related to the spacecraft's health and operating performance and 2) science data, which pertained to the various experiments that Galileo was conducting. Mission support personnel studied the former carefully and passed on the latter to the various science teams around the country and in Europe. Mission support sent on science data packets without ever actually looking at the data they contained. Such data were very sensitive and carefully protected by the science team. That team might have been waiting for the data for many years and had likely spent a large sum of money building the instrumentation for its experiment. The team wanted to be sure that it controlled access to its own data.[21]

Navigating and Orienting Galileo

For mission engineers to control the Orbiter's trajectory accurately, they needed to analyze both radio tracking data and optical data (positions of stars and satellites used as reference points). This information was then used to make the needed modifications in spacecraft flight direction and orientation.[22]

One of the critical "housekeeping" chores of the mission was to keep the spacecraft's low-gain antenna (LGA) pointed within 4 degrees of Earth in order to maintain radio contact. For optimal signal reception on Earth, it was better to maintain the LGA

---

[20] Gerry Snyder interview.

[21] Ibid.

[22] JPL, "The Tour," p. 62.

direction within 1 degree of Earth. Two of the data streams that mission support personnel watched related to spacecraft attitude control and indicated how far off the spacecraft's antenna was from where it had been commanded to point.[23]

Spacecraft attitude data were also used in turning the Orbiter to allow its instruments to view different parts of the sky that might otherwise be blocked by the science boom or the main body of the spacecraft. Such reorientations of the Orbiter were called "spacecraft inertial turns" (SITURNS) or, more commonly, "science turns." The number of science turns that could be made was severely limited by the amount of propellant aboard. Only 20 kilograms of propellant were budgeted for science turns during the mission; a 90-degree turn and return used about 3 kilograms.[24]

Preventing Spacecraft Problems

Mission support engineers such as Gerry Snyder had to keep tabs on many different data streams and quickly spot potential spacecraft trouble or conditions that would interfere with science experiments. One data stream relayed spin error, which was the difference between the spacecraft's actual and intended spin rates. The "star code" record gave an indication of what Galileo's star scanner navigation sensor was doing, while the "running count"—the number of background pulses per unit of time that the star scanner was seeing—related to radiation levels around the spacecraft. In low-radiation regions, normal background counts ran about 170 to 270 per second, allowing the star scanner to track both faint and bright stars. But when the counts exceeded 1,000 per second, many faint stars were masked by the background and the scanner could track only bright stars reliably.[25]

The "slew count" was a measure of the times that the spacecraft's scan platform, on which the remote sensing instruments were mounted, had been reoriented to allow an instrument to make a particular observation. Other data indicated whether it would be necessary to perform a "repulsive maneuver," in which one or more of the science instruments had to be turned off or put into safe modes because, for instance, fumes from engine firing might present risks to them. The spacecraft's central data system had fixed responses for such occurrences, like turning off an instrument's high voltage in order to prevent arcing.[26]

Another data stream provided the "Doppler residuals," important to the navigation team in determining how close Galileo was to its aim point for the flyby. There was also a sequence of events (SOE) log that listed all spacecraft commands and ground events (for instance, actions that the DSN stations took).[27]

Using Gravity-Assists To Minimize Propellant Requirements

Project engineers had the task of "steering" the spacecraft along its satellite tour so that it could fly by one moon after another with a minimum expenditure of fuel. The better

---

[23] Gerry Snyder interview.

[24] JPL, "The Tour," p. 65.

[25] Galileo mission support crew interview.

[26] Gerry Snyder interview.

[27] Brad Compton, Galileo mission control team chief, interview, JPL, 5 August 2001.

the engineers were at their job, the longer the craft could operate (assuming that its other systems held together). To minimize propellant use, each satellite encounter was designed to make optimal use of the moon's gravitational force to alter the course of the spacecraft toward its next target. Only a small amount of propellant was needed to fine-tune the Orbiter's trajectory. The entire Prime Mission satellite tour required the thrusters to supply a total change in speed of only about 100 meters per second, about 60 times less than if gravity-assists had not been used. Galileo staff viewed this trajectory design as a "10-cushion shot in a celestial billiard ball game," with the caveat that in this case, small corrections in billiard ball direction could be made along the way.[28]

The Prime Mission

The trip from Earth to Jupiter, the Probe's exploration of the Jovian atmosphere, and an Orbiter tour consisting of 11 orbits of Jupiter constituted Galileo's Prime Mission. Table 9.1 lists major milestones of the Prime Mission.

Table 9.1. **Overview of Galileo's Prime Mission.**[29]

Launch	18 October 1989
Probe Release	12 July 1995 (PDT)
Jupiter Arrival Day	7 December 1995
Probe atmospheric penetration	7 December 1995
Prime Mission Orbiter tour (11 orbits of Jupiter)	7 December 1995 to 7 December 1997

On Jupiter Arrival Day (7 December 1995), the Galileo spacecraft was given a gravity-assist from Io and then subjected to the Jupiter orbit insertion (JOI) maneuver, which slowed the spacecraft down and let it be "caught" by the planet. These two actions placed the Orbiter on its proper trajectory to tour the Jovian moons. The Jupiter orbit insertion maneuver involved an orbit around the planet, which is referred to as the spacecraft's "zeroth" orbit. The spacecraft's "first," and by far longest, orbit around Jupiter followed the JOI and lasted nearly seven months. On 27 June 1996, this initial orbit culminated in a close encounter with Ganymede, the largest of the four Galilean satellites.

Figure 9.1 depicts the "flower petal" pattern of the 11 Jovian orbits on Galileo's Prime Mission. This tour included four close encounters with Ganymede, three with Europa, and three with Callisto. No Io encounters were planned for the Prime Mission (besides the flyby on Arrival Day) because mission scientists feared that the high radiation levels so close to Jupiter could damage the spacecraft and possibly end the project.

---

[28] JPL, "The Tour," p. 62.

[29] Project Galileo Outreach Office, "The Galileo Mission at Jupiter—Fact Sheet," *Bringing Jupiter to Earth*, revision 3, November 1998, *http://www.jpl.nasa.gove/galileo*.

After the first Jupiter orbit of seven months, subsequent orbits were much shorter, ranging from one to two and a half months. As figure 9.1 shows, orbits 2 through 11 followed much less eccentric (elongated) elliptical paths than the first orbit. The Prime Mission ended in December 1997, two years after Jupiter arrival.[30]

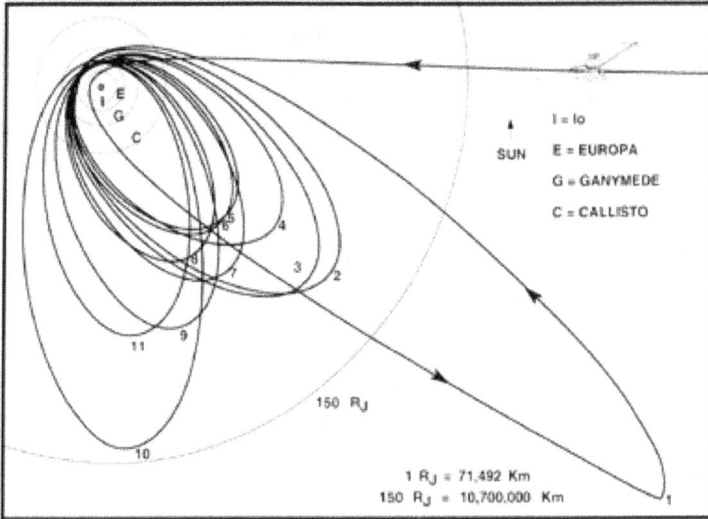

Figure 9.1. **"Flower petal" plot of the Prime Mission Orbiter tour, depicting close encounters with satellites. Note the eccentricity of the first orbit as compared to the others. Due to a solar conjunction, the spacecraft did not have a close encounter with any satellites on its fifth orbit. (Adapted from JPL image number P45516B)**

Naming the Orbits of Jupiter

The Galileo mission used a two-character code to specify each orbit. The first character was the first letter of the name of the moon that would receive a flyby on the orbit, while the second character indicated the number of the orbit. Table 9.2 lists the code name for each Jupiter orbit on the Prime Mission satellite tour, the satellite encountered during the orbit, and the date of closest approach to the satellite. Note that on the fifth orbit (J5), no close satellite encounter occurred.[31] Galileo was out of communication during part of J5 due to a solar conjunction, when the Sun was between Jupiter and Earth, so mission engineering staff did not plan a close flyby for this orbit (figure 9.2 illustrates a solar conjunction situation).[32]

---

[30] JPL, "The Tour," p. 59.

[31] Ibid., p. 60.

[32] "Galileo Highlights of Jupiter Orbital Tour 92-14a," *Project Galileo Homepage, http://www.jpl.nasa.gov/galileo/ tourhilites.html* (accessed 19 May 2001).

Table 9.2. **Prime Mission satellite tour. Includes the naming convention for Jupiter orbits and dates of closest approach to the designated satellite on each orbit.**[33]

Orbit	Satellite Encounter	Closest Approach
G1	Ganymede	27 May 1996
G2	Ganymede	6 September 1996
C3	Callisto	4 November 1996
E4	Europa	19 December 1996
J5	solar conjunction	no close flyby
E6	Europa	20 February 1997
G7	Ganymede	5 April 1997
G8	Ganymede	7 May 1997
C9	Callisto	25 June 1997
C10	Callisto	17 September 1997
E11	Europa	6 November 1997

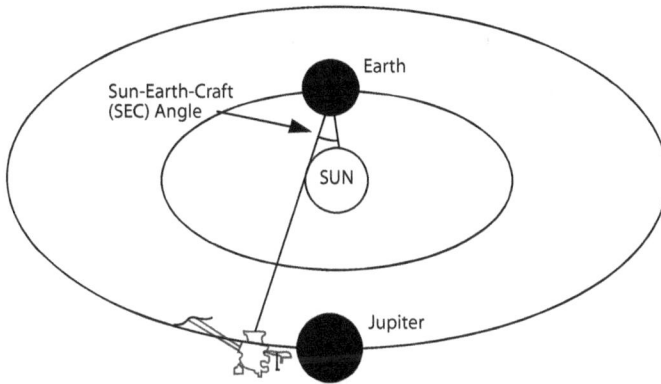

Figure 9.2. **Solar conjunction occurs during the Galileo mission when the Sun lies directly between Jupiter and Earth. The Sun is a strong source of electromagnetic activity, and it wreaks havoc with the spacecraft's radio signal, essentially reducing the spacecraft's data-transmission rate to Earth to almost zero for the two and a half weeks spanning the conjunction. Mission planners and telemetry engineers define this problem time as occurring when the Sun-Earth-Craft (SEC) angle is less than 7 degrees (see figure), although a relatively "quiet" Sun can allow data to be successfully transmitted to Earth at SEC angles as small as 3 to 5 degrees. During Galileo's primary mission, solar conjunction periods occurred from 11 to 28 December 1995 and 11 to 28 January 1997. During the Galileo Europa Mission (GEM), solar conjunctions occurred from 14 February 1998 to 4 March 1998 and from 22 March 1999 to 10 April 1999. A Galileo Millennium Mission (GMM) solar conjunction period ran from 28 April 2000 to 17 May 2000.**

[33] JPL, "The Tour," p. 60.

The Galileo Europa Mission

The Galileo project would have been considered a success even if the spacecraft had stayed operational only through the end of the Prime Mission on 7 December 1997, two years after Jupiter Arrival Day. The Orbiter was an extremely robust machine, however, with many backup systems. It showed no sign of quitting at the end of the Prime Mission, so it was given a highly focused set of new exploration objectives, defined in part by the findings of the Prime Mission. These new objectives centered on investigating Europa in great detail, and the new mission was, appropriately, called the "Galileo Europa Mission" (GEM). Mission objectives were not limited to Europa, however; they included analyses of other satellites, as well as of Jovian fields and particles and atmospheric characteristics. During GEM, some of the most important and spectacular observations of the volcanic moon Io were taken.[34]

GEM ran for slightly over two years, from 8 December 1997 to 31 December 1999. It was a low-cost mission with a budget of only $30 million.[35] At the end of the Prime Mission, most of the 200 Galileo staff members, including Project Manager Bill O'Neil, left for other assignments. The remaining bare-bones crew, about one-fifth the size of the Prime Mission, was left to run GEM and achieve the objectives of four separate studies:

- Europa campaign.

- Io campaign.

- Io plasma torus study.

- Jupiter water study.

Europa Campaign

The spacecraft analyzed Europa's crust and atmosphere and searched for further evidence of a past or present ocean beneath the moon's ice crust. The Galileo team compared GEM image data with previous images and identified surface changes and signs of spewing ice volcanoes that might indicate the presence of a subsurface ocean. Because a flowing, salty ocean can generate a magnetic field, the Orbiter needed to determine whether the magnetic signals near Europa were generated from within the moon. By measuring variations in the pull of the moon's gravity, the thickness of the ice crust and depth of a possible ocean were estimated.

The Europa campaign involved a one-year study of the moon that included eight consecutive close flybys. During these encounters, the Orbiter conducted extensive

---

[34] "Galileo Europa Mission (GEM) Fact Sheet," *JPL Galileo Home Page, http://www.jpl.nasa.gov/galileo/gem/fact.html* (accessed 31 March 2000).

[35] "Ice, Water, and Fire: The Galileo Europa Mission," extracted from *The Planetary Report* (January/February 1998, courtesy of the Planetary Society), *Galileo Europa Mission (GEM) Home Page, http://www.jpl.nasa.gov/galileo/gem/gem1.html* (accessed 31 May 2001).

remote sensing and fields-and-particles observations. A timeline for the Europa campaign and other parts of GEM is included in table 9.3.[36]

Table 9.3. **Galileo Europa Mission key events.**

Event	Dates	Orbits	Closest Approach in Kilometers (Miles)
Europa campaign	16 December 1997 to 4 May 1999	8 orbits: E12–E19	201 km (125 mi), 16 December 1997 (E12)
Jupiter water/Io torus studies	5 May to 10 October 1999	4 orbits: C20–C23	467,000 km (290,000 mi), 14 September 1999 (C23)
Io campaign	11 October to 31 December 1999	2 orbits: I24–I25	300 km (186 mi), 26 November 1999 (I25)

The mission's engineering staff expected to collect Europa images with resolutions better than 50 meters. The spacecraft flew close enough to the moon that it was able to take in situ sampling of its very tenuous atmosphere.

Jupiter Water Study

GEM's agenda included making detailed observations of storm and wind patterns in Jupiter's atmosphere and mapping the water distribution. Besides contributing to our knowledge of Jovian meteorology, this study also helped scientists better understand Earth's weather system.[37] The Orbiter flew the closest to Jupiter that it had since Arrival Day. In particular, it sought to examine the billowing thunderstorms that grow to heights several times those on Earth. The craft studied the water circulation patterns in the Jovian atmosphere's top layers that cause large areas of the surface to be drier than our Sahara Desert while other parts receive rainfall similar to that in Earth's tropical regions.

Io Plasma Torus Study

The Orbiter explored and mapped the density of the donut-shaped cloud of charged particles that surrounds Io's orbit. During passes through this cloud, the spacecraft mapped the density of the sulfur streaming from Io's erupting volcanoes. The spacecraft also measured the sodium and potassium that particles accelerated by Jupiter's rotating magnetic field "sand-blast" off the moon's surface.[38]

---

[36] "GEM Fact Sheet"; "Ice, Water, and Fire: The Galileo Europa Mission"; JPL, "The Europa Mission—Exploring Through 1999," *JPL Galileo Home Page, http://www.google.com/search?q=cache:KSWkENJMpCQJ:www2.jpl.nasa.gov/galileo/Page_2_GEM_Fact_Sheet.pdf+The+Galileo+Europa+Mission+%E2%80%93+Exploring+Through+1999&hl=en&ie=UTF-8* (accessed 20 November 2004) (available in folder 18522, NASA Historical Reference Collection, Washington, DC).

[37] "GEM Fact Sheet"; "The Galileo Europa Mission—Exploring Through 1999."

[38] "GEM Fact Sheet"; "Ice, Water, and Fire: The Galileo Europa Mission."

Io Campaign

Now that the Prime Mission was successfully completed and the Orbiter still operated well, mission engineers decided to attempt close flybys and a detailed examination of Io. During the first Jupiter orbit of the Prime Mission, the Orbiter's perijove (its point of closest approach to Jupiter) had been raised to keep the craft out of the most intense parts of Jupiter's radiation belts (through which Io orbits). The radiation intensity in those belts exceeds human fatality levels, and there was a high risk that traveling through the belts would damage the Orbiter's equipment and possibly end the mission. For the spacecraft to attempt close flybys of Io at this point, its perijove had to be reduced. This was done through a series of maneuvers that involved four Callisto encounters. Thus, the Galileo Europa Mission provided opportunities to study at close range not only Europa and Io, but Callisto as well. For six months in 1999, Galileo used gravity-assists from Callisto, as well as fine tuning from the craft's thrusters, to gradually halve its closest approach to Jupiter.[39]

The first Io flyby occurred in October 1999. The spacecraft was still operating after passing through the high-radiation region near the moon, and so it flew by Io again six weeks later. The Io campaign conducted a comprehensive examination of the satellite's volcanism, atmosphere, and magnetosphere. Project planners obtained high-resolution images and a compositional map of the moon. The Orbiter was able to sample and analyze volcanic particulates and, as at Europa, particles from Io's tenuous atmosphere. The spacecraft also made measurements to determine whether Io generated its own magnetic field.[40]

GEM Power, Propellant, and Data-Recording Issues

Project engineers predicted that throughout GEM, the Orbiter's radioisotope thermoelectric generators would be able to supply ample power for spacecraft functions and that sufficient propellant would remain to power the craft's thrusters. However, they thought that by GEM, the Orbiter's tape recorder would have surpassed its design limit for starts and stops. GEM's scope would be significantly reduced if the tape recorder failed. In such an event, project engineers would load the Orbiter's on-board computer with a program that would allow the craft's instruments to receive and transmit a severely limited amount of data in real time.[41]

Organizational and Operational Changes During GEM

To meet its scientific goals with a bare-bones budget of $15 million per year, GEM employed only 20 percent of the original personnel to operate the Orbiter and analyze its reduced data stream. Engineering and science teams streamlined and automated their operations and software. On each flyby, the spacecraft took only two days of data versus the seven days it had taken during the Prime Mission. Minimal Jovian magnetic field data were collected. Orbiter turning maneuvers were kept to a minimum. The GEM team did not include the expertise to deal with unexpected problems, as the Prime Mission had.

---

[39] "GEM Fact Sheet"; "Ice, Water, and Fire: The Galileo Europa Mission."

[40] "The Galileo Europa Mission—Exploring Through 1999"; "GEM Fact Sheet."

[41] "Ice, Water, and Fire: The Galileo Europa Mission."

When issues arose, specialists who had gone on to other missions were temporarily brought back and placed on "tiger teams" to work through the problems quickly.[42]

The Galileo Millennium Mission

Because the Orbiter spacecraft continued to operate so well, a further extension to the original project, the Galileo Millennium Mission (GMM), was added to pursue answers to key questions raised during GEM. The original GMM schedule ran from January 2000 through March 2001, but it was then extended to the end of mission operations in January 2003.[43] The spacecraft met its demise in September 2003, when its trajectory took it on a collision course toward Jupiter and it burned up in the planet's atmosphere.[44]

GMM conducted additional investigations of Europa, including a magnetic field measurement key to detecting the presence of liquid water. GMM also added to our knowledge of Io, studied the dynamics of Ganymede's unique magnetosphere, determined particle sizes in Jupiter's rings, and performed a joint investigation with the Cassini spacecraft, whose closest approach to Jupiter was on 30 December 2000. Primary science goals of the joint Galileo-Cassini study were to accomplish the following:

- Analyze how interactions with the solar wind affect Jupiter's magnetosphere and, in particular, its auroral regions.

- Improve our understanding of Jupiter's atmosphere, particularly its active storm regions.

- Analyze dynamics of Jovian dust streams.

- Observe Io in eclipse in order to understand airglow phenomena and monitor hotspots.[45]

Earth-based telescopes joined Galileo and Cassini in studying Jupiter during GMM. Table 9.4 gives dates for GMM's close encounters and Jovian magnetosphere–solar wind study.[46] Some of Galileo's instruments were not operating at full performance during GMM because exposure to Jupiter's intense radiation belts had damaged them. This was not surprising; the total radiation that the spacecraft had received was three times the amount that its systems had been built to withstand. But even with its impaired systems, Galileo continued to make valuable observations and generate important scientific data.

---

[42] Ibid.

[43] Eilene Theilig e-mail message, "Re: Definition of GMM," 20 September 2001.

[44] Guy Webster, "Galileo Gets One Last Frequent-Flyer Upgrade," JPL Media Relations Office news release, 15 March 2001.

[45] J. K. Erickson, Z. N. Cox, B. G. Paczkowski, R. W. Sible, and E. E. Theilig, "Project Galileo Completing Europa, Preparing for Io" (paper number IAF-99-Q.2.02 of the International Astronautical Federation, Paris, France, from the 50th International Astronautical Congress, Amsterdam, The Netherlands, 4–8 October 1999), p. 15; J. K. Erickson, D. L. Bindschadler, E. E. Theilig, and N. Vandermey, "Project Galileo: Surviving Io, Meeting Cassini" (paper IAF-00-Q.2.01 of the International Astronautical Federation, Paris, France, from the 51st International Astronautical Congress, Rio de Janeiro, Brazil, 2–6 October 2000), pp. 14–15.

[46] Erickson et al., "Project Galileo Completing Europa, Preparing for Io."

Table 9.4. **GMM activities. During GMM, the spacecraft conducted a study of the Jovian magnetosphere–solar wind interaction in conjunction with the Cassini spacecraft (which was passing by Jupiter on its way to Saturn) and with Earth-based telescopes.**[47]

Orbit	Date of Closest Encounter
E26	3 January 2000
I27	22 February 2000
G28	20 May 2000
Magnetosphere–solar wind interaction	15 June 2000 to 15 November 2000
G29	28 December 2000
C30	25 May 2001
I31	5 August 2001
I32	16 October 2001
I33	17 January 2002
Amalthea 34	5 November 2002
Jupiter 35 (impact)	21 September 2003

Fundamental Questions Addressed by the Orbiter Tour

Certain achievements of the Orbiter tour emerge as particularly memorable because they addressed highly compelling questions that cut across the boundaries of different disciplines. Among these questions were the following:

- *What did our own planet look like during its early years, billions of years before humans appeared?* The Galileo mission's study of the moon Io helped answer this question.

- *Are we alone in the universe? In other words, does life exist on any other planet than Earth? And if so, how rare is its occurrence?* Although Galileo could not answer this question, data sent back from encounters with Europa raised tantalizing possibilities that will influence planetary exploration policy for many decades to come.

- *What is the impact of Jupiter's intense, complex magnetic field and its massive gravitational field on its moons and rings?* This question probes at the basic nature of the outer planet satellite environment and was addressed by many of the Orbiter's moon encounters, especially those with Io.

---

[47] Ibid.; Eilene Theilig e-mail message, 24 September 2001.

- *How does the Jovian system interact with surrounding space? In particular, what effect does the solar wind have on the shape and characteristics of Jupiter's magnetosphere and particle fluxes, and what were the results of collisions between Jupiter's satellites and interplanetary meteoroids and comets?*

- *What is the nature of Jupiter's weather, especially its patterns of water distribution and its dynamic, long-lasting storms (so much more persistent than Earth's weather systems)?*

Although many Orbiter tour activities addressed the above questions, other actions taken by the spacecraft could not be so readily categorized. However, what can be said about all of the Orbiter tour's observations is that they strove to answer the question, what is out there? If Galileo's scientific activity had been limited to the pursuit of well-defined, preconceived objectives, unexpected phenomena may have been missed. The Orbiter tour resulted in discoveries that addressed the questions above, but it also repeatedly brought to light the unexpected.

The following sections attempt to capture the excitement and importance of the Orbiter tour's discoveries, including those that came as complete surprises. They also spotlight what an Earth-based flight team was able to do to keep an aging, problem-ridden spacecraft operational from a distance of nearly half a billion miles. Again and again, Galileo's multitalented team of engineers and scientists found ways to save the mission when the Orbiter's electronic or mechanical systems shut down. The flight team's innovations were among the most impressive mission achievements. The team's efforts enabled Galileo to become more than just a robot spacecraft: Galileo became a virtual extension of human senses, allowing those on Earth to ride along to Jupiter and observe what was there.

Io: Volcanoes, Mountains, and Magnetism

Io's Volcanism

Io's fiery surface has provided scientists with a window on dynamic volcanism and colossal lava eruptions such as those that raged on Earth many eons ago. Galileo data revealed that Io was the most geologically active body in the solar system, with more than 100 erupting volcanoes. Dr. Alfred McEwen, a member of Galileo's imaging team from the University of Arizona, spoke of Io's "gigantic lava flows and lava lakes, and towering, collapsing mountains. Io makes Dante's Inferno seem like another day in paradise."[48]

The last comparable lava eruptions on Earth occurred 15 million years ago. It has been over 2 billion years since lava as hot as the 1,500°C (2,700°F) molten material found on Io has flowed on our planet. "No people were around to observe and document these past events," said Dr. Torrence Johnson, Galileo project scientist. "Io is the next best thing

---

[48] Jane Platt, "Jupiter's Moon Io: A Flashback to Earth's Volcanic Past," JPL news release no. 99-097, 19 November 1999, *http://www.jpl.nasa.gov/releases/99/ioishot.html.*

to traveling back in time to Earth's earlier years. It gives us an opportunity to watch, in action, phenomena long dead in the rest of the solar system."[49]

Table 9.5 lists the Io encounters that Galileo made during its Prime, Europa, and Millennium Missions. On 11 October 1999, during GEM's I24 encounter, Galileo's camera captured a closeup at Io's Pele volcano of hot lava, at most a few minutes old, breaking through the surface crust. The lava formed a thin, curving line 10 kilometers (6 miles) long and up to 50 meters (150 feet) wide. Galileo scientists hypothesized that, under the surface crust at Pele, there might be an extremely active lava lake that constantly exposes new molten material. The curved line may mark the edge of the molten lake where it presses against a rock wall and breaks through to the surface. The Kilauea volcano in Hawaii has such a structure and behavior, although Pele's lava lake appears to be 100 times larger than those found in Hawaii.[50]

Table 9.5. **Io encounters.**[51]

ENCOUNTER	CLOSET APPROACH IN KILOMETERS (MILES)	DATE IN PACIFIC TIME	MISSION	COMMENTS
Jupiter Arrival Day	1,000 (600)	7 December 1995	Prime Mission	Spacecraft did no imaging due to tape-recorder problems.
I24	611 (379)	11 October 1999	Galileo Europa Mission	Spacecraft safed at 19 hours before encounter due to radiation memory hit. Obtained valuable imaging of Io volcanism. Observed a 10-kilometer-long eruption of Pele volcano.
I25	300 (186)	25 November 1999	Galileo Europa Mission	Spacecraft safed at 4 hours before encounter due to software problem. Collected dramatic pictures of Io volcanic activity. Observed mile-high lava fountain.
I27	198 (123)	22 February 2000	Galileo Millennium Mission	Discovered volcanoes that change from hot to cool in several weeks. Spacecraft safed due to transient bus reset. Some Io 27 data played back during Ganymede 28.

[49] Ibid.

[50] Ibid.; Ron Cowen, "Close Encounter: Galileo Eyes Io," *Science News* (11 December 1999): 382.

[51] Erickson et al., "Project Galileo: Completing Europa, Preparing for Io," p. 15; Cowen, "Close Encounter: Galileo Eyes Io"; JPL, "Galileo Sees Dazzling Lava Fountain on Io," JPL Media Relations Office news release, 17 December 1999; Theilig e-mail message, 24 September 2001.

Table 9.5. **Io encounters.** (continued)

ENCOUNTER	CLOSET APPROACH IN KILOMETERS (MILES)	DATE IN PACIFIC TIME	MISSION	COMMENTS
I31	200 (120)	5 August 2001	Galileo Millennium Mission	Magnetic measurements indicated a weak or absent internally generated field. Spacecraft directly sampled fresh sulfur dioxide "snow-flakes" from a volcanic vent.
I32	181 (112)	16 October 2001	Galileo Millennium Mission	Galileo observed the Loki volcano (largest in the solar system) and a new eruption in the southern region of the moon.
I33	102 (63)	17 January 2002	Galileo Millennium Mission	This was the closest of all the flybys. The moon provided a gravity-assist necessary for Galileo's ultimate collision course with Jupiter. A safing event prevented most of the planned data from being collected.

Io has an even more active volcano than Pele: Loki, the most powerful volcano in the solar system. Galileo's photopolarimeter radiometer and near infrared mapping spectrometer have constructed detailed temperature maps of Loki; these maps show that the volcano generates more heat than all of Earth's active volcanoes combined.

Subsurface volcanic mechanisms on Loki are somewhat different from those on Pele. According to Dr. Rosaly Lopes-Gautier, a member of the spectrometer team, "Unlike the active lava lake at Pele, Loki has an enormous caldera [a crater at the top of a volcano] that is repeatedly flooded by lava, over an area larger than the state of Maryland."[52]

On Thanksgiving night, 25 November 1999 (during the I25 encounter), the Orbiter's solid state imaging camera and its near infrared mapping spectrometer captured images of a dazzling volcanic display. The craft observed a fountain of sparkling lava that extended more than a mile above the satellite's surface. As a means of comparison, lava fountains on Earth rarely exceed a few hundred yards in height. The Orbiter's images depicted a curtain of fiery lava erupting from a massive crater. Dr. Torrence Johnson commented that the "lava was so hot and bright, it over-exposed part of the camera picture and left a bright blur in the middle."[53] The lava fountains were hot enough and tall enough that they were seen by NASA's infrared telescope atop Mauna Kea in Hawaii.

Because lava fountains appear infrequently and briefly, it is hard to capture images of them. According to Galileo scientist Dr. Alfred McEwen, "Catching these fountains was a one-in-500 chance observation."[54]

---

[52] Platt, "Jupiter's Moon Io"; *Encarta World English Dictionary* (North American Edition), 2001.

[53] JPL, "Galileo Sees Dazzling Lava Fountain on Io," JPL Media Relations Office news release, 17 December 1999.

[54] Ibid.

Io's Mountains

Although Io is far smaller than Earth, the satellite has mountains soaring up to 16 kilometers (52,000 feet) high—significantly taller than those on our planet. Scientists are not sure how these huge mountains formed. They do not appear to be volcanic in origin. Galileo images have captured evidence of how they die: concentric ridges covering the mountains and surrounding plateaus suggest that immense landslides are triggered as the mountains collapse under the force of their own weight. These ridges strongly resemble the terrain surrounding Olympus Mons, the largest peak on Mars.[55]

The Cause of Io's Geologic Activism

Io's geological phenomena may provide important analogs of early Earth events, but the mechanism driving the moon's tectonics appears to be far different from anything our planet has experienced. Galileo scientists believe that Io's geologic activism is related to intense frictional heating generated by the constant deformation of the moon by Jupiter's powerful gravitational forces. "Jupiter's massive gravity field distorts the shape of Io in the same way that tides are raised in Earth's oceans by the gravitational tugs of the Sun and Moon," said Dr. Torrence Johnson. The face of Io closest to Jupiter is pulled harder than its other, outer face; this difference in force results in a stretching effect on the moon. Irregularities in the moon's orbit around Jupiter cause "body tides" to rise and fall. The orbital perturbations are, in turn, caused by gravitational nudges from the moons Europa and Ganymede. As a result of the varying tides, Io is stretched and released like a rubber ball. Friction created by this process heats and melts rock and produces the volcanoes and lava flows seen all over its surface, as well as the huge geysers that spew sulfur dioxide onto Io's landscape. Though it is less than a third of Earth's size, Io generates twice as much heat.[56]

Summary of Io Statistics

Table 9.6 summarizes some of the facts known about Io. Note that although its distance from Jupiter is roughly the distance of our Moon from Earth, Io moves far quicker, orbiting Jupiter in less than two days, while our Moon takes a month to complete its orbit. Also, note that Io's orbital period and its rotational period (the length of its day) are the same. This means that, like our Moon, Io always presents the same face to Jupiter. Just as we on Earth never see the far side of our Moon, an observer on Jupiter will never see the far side of Io. In fact, all four of the Galilean satellites have this property. The mechanism that causes it is related to the tides induced in satellites by the gravitational pull of the parent planet. Although we normally think of tides as occurring in an ocean, they also occur in the solid material of a planet or moon, making the body slightly elongated. It is

---

[55] Don Savage and Guy Webster, "Jupiter Millennium Mission: The Galileo and Cassini Encounter at the Fifth Planet," NASA press kit; Platt, "Jupiter's Moon Io."

[56] Douglas Isbell (NASA Headquarters) and Mary Beth Murrill (JPL), "Galileo Finds Giant Iron Core in Jupiter's Moon Io," NASA press release 96-89, 3 May 1996, *http://www.solarviews.com/eng/galpr4.htm*.

the gravitational pull of the parent body on these elongations that can "tidally lock" the satellite so that it always shows the same side to the planet.[57]

Table 9.6. **Io statistics.**[58]

Discovery	7 January 1610, by Galileo Galilei
Diameter	3,630 kilometers (2,256 miles)
Mass	8.94×10^{22} kilograms (2×10^{23} pounds)
Mass relative to that of Earth	0.014960 times the mass of Earth
Surface gravity (Earth = 1)	18.3 percent of Earth's gravity
Mean distance from Jupiter	422,000 kilometers (261,000 miles)
Mean distance from Jupiter in radii	5.905 R_J
Mean distance from the Sun	5.20 AU (1 AU = distance from Sun to Earth)
Orbital period	1.77 days
Rotational period	1.77 days
Density	3.57 grams per cubic centimeter
Typical subsolar temperature	$\approx -135°C$ ($-211°F$)
Surface composition	Sulfur and frozen sulfur dioxide

Discovery of a Possible Iron Core and Magnetic Field

On 3 May 1996, JPL and *Science* magazine reported that Galileo might have found a giant iron core taking up half the moon's diameter. The spacecraft also observed a large "hole" in Jupiter's magnetic field near Io, where the strength of the magnetic field took a sudden drop of about 30 percent. This suggested that Io might be generating its own field. At the time of this discovery, no other planetary moons in our solar system were known to generate their own magnetic fields (although many have fields induced by the magnetism of the parent planet). Only Earth, Mercury, and the outer gas-giant planets were known to generate their own fields.

"It's an astonishing result and completely unexpected," said Dr. Margaret Kivelson of UCLA and the leader of Galileo's magnetic fields investigation team. "The data suggest that something around Io—possibly a magnetic field generated by Io itself—is creating a bubble or hole in Jupiter's own powerful magnetic field. But it's not clear to us just how Io can dig such a deep and wide magnetic hole."[59]

The existence of a large, dense core within Io was deduced from Galileo data taken during the spacecraft's flyby within 900 kilometers (560 miles) of the moon on

---

[57] Ron Baalke, *Today on Galileo—May 23, 2001,* JPL e-mail newsletter, 23 May 2001.

[58] JPL, "A Summary of Facts About Io," *Io, A Continuing Story of Discovery,* 22 March 1999, *http://www.jpl.nasa.gov/galileo/io/io_summary.html.*

[59] Isbell and Murrill, "Galileo Finds Giant Iron Core."

Jupiter Arrival Day, 7 December 1995. In particular, Doppler data generated with the spacecraft's radio carrier wave were analyzed to measure Io's external gravitational field. A metallic core composed of iron and iron sulfide with a radius of about 900 kilometers (560 miles), or 52 percent of the moon's mean radius, would be consistent with the spacecraft's observations. Scientists think that the core was formed as a result of interior heating of the moon into a molten state, causing the moon's heavier components, such as iron, to sink toward its center. The internal heat source of the moon could have been either its original heat of formation or frictional heating resulting from perpetual distortions of the moon due to Jupiter's strong gravitational field.[60]

The Rapid Timescale of Io's Geological Changes

On Earth, we think of geological changes as happening ever so slowly, over periods of thousands or millions of years. Large alterations to Earth's surface rarely occur quickly, unless there is a cataclysmic event such as a massive volcanic eruption. But Io is a moon where cataclysmic events are the norm. For instance, a Galileo observation of a section of Io's surface taken only five months after the previous observation revealed that a new dark spot the size of Arizona had appeared, indicating dramatic volcanic activity.[61]

The visible change, almost 250 miles in diameter, occurred during the five months between Galileo's 7th (G7) and 10th (C10) orbits of Jupiter. The observations were not taken during an Io close encounter, but from much further away. The alteration in Io's surface occurred around a volcanic center named Pillan Patera (after the South American god of thunder, fire, and volcanoes). Pictures revealing the changes were captured by the Orbiter's SSI camera. The pictures showed marked differences between the Pillan Patera region on 4 April 1997 and 19 September 1997. In June of 1997, both Galileo and the Hubble Space Telescope observed an active plume over Pillan 120 kilometers (75 miles) in height, and both Galileo and ground-based astronomers observed an intense hotspot.[62]

The composition of the rapidly created dark spot appeared to be different from that of typical deposits. "Most of the volcanic plume deposits on Io show up as white, yellow, or red due to sulfur compounds. However, this new deposit is gray, which tells us it has a different composition, possibly richer in silicates than the other regions," said Galileo imaging team member Dr. Alfred McEwen, a research scientist at the University of Arizona in Tucson. "While scientists knew that silicate volcanism existed on Io from high temperatures, this may provide clues as to the composition of the silicates, which in turn tells us about Io's evolution. Io is probably primarily composed of silicates, which is the type of volcanic rock found on Earth," McEwen said. "But the extreme volcanism of Io may have led to the creation of silicate compositions that are unusual on Earth."[63]

---

[60] J. D. Anderson, W. L. Sjogren, and G. Schubert, "Galileo Gravity Results and the Internal Structure of Io," *Science* 272, no. 5262 (3 May 1996): 709; Isbell and Murrill, "Galileo Finds Giant Iron Core."

[61] Douglas Isbell (NASA Headquarters) and Jane Platt (JPL), "Galileo Finds Arizona-Sized Volcanic Deposit on Io," NASA news release 97-257, 5 November 1997, *http://www.jpl.nasa.gov/galileo/status971105.html*.

[62] Ibid.

[63] Ibid.

A dark region southwest of the Pele volcano, which was first seen during the Voyager flybys in 1979, is similar in appearance to the Pillan deposits. The Io images showing the changes in Pillan Patera also revealed alterations in a plume deposit at the Pele volcano, which may indicate that both plumes were active at the same time and interacted with one another.[64]

Io's Ionosphere

In analyzing Galileo data, scientists discovered that in December 1995, during the spacecraft's first encounter with Io, the craft may have flown through a dense, high-altitude ionosphere of the moon. An ionosphere is a region of electrically charged gas. Such a layer occurs at the top of some planetary atmospheres, including that of Earth. The scientists thought that the ionosphere consisted of volcanic gases that rose to very high altitudes.[65]

According to Dr. Louis A. Frank of the University of Iowa, principal investigator on Galileo's plasma science experiment, the spacecraft's sensors registered a dense region of ionized oxygen, sulfur, and sulfur dioxide 900 kilometers (560 miles) above Io. No one expected an ionosphere at such an altitude because 1970s data from the Pioneer and Voyager spacecraft only saw particulate layers as high as a few hundred kilometers. Scientists also were surprised that the ionized gases found by Galileo stayed with Io rather than being swept away by forces from Jupiter's rotating magnetosphere.

The difference between what the Pioneer 10 spacecraft's 1973 radio occultation experiments revealed—an ionosphere only 50 to 100 kilometers (30 to 60 miles) high—and what Galileo saw indicates that Io's ionospheric activity is highly variable. The moon's ionosphere may grow and shrink with volcanic activity. The Galileo results gave credence to the "stealth plume" theory of project scientist Torrence Johnson. He proposed that because of the moon's weak gravity, invisible gases from Io's volcanoes rose to great heights far beyond those achieved by dust and other visible volcanic ejecta.[66]

Arrival Day Io Flyby: The Great Disappointment

Galileo flybys of the satellite Io were filled with surprises, drama, and sometimes disappointment. Since the beginning of the Galileo program in the late 1970s, its scientists had been looking forward to the spacecraft's Io flyby on Jupiter Arrival Day (7 December 1995). Because of the high radiation levels around Io and their effect on sensitive instruments, the scientists knew that Arrival Day might be the only time they would get a detailed look at Io—the only time they could obtain high-resolution, closeup images. But concerns regarding Galileo's malfunctioning tape recorder made it necessary for the Project Manager, Bill O'Neil, to cancel all remote sensing operations during the flyby. Only fields-and-particles instruments were allowed to take and record data. Those

---

[64] Ibid.

[65] Douglas Isbell (NASA Headquarters) and Mary Beth Murrill (JPL), "High Altitude Ionosphere Found at Io by Galileo Spacecraft," NASA news release 96-216, 23 October 1996.

[66] Ibid.

instruments needed the tape recorder to run continuously, but at low speeds, and project staff thought that the recorder could handle this. Remote sensing instruments such as the SSI camera required the recorder to run discontinuously, with abrupt starts and stops, and at high speeds. That kind of operation might well have permanently crippled the recorder and, with it, the entire mission. It is hard to imagine the frustration felt by scientists who had waited much of their careers for the Io flyby when they learned that they would not receive the data for which they had hoped.[67]

I24 and I25: The Great Flybys That Almost Weren't

Much to the delight of scientists around the country, Galileo survived its Prime Mission and continued to perform beautifully. Two more Io missions were planned during the Galileo Europa Mission, after the spacecraft had had a chance to study Europa in detail. The missions were much anticipated, but the observations that were planned almost did not happen. During the I24 and I25 flybys, the first Io encounters since Jupiter Arrival Day, problems arose that once again threatened to ruin the chances of close-flyby Io observations.

I24. After the extremely limited flyby on Jupiter Arrival Day, the next Io encounter, I24, occurred during GEM. Its closest approach took place on 11 October 1999. The entire encounter sequence lasted four days, beginning on 10 October and running to 14 October. Mission scientists were tremendously interested in the first images of Io taken from a low-altitude flyby skimming only 611 kilometers (379 miles) above the surface. The SSI camera and NIMS were going to observe numerous volcanic features, including the terrain around Pele and Pillan Patera. NIMS would also produce a regional map and a plume image, in addition to searching for undiscovered hotspots. Other remote sensing instruments would take key atmospheric and radiometric measurements. But all this almost did not happen.[68]

Io is the Galilean moon that is closest to Jupiter. It lies in a region of intense radiation from the parent planet that "can wreak havoc with spacecraft instruments."[69] During the flyby, this radiation interfered with the Orbiter computer's memory. The Orbiter's on-board flight software was programmed to detect anomalous situations in the spacecraft's many subsystems. Any time the spacecraft computer executed a portion of memory that had an error in it, the craft's flight software was designed to detect it and put the Orbiter into a "safe mode." This meant that all nonessential functions were turned off to make the spacecraft as safe as possible from a power, thermal, and communications perspective until mission staff back on Earth could figure out what had happened and how to get the craft functioning again without damaging it. A safing incident occurred at

---

[67] J. K. Erickson, D. L. Bindschadler, E. E. Theilig, and N. Vandermey, "Project Galileo: Surviving Io, Meeting Cassini" (paper IAF-00-Q.2.01 of the International Astronautical Federation, Paris, France, from the 51st International Astronautical Congress, Rio de Janeiro, Brazil, 2–6 October 2000), p. 7; Bill O'Neil, former Galileo Project Manager, interview, tape-recorded conversation, 15 May 2001. For more detailed information on the tape-recorder anomaly, see chapter 8.

[68] J. K. Erickson, Z. N. Cox, B. G. Paczkowski, R. W. Sible, and E. E. Theilig, "Project Galileo: Completing Europa, Preparing for Io" (paper IAF-99-Q.2.02 of the International Astronautical Federation, Paris, France, from the 50th International Astronautical Congress of the International Astronautical Federation, Amsterdam, The Netherlands, 4–8 October 1999), pp. 13–14.

[69] JPL, "Galileo Succeeds in Historic Flyby of Jupiter's Volcanic Moon," JPL Media Relations Office news release, 11 October 1999.

3:09 a.m. Pacific time on Sunday, 11 December 1999. The closest approach to Io was only 19 hours away.[70, 71]

According to Dr. Duane Bindschadler, Team Chief of the spacecraft and sequence team, what occurred on I24 was that high-energy electrons penetrated the Galileo spacecraft, hit a memory chip, and physically altered a single bit of memory. This caused the spacecraft to "safe."[72]

Bringing Galileo out of safe mode was orders of magnitude more complex than restoring a personal computer that has crashed. For one thing, the travel time for radio transmissions from Earth to Jupiter was 35 to 55 minutes one way (depending on the relative positions of the two planets). Querying the on-board computer, analyzing what had gone wrong, and responding to the situation constituted a very slow, multistep process. The early "safings" aboard Galileo typically took a week to 10 days to bring the spacecraft gradually back to full operation, although as more was learned about the craft's limitations and operating characteristics, the recovery time grew shorter. But 19 hours was a very short time in which to bring the craft out of safe mode. Once again, it seemed, Galileo might pass close by Io and not be able to send the imaging data that scientists so desperately wanted.[73] Scientists had been waiting for 20 years for close-encounter images from Io. It was critical not just for JPL, but also for NASA and the U.S. interplanetary space program, that Galileo do well.

The Galileo engineering staff had to develop and implement a recovery plan on the fly, altering the spacecraft's code sequences so that it would obtain the necessary data. Late Sunday afternoon, the staff sent out its code modifications, and they worked as hoped, getting the spacecraft fully operational by 8 p.m. Pacific time, a mere 2 hours before the closest approach. Nagin Cox, the Deputy Team Chief for the spacecraft and sequence team, said, "I think I will be proud of the Galileo team and those few days for the rest of my life . . . . Without them, there [would have been] no recovery."[74]

"It was a heroic effort to pull this off," said Jim Erickson, Galileo Project Manager. "The team diagnosed and corrected a problem we'd never come across before, and they put things back on track." Because of this effort, his staff could look forward to seeing the closest pictures of Io ever taken, as well as observations of the moon's surface chemistry, thermal characteristics, gravitational field, and magnetic properties.[75]

The Orbiter was not able to carry out and record all of its planned observations. Some photopolarimeter radiometer (PPR) observations of Jupiter and Io's night side were lost, as well as some fields-and-particles data of Io's plasma. But the spacecraft did obtain numerous high-resolution images of Io's surface, including the Pillan Patera, Zamama, Prometheus, and Pele volcanic eruption centers. The Orbiter also obtained global-scale color images and an observation of Io while in eclipse, as well as atmospheric airglow and auroral emissions. In addition, excellent infrared measurements were

---

[70] Nagin Cox and Bill O'Neil interviews, tape-recorded telephone conversations, both conducted on 15 May 2001. Nagin Cox was the Deputy Team Chief for the spacecraft sequence team at the times of Io 24 and 25. Bill O'Neil is a former Galileo Project Manager.

[71] JPL, "Galileo Succeeds in Historic Flyby."

[72] Duane Bindschadler, Team Chief of the Galileo spacecraft and sequence team, interview, JPL, 5 August 2001.

[73] Bill O'Neil interview, 15 May 2001.

[74] Nagin Cox interview, 15 May 2001.

[75] JPL, "Galileo Succeeds in Historic Flyby."

taken of various volcanic centers and mountains, and two infrared regional maps were constructed.[76]

I25. Galileo encounters, according to mission staff, seemed to occur again and again over holidays, when normal people are home enjoying their families and taking it easy. I25, the second and last Io encounter during GEM, occurred on Thursday, 25 November 1999, at 8:40 p.m. Pacific time—Thanksgiving evening. Project staff ate their Thanksgiving dinners in Galileo's Mission Control Center while waiting for the closest approach, which was to be a mere 300 kilometers (186 miles) above the moon's surface. Remembering the problems that had happened on I24, engineering staff had been counting down the hours. At 12 hours from closest encounter, everything was fine. Things remained so as the time grew shorter: 10 hours . . . 8 . . . 7 . . . . Then, at 4 hours out, the spacecraft safed again. A software problem that had been present since launch interacted with the software patch that had been loaded into the Orbiter's computer to help bring it out of safe mode during the I24 encounter. The spacecraft safed once more but did not shut down to as deep a level as on the last Io encounter. Nevertheless, trying to pull the Orbiter out of safe mode at 4 hours from the closest approach was very ambitious, if not outright futile. "With so little time to spare, it would have been easy to think, no way can we do this," said Galileo Project Manager Jim Erickson. "But our team members jumped to the challenge, in some cases leaving behind half-eaten Thanksgiving dinners."[77]

"I have very little memory of those four hours," said Nagin Cox, "because [we were] so incredibly focused and intense. With that amount of time to recover, everybody simply had to do their jobs as they knew them in their heads. There was not enough time to rely upon the automated tools that normally help us make sure we are doing the right thing." The team formulated command sequences, got them to Jupiter, and recovered the spacecraft, 500 million miles away, only 3 minutes before the closest encounter with Io. "There was clapping and cheering and smiles all around. There was a lot of adrenalin flowing. We were all thrilled and a bit surprised at having recovered so quickly." But they could not celebrate for long because the flyby was upon them. When the team had gotten I24 back online, they had had 2 hours to catch their breath before the flyby. On I25, they had all of 3 minutes.[78]

Dr. Eilene Theilig, who worked on Galileo for 11 years and became its Project Manager in January 2001, said that the recovery of I25 was her most exciting moment on the mission. "Just working through that with the people involved and the teamwork, and the fact that everybody knew just what to do and was right on top of it was an exciting evening. And then to see an image that we got back, which was our first image of an active volcano. I mean, we caught the curtain of fire. All of that tied together had to be the highlight."[79]

The literally last-minute recovery of I25 allowed stunning observations to be taken and sent to Earth. The encounter's high-resolution imaging data and spectral maps met many of the flyby's ambitious remote sensing goals. Galileo's SSI and NIMS systems caught one of the most impressive events ever recorded anywhere in the

[76] Erickson, Bindschadler, Theilig, and Vandermey, "Project Galileo: Surviving Io, Meeting Cassini," p. 9.

[77] Nagin Cox interview, 15 May 2001; JPL, "Galileo Mission Status: November 25, 1999," JPL Media Relations Office news release, 25 November 1999; Erickson, Bindschadler, Theilig, and Vandermey, "Project Galileo: Surviving Io, Meeting Cassini," p. 4.

[78] Nagin Cox interview.

[79] Eilene Theilig, former Galileo Project Manager, interview, tape-recorded conversation, 8 May 2001.

solar system—an actively erupting volcanic fountain that generated a curtain of lava 20 kilometers (12 miles) long. Capturing such a massive eruption was "extremely fortuitous," based on similar events on Earth, which typically last only a few hours to one to two days. Even considering the more frequent volcanic activity found on Io, Galileo was very lucky to have seen such a massive eruption.[80]

The I25 encounter was a smashing success, due largely to the frenetic efforts of the project staff who recovered it. But some important observations were never made because of the 4 hours the spacecraft spent in safe mode. Among the most significant scientific data that were lost were the fields-and-particles observations that were to address the question of whether Io has an internally generated magnetic field, which would have bearings on the internal structure of the moon. The observations were to have helped determine the extent to which electrical conductivity within Io and its atmosphere could have generated the moon's observed magnetic signature. In fact, the polar path of the flyby had been chosen largely in order to help answer this question.[81]

Galileo Millennium Mission Observations of Io

I27. The first Io flyby of the Galileo Millennium Mission (GMM) occurred on 22 February 2000, when the Orbiter skimmed above the satellite surface at an altitude of only 198 kilometers (123 miles). The plan for this encounter was influenced to a large extent by the data returned from I24 and I25 but had to be very limited in scope due to a solar conjunction that occurred during the latter part of the I27 cruise period. The solar conjunction interfered with data transmission by reducing the rate at which the data could be sent to Earth. The spacecraft was also near its maximum distance from Earth at this time, which further reduced the data transmission rate. Both of these factors significantly reduced the volume of information that could be sent before the next satellite encounter (G28, a flyby of Ganymede) required the space on the recording tape. With these constraints, Galileo staff focused on the time of closest approach to Io as yielding the most valuable data. Approximately 75 percent of the tape-recorded observations were taken during a 3-hour period beginning 40 minutes before closest approach.[82]

The spacecraft safings in I24 and I25 caused data gaps in the survey of the Io plasma torus. The observations during I27 filled in some of those gaps. The Orbiter also measured the global distribution of daytime temperatures on Io and took extensive visible and infrared images of the moon. Results from previous Io encounters indicated that images taken at low-Sun-angle conditions (when the Sun was close to the satellite's horizon) were especially helpful in establishing the shapes and geology of the satellite's features. Radio science measurements of the moon's gravitational field were made to improve characterizations of the field and to refine Io interior structure models.[83]

The high-resolution images taken by Galileo's NIMS during I27 revealed 14 volcanoes in a region that was previously thought to contain only 4. The region covers

---

[80] Erickson, Bindschadler, Theilig, and Vandermey, "Project Galileo: Surviving Io, Meeting Cassini," p. 11.

[81] Ibid., p. 11.

[82] Ibid., pp. 12-13.

[83] Ibid., pp. 13–14.

about 5 percent of Io's surface and is approximately three times larger than Texas. Integrating data from I24, I25, and I27 revealed that some of the moon's smaller, fainter volcanoes turned rapidly on and off, altering from hot and glowing objects to cool and dim ones in the space of a few weeks. On the other hand, the moon's larger, brighter volcanoes tended to remain active for years and possibly decades, based on Galileo data and the images taken by Voyager when it flew by the Jovian system in 1979.[84]

During I27, Galileo noted dramatic changes in Loki, the most powerful known volcano in the solar system. According to Dr. John Spencer of Lowell Observatory in Arizona and a co-investigator for Galileo's radiometer experiment, most of Loki's caldera, a region half the size of Massachusetts (about 10,000 square kilometers or 4,000 square miles), appeared to have been covered by fresh lava in the four and a half months between I24 (11 October 1999) and I27 (22 February 2000).

Images of the Chaac Patera volcanic region revealed a caldera wall 2.8 kilometers (1.7 miles) in height, with a slope of 70 degrees—twice as high and far steeper than Earth's Grand Canyon. This finding indicates that the rocks around Chaac Patera are very strong to be able to support such an extreme topography. The spacecraft also imaged a caldera filled with bright white deposits containing sulfur dioxide of a purity higher than at any other place on Io. Mission scientists believed that it might be a layer of sulfur dioxide ice. "There are processes on Io for which we have no terrestrial experience," said Dr. Alfred McEwen of the University of Arizona and a member of Galileo's imaging team. "Strange new observations like these will provide fodder to current and future scientists for understanding the processes that have shaped this fascinating world."[85]

A series of extreme ultraviolet (EUV) scans of the Io plasma torus and Jupiter, planned as a final observation for I27, had to be curtailed because of yet another space-craft safing event. Project staff believe that this safing was also induced by radiation exposure, but not of the spacecraft's memory, as was the case in I24. This radiation exposure caused a "transient bus reset," which was triggered by inadvertent short circuits across the spin bearing assembly that connected the Orbiter's spun and despun sections. Transient bus resets continued to occur up to several times during each encounter. A soft-ware patch loaded into the Orbiter's computer after a Europa encounter (E19) allowed the spacecraft to recognize a bus reset and refrain from going into safe mode. This patch was typically enabled when the Orbiter flew within 15 R_j of Jupiter, where the most intense radiation occurs. On I27, however, a bus reset unexpectedly occurred at 29 R_j, when the patch had been disabled. As a result, the spacecraft safed.[86]

Besides interrupting the extreme ultraviolet scan, the safing event also impacted the return of other Io data to Earth. As mentioned above, the amount of data returnable to Earth was already severely limited because of a solar conjunction and because of the extreme distance of the spacecraft from Earth. Loss of playback time due to the safing event further impacted data return. To help remedy the situation, Galileo managers decided to carry some I27 data into the next orbit (G28, a Ganymede flyby) and play it back at that time. This would reduce the amount of tape that could be used to record

---

[84] Susan Mitgand and Mary Beth Murrill, "Dynamic Terrain and Volcanoes Galore on Io," JPL Media Relations Office news release, 31 May 2000.

[85] Ibid.

[86] Erickson, Bindschadler, Theilig, and Vandermey, "Project Galileo: Surviving Io, Meeting Cassini," p. 14; Nagin Cox interview.

G28 data, but Galileo management perceived Io data as having a high enough science value to justify this.[87]

I31. GMM activities included a Callisto encounter on 25 May 2001 (C30), on which the moon's gravitational field was used to set up two encounters with Io (I31 and I32). During I31, which occurred on 5 August 2001 at about 10 p.m. PDT, Galileo skimmed as close as 200 kilometers (120 miles) above Io's surface. A north polar path was picked for the I31 flyby because magnetic readings above the pole might help to reveal whether Io generated its own magnetic field. According to Torrence Johnson, determining that was important for understanding the processes taking place in Io's hot interior. "All of our previous magnetic measurements at Io have been on equatorial passes, and from those we can't tell whether the field at Io is induced by Jupiter's strong magnetic field or produced by Io itself," said Johnson. Initial interpretations of I31 data suggested that either the moon did not have an internally generated field or it was extremely weak, according to the principal investigator for the magnetometer experiment, Margaret Kivelson.[88]

As a side benefit of the flyby, Galileo sped through a region that had been occupied by a giant gas plume from the volcano Tvashtar seven months earlier. This was the first time that the spacecraft had an opportunity to sample a plume's constituents directly. Io's polar plumes, however, appeared to be short-lived, according to Project Manager Eilene Theilig. It was not clear whether remnants of the Tvashtar plume would still be there when Galileo passed through.[89]

Galileo's investigations have often turned up the unexpected, and the I31 flyby was no exception. The Tvashtar volcano was quiet, but the spacecraft spotted a new eruption from a previously unknown volcano 600 kilometers away. The plume this new volcano spewed out was the tallest yet observed, soaring 500 kilometers (300 miles) above the moon's surface. Fortuitously, Galileo's flightpath took it through the outskirts of this plume, and the craft's plasma science instrument detected particles that had erupted from Io's interior only minutes earlier. Louis Frank, leader of the plasma experiment, commented that although Galileo had taken wonderful images of Io's volcanoes, it had never before "caught the breath" of one. The particles Galileo detected were not hot volcanic embers, but rather tiny sulfur dioxide snowflakes of 15 to 20 molecules each. The snowflakes formed as hot gases from the vent rose through Io's frigid, thin atmosphere and condensed.[90]

I32. On 18 September 2001, Galileo performed an "orbit trim" maneuver in which the spacecraft's thrusters were pulsed briefly three times to "gently nudge" the vehicle toward its 16 October encounter with Io. Up to this point, this was the smallest maneuver of the mission, using only 15 grams of propellant. But it was necessary because the amount of fuel remaining in Galileo's tanks was "getting so low that spending 15 grams today may save us from spending 20 grams next week to reach the same place." Project staff were painstakingly trying to get the maximum science results out of the

[87] Erickson, Bindschadler, Theilig, and Vandermey, "Project Galileo: Surviving Io, Meeting Cassini," p.14.

[88] Guy Webster, "Spacecraft to Fly Over Source of Recent Polar Eruption on Io," JPL news release 2001-161, 2 August 2001; Guy Webster, "Galileo Millennium Mission Status," JPL news release 2001-166, 6 August 2001; Margaret G. Kivelson, Krishan K. Khurana, Christopher T. Russell, and Raymond J. Walker, "Magnetic Signature of a Polar Pass Over Io," Abstract P11A-01, Eos Transactions, American Geophysical Union 82, no. 47, Fall Meeting Suppl., 2001.

[89] Webster, "Spacecraft to Fly Over Source."

[90] "Dashing Through the Snows of Io," NASA Science News (16 October 2001).

space vehicle's activities and still attain the end-of-mission goal of impact with Jupiter in September of 2003.[91]

During the 15 October 2001 flyby, Galileo passed closer to Io than ever before— only 181 kilometers (112 miles) above the moon's south pole. The spacecraft made observations of several volcanic areas on the moon's surface, including Loki, the most powerful volcano in the solar system, as well as a new hotspot and plume eruption in the moon's southern region that had been discovered on the I31 flyby in August.[92]

Magnetic field measurements had a high priority during I32, as they did on I31, because they could help determine whether Io generated an intrinsic magnetic field of its own within the greater magnetic field of Jupiter. Analysis of the measurements indicated that "an internal magnetic field is negligibly small and probably absent."[93] This finding implied that Io's molten iron core did not have the same type of convective overturning by which Earth's molten core generates a magnetic field. This result fits a model in which Io's core is heated from the outside, by tidal flexing of the layers around it, rather than being heated from its center.[94]

I33. This was the closest, and last, flyby that Galileo performed past any of Jupiter's four major moons. The spacecraft cruised to within 102 kilometers (63 miles) of Io's surface on 17 January 2002. The close encounter gave Galileo the gravity-assist necessary for it to end its multiyear mission. "The reason we're going so close," according to Eilene Theilig, "is to put Galileo on a ballistic trajectory for impact into Jupiter in September 2003."[95] The propellant needed to steer the spacecraft and keep its antenna pointed toward Earth was nearly exhausted. Mission scientists wanted to avoid the small chance that Galileo might crash into Europa after its mission ended and possibly contaminate it with Earth microorganisms. If that happened, it would not be clear whether any life-forms that might ultimately be discovered on Europa originated there or on Earth.

On I33, scientists hoped to observe how several of Io's regions had changed over the years. The spacecraft was also supposed to make direct measurements of charged particles and magnetic fields around the moon, as well as using its camera and other instruments for infrared and thermal imaging. Plans for science observations did not go as hoped, however. While approaching the moon, 28 minutes before closest approach, the spacecraft placed itself in safe mode. In this mode, on-board fault protection software instructed the craft's cameras and science instruments to stop taking data and await further instructions from Earth. Mission scientists thought that the safing incident probably had resulted from the radiation environment near Jupiter. The flight team

[91] Ron Baalke, "This Week On Galileo—September 17–23, 2001," 18 September 2001.

[92] JPL, "Galileo Millennium Status," news release, 16 October 2001.

[93] Margaret G. Kivelson, Fran Bagenal, William S. Kurth, Fritz M. Neubauer, Chris Paranicas, and Joachim Saur, "Magnetic Interactions With Satellites," in *Jupiter*, ed. Fran Bagenal, Tim Dowling, and Bill McKinnon (Port Melbourne, Australia: Cambridge University Press, 2004).

[94] JPL, "Jupiter's Io Generates Power and Noise, But No Magnetic Field," JPL news release, 10 December 2001, *http://galileo.jpl.nasa.gov/news/release/press011210.html*.

[95] Donald Savage and Guy Webster, "Farewell, Io: Galileo Paying Last Visit to a Restless Moon," NASA news release 02-10, 15 January 2002.

worked through the day and evening to restore spacecraft operations. Unfortunately, three of the tracks of data that were to be put on Galileo's tape recorder were lost.[96]

Eilene Theilig said of this last Io flyby, "As expected, visiting Io has proved to be a challenging and risky endeavor. It's disappointing not to get the observations of Io that were planned for this encounter, but I am very proud of the flight team that has kept Galileo functioning in orbit more than three times longer than originally planned and revived it once more."[97]

Galileo's observations revealed the incredibly dramatic nature of Io geology, but there is still much more about the moon to study. Dr. Rosaly Lopes-Gautier, a research scientist at JPL and member of the Galileo spectrometer team, estimated the total quantity of volcanoes on Io. She based her estimate on the volcanoes seen so far and on the fact that volcanism appears to be uniformly distributed across the surface. According to Lopes-Gautier, "we can expect Io to have some 300 active volcanoes, most of which have not been discovered."[98]

Europa: A Possible Ocean Under the Ice Crust

Europa looks like broken glass that is repaired by an icy glue oozing up from below.[99]

Frozen water appears to cover Europa's entire surface, resembling "a broken pane of glass, its shards repaired by an icy glue from below."[100] The "shards" also look like the fragmented chunks of ice seen in Earth's polar regions during springtime thaws (see figure 9.3) and are separated by a network of low ridges. The surface of Europa also contains gigantic ice rafts, cold-water volcanoes, and probably an underwater ocean.[101]

Table 9.7. **Europa data.**[102]

Discovery	7 January 1610, by Galileo Galilei
Diameter	3,138 kilometers (1,946 miles)
Mass	4.8×10^{22} kilograms (1.1×10^{23} pounds)
Mass relative to that of Earth	0.0083021
Surface gravity (Earth = 1)	13.5 percent of Earth's gravity
Mean distance from Jupiter	670,900 kilometers (416,000 miles)

[96] JPL, "Galileo Millennium Mission Status," news release, 17 January 2002; JPL, "Farewell to Io!" *Today on Galileo*, news release, 18–20 January 2002; Donald Savage and Guy Webster, "Farewell, Io: Galileo Paying Last Visit to a Restless Moon," NASA news release 02-10, 15 January 2002.

[97] JPL, "Galileo Millennium Mission Status," news release, 18 January 2002, *http://www.jpl.nasa.gov/galileo/news/release/press020118.html*.

[98] Mitgand and Murrill, "Dynamic Terrain and Volcanoes Galore on Io."

[99] JPL, "Europa: Another Water World?" *Moons and Rings of Jupiter—Galilean Moons: Europa*, 1 October 2001, *http://www2.jpl.nasa.gov/galileo/moons/europa.html* (available in folder 18522, NASA Historical Reference Collection, Washington, DC).

[100] JPL, "Why Europa?" *Galileo Solid State Imaging (SSI) Team Education and Public Outreach* Web site, 17 September 1999, *http://www2.jpl.nasa.gov/galileo/sepo/education/europa/whyeuropa.html* (accessed 22 November 2004) (available in folder 18522, NASA Historical Reference Collection, Washington, DC).

[101] "Why Europa?" and "Diving Water on Europa," Science@NASA, 9 September 1999, *http://science.nasa.gov/newhome/headlines/ast09sep99_1.htm*.

[102] "A Summary of Facts about Europa," JPL Galileo Home Page, *http://www.jpl.nasa.gov/galileo/europa/e-summary.html*.

Table 9.7. **Europa data.** (continued)

Mean distance from Jupiter in radii	9.5 R$_J$
Mean distance from the Sun	5.203 AU
Orbital period	3.551181 days
Rotational period	3.551181 days
Density	3.01 grams per cubic centimeter
Orbit speed	13.74 kilometers per second (30,740 mph)
Visual albedo	0.64
Surface composition	Water ice

Figure 9.3. **Europa's surface is covered with water ice. Straight and curved low ridges form the boundaries of ice fragments that resemble those seen in Earth's polar regions during springtime thaws. The white and blue colors outline areas that have been blanketed by a fine dust of ice particles. The unblanketed surface has a reddish-brown color that has been painted by mineral contaminants carried and spread by water vapor released from below the crust when it was disrupted. The original color of the icy surface probably was a deep blue seen in large areas elsewhere on the moon. The colors in this picture have been enhanced for visibility. (NASA image number PIA01127)**

Voyager's View of Europa

In 1979, Voyager's twin spacecraft provided Earth with its first closeup view of Europa. The pictures showed pale yellow, icy plains with mottled red and brown areas. Long cracks extended over the surface for thousands of kilometers. None of these features rose more than a few kilometers above the surface, which placed Europa among the smoothest bodies in our solar system.[103]

---

[103] "Europa: Another Water World?"

Galileo Prime Mission Europa Encounters

Galileo passed above Europa's south pole at a distance of 32,000 kilometers (20,000 miles) when it first reached the Jovian system. Due to tape-recorder limitations, observations had to be extremely restricted to ensure the satisfactory storage of data from the Probe's plunge into Jupiter's atmosphere. After the Probe data were received, stored, and sent to Earth, the tape recorder could store more satellite data. The Orbiter's remote sensing observations (including extensive imaging activities) were initiated shortly before the Ganymede encounter (G1) in June 1996. On each orbit, the spacecraft made observations not only of the satellite being encountered, but also of other satellites, as well as Jupiter and its fields and particles.[104]

Europan observations taken during the G1 and G2 encounters revealed areas that appeared to be ice floes that had once moved and rotated, indicating water or soft ice below the surface. The images also showed many long lines on the moon, which appeared in groups of three—two dark lines with a lighter line between them. Scientists thought that these were cracks in Europa's icy surface due to upwellings of water or soft ice from below. Debris carried in the upwellings might have caused the dark lines. Deposits observed along the lines may have been caused by explosive geyser-like activity.[105]

The spacecraft's C3 orbit, in which it carried out a close Callisto encounter on 4 November 1996, also included a "nontargeted" pass by Europa on 6 November at an altitude of 34,800 kilometers (21,600 miles). Nontargeted encounters are secondary flybys on a given orbit, up to a distance of 100,000 kilometers (60,000 miles). Nontargeted encounters such as this Europa flyby are not accidental, but are designed into the orbit. They are untargeted in the sense that no effort is made to control the encounter conditions. With the limited amount of propellant available, it would be very difficult, if not impossible, to control the close flybys past two satellites on the same orbit and still achieve the necessary gravity effects to reach the next encounter on the tour. One of the primary objectives of nontargeted encounters is to take global observations of the satellite, which are possible because of the greater flyby distances.[106]

E4. Galileo's first close encounter with Europa (E4) occurred on 19 December 1996 Greenwich mean time (GMT), about a year after Jupiter Arrival Day. Galileo's primary science objectives during E4 were to conduct remote sensing observations of Europa's surface, collect data on the moon's interactions with Jupiter's magnetosphere, and analyze Jovian atmospheric features. During E4, the Orbiter flew as close as 692 kilometers (429 miles) above the moon's surface. The encounter sequence ran from 15 to 22 December 1996 GMT. It included occultations of the Sun and Earth by both Jupiter and Europa, which provided an opportunity to search for indications of an ionosphere and atmosphere on the moon. The return of data from E4 was limited by a solar conjunction

[104] W. J. O'Neil, N. E. Ausman, J. A. Gleason, M. R. Landano, J. C. Marr, R. T. Mitchell, R. J. Reichert, and M. A. Smith, "Project Galileo at Jupiter" (paper number IAF-96-Q.2.01 of the International Astronautical Federation, Paris, France), pp. 27–31. Taken from a compilation of papers, *The Flight of Project Galileo as Reported Annually to the IAF/AIAA*, with the number 1625-592, and published by NASA in May 1997. All papers in the compilation were presented at the 47th International Astronautical Congress, Beijing, China, 7–11 October 1996.

[105] O'Neil et al., "Project Galileo at Jupiter," pp. 27–31.

[106] W. J. O'Neil, J. K. Erickson, J. A. Gleason, G. R. Kunstmann, J. M. Ludwinski, R. T. Mitchell, and R. J. Reichert, "Project Galileo Completing Its Primary Mission" (paper number IAF-97-Q.2.02 of the International Astronautical Federation, Paris, France, from the 48th International Astronautical Congress, Turin, Italy, 6–10 October 1997), pp. 3, 5–6.

on 19 January 1997, occurring approximately midway between the E4 and E6 encounters. During this time, the Sun's proximity to the signal path from Galileo to Earth completely blocked the return of data for about 10 days and kept the data-transmission rate low for many days before and after this period.[107]

J5. No close encounter to a Jovian moon was designed into the spacecraft's fifth orbit around Jupiter because Earth and Jupiter were in solar conjunction about the time that the closest approach would have occurred, and there would have been minimal communication capability between the spacecraft and Earth. There also would have been insufficient capability during the cruise part of the orbit, after the close encounter, to return collected data to Earth.[108]

E6. The Orbiter's next close encounter with Europa took place on 20 February 1997 at an altitude of 586 kilometers (363 miles). The main scientific objective was to conduct high-resolution coverage of Europa. This was a similar objective to E4, but with some new Europa surface terrain observed. Monitoring of Io was also conducted. During E6, the spacecraft would fly as close as about 400,000 kilometers (250,000 miles) to Io. Jupiter atmospheric observations during E4 involved a coordinated effort by all of the Orbiter's remote sensing instruments to analyze white oval features in the infrared and ultraviolet regions of the spectrum. Four occultations of Earth occurred during E6—two by Europa, one by Io, and one by Jupiter. The radio science occultation measurements made during these events provided data on atmospheric profiles of the moons and Jupiter, and also on Europa's gravitational field.[109]

E11. The Orbiter's last close encounter of its Prime Mission took place on 6 November 1997 at an altitude of 2,042 kilometers (1,266 miles). The encounter sequence lasted seven days, from 2 to 9 November 1997 GMT. The encounter included the longest recording to date, lasting almost 3 hours, of Jovian magnetospheric data close to Europa. The data were helpful not only in the study of Europa, but also for analyzing Io's plasma torus, whose charged particles are strongly influenced by the magnetic fields they encounter. Primary science objectives of E11 included more remote sensing of the moon's surface and more Jovian atmospheric observations. Another objective was to obtain the highest resolution images yet taken of four small, inner Jovian satellites: Thebe, Metis, Amalthea, and Adrastea.[110]

Galileo Europa Mission (GEM) Encounters

The first eight close encounters on GEM, orbits E12 through E19, were with the satellite Europa. Due to reduced resources at the start of GEM (a smaller project budget and only a fraction of the project staff), encounter sequences were shortened from typically seven

---

[107] Ibid.

[108] Ibid., p. 7.

[109] Ibid., p. 7.

[110] R.T. Mitchell, Z. N. Cox, J. K. Erickson, B. G. Paczkowski, and E. E. Theilig, "Project Galileo: The Europa Mission" (paper number IAF-98-Q.2.01 of the International Astronautical Federation, Paris, France, from the 49th International Astronautical Congress, Melbourne, Australia, 28 September to 2 October 1998), p. 8.

days during the Prime Mission to about two days. During the Prime Mission, Galileo's Europa observations had to compete with many other science objectives, including observations of Jupiter's atmosphere and other satellites. During GEM, Europa encounters got the lion's share of Orbiter resources. Details of these Europa encounters are included in table 9.8. Note the range of regions and surface features that Galileo observed.[111]

Table 9.8. **GEM Europa encounters.**[112]

ORBIT	CLOSEST APPROACH IN KILOMETERS (MILES)	DATE	OBSERVATIONS
E12	196 (122)	16 December 1997	The instruments observed the Conamara ice raft region and took stereo images of the Pwyll crater region. Stereo imaging discerned the topography of a region.
E13	3,562 (2,212)	10 February 1998	No remote sensing or magnetospheric data were collected because of the solar conjunction, which reduced the capacity to transmit science data to Earth. Radio science data for studying Europan gravitational field and internal structure were taken.
E14	1,645 (1,022)	28 March 1998	Stereo imaging of Mannann'an crater and Tyre Macula dark spot was accomplished. The spacecraft observed banded terrain, bright plains, and ice rafts.
E15	2,515 (1,562)	31 May 1998	The spacecraft carried out stereo and color imaging of Cilix massif, previously believed to be the largest mountain on Europa (but E15 data revealed that it was an impact crater). Created near-terminator maps of unexplored mottled terrain. (The terminator is the boundary between the part of the moon that is illuminated and that which is dark.) Because of the low Sun angles near the terminator, shadows cast by uneven terrain are more measurable, and the heights of mounds, ridges, and ice rafts can be determined.
E16	1,830 (1,136)	21 July 1998	A spacecraft safing event prevented Europan science observations. The cause of the event was believed to be debris generated in the slip rings between the spun and despun sections of the Orbiter. The spacecraft passed over the Europan south pole.

---

[111] Ibid.

[112] Ibid.; J. K. Erickson, Z. N. Cox, B. G. Paczkowski, K. W. Sible, and E. E. Theilig, "Project Galileo Completing Europa, Preparing for Io," paper number IAF-99-Q.2.02 in *Proceedings of 50th International Astronautical Congress of the International Astronautical Federation* (held in Amsterdam, The Netherlands, 4–8 October 1999); UCLA Institute of Geophysics and Planetary Physics, "Mission Data," *Galileo Magnetometer Team's Home Page*, 27 October 2000, *http://www.igpp.ucla.edu/galileo/qlook/mission.htm* (accessed 24 November 2004).

Table 9.8. **GEM Europa encounters.** (continued)

ORBIT	CLOSEST APPROACH IN KILOMETERS (MILES)	DATE	OBSERVATIONS
E17	3,582 (2,224)	26 September 1998	A south polar pass (like that of E16) allowed observations of many of the targets missed during E16. The spacecraft searched for evidence of large-scale shifting of surface features, which would indicate a possible liquid sublayer. The spacecraft's instruments took images of the Agenor Linea-Thrace Macula region, Libya Linea, a strike-slip fault zone, Rhiannon Crater, Thynia Linea, and south polar terrain (for comparison with E4 and E6 equatorial terrain images). Thermal maps of Europa were generated. Radio-science analyses of the Europan gravity field were made over a 20-hour period. The instruments also made ultraviolet observations of Europa outgassing and atmospheric emissions.
E18	2,273 (1,412)	22 November 1998	A safing event terminated science observations 6 hours before the Europan closest approach. The primary collection was of radio-science Doppler data.
E19	1,439 (894)	1 February 1999	The instruments carried out global- and regional-scale mapping, along with imaging of Tegid Crater, Rhadmanthys Linea volcanic features, mottled terrain, and a dark spot. The ultraviolet instruments also made observations of atmospheric emissions and possible outgassing. A safing event terminated science observations 4 hours after the Europan closest approach. Outbound distant observations of Europa (as well as Jupiter and Io) were lost.

Galileo Millennium Mission Europa Encounter

The Europa satellite encounter on 3 January 2000 (E26) marked the beginning of the Galileo Millennium Mission (GMM). During this encounter, the spacecraft flew as close to Europa as 351 kilometers (218 miles) and passed slightly south of the prominent Pwyll impact crater. Only limited observations were made during E26 due to factors such as the decreasing periods of Galileo orbits (allowing less time to develop orbital sequences), a smaller workforce and budget than during GEM, and reduced downlink resources.

The E26 encounter with Europa began with real-time data collection[113] by Galileo's six fields-and-particles instruments, which continued the survey of Jupiter's magnetosphere that had been carried out during previous encounters. The fields-and-particles instruments

---

[113] During real-time data collection, data were sent to Earth as they were gathered, rather than being stored on the spacecraft's tape recorder and sent to Earth at a later time.

began their observations on 1 January 2000 at a distance from Jupiter of 29 R_J, and they continued it until 5 January, when the spacecraft was only about 15 R_J from the planet.[114]

Recorded observations during E26 included high-resolution pictures near the Europan terminator,[115] images of three of the four Jovian inner moons (Thebe, Amalthea, and Metis), and observations of the Loki volcanic region on Io. Near the terminator line, the Sun angle is low, and shadows are cast that can help determine the heights of surface features. The spacecraft took low-Sun-angle images of Europan surface features that included possible volcanic flows, ejecta from the Callanish impact basin, and the point of intersection of two ridges. These features were chosen because they related to issues of subsurface water inside the moon.[116]

The E26 flyby was also designed to better characterize Europa's magnetic field signature and to help determine whether the moon either generated its own magnetic field through some internal process or had only a magnetic field induced by Jupiter's magnetism. Scientists hoped that E26 data would show whether the Europan magnetic field was constant, indicating an intrinsic, internally generated field, or time-varying, suggesting an induced field whose characteristics were affected by Europa's location within Jupiter's magnetosphere.[117]

Europa Scientific Findings

Lack of Craters. Europa has relatively few craters, as compared to Ganymede and Callisto, or, for that matter, to Earth's Moon. Models of current cratering rates due to impacts with comets and meteoroids indicate that our Moon's surface has been geologically inactive for more than a billion years. Geological activity during this period would have obliterated many of the craters. Earth has been impacted at least as many times as the Moon during the last billion years, but geological processes such as plate tectonics and volcanic flows, as well as continuous weathering, have smoothed over most of our cratered areas.[118]

Like our Moon, the Jovian satellites Ganymede and Callisto are heavily cratered, suggesting that they too have very old and inactive surfaces. It is unlikely that Europa has somehow avoided meteoroid and comet impacts when Ganymede and Callisto, whose orbits are not far outside Europa's, show evidence of such collisions. Since Jupiter was hit only a few years ago with numerous fragments of the Shoemaker-Levy comet, it is likely that the planet, as well as its moons, sustained similar impacts in the recent past. If Europa has indeed been subjected to many collisions in the past, it suggests that the moon has an active geology capable of rapidly obliterating the impact craters.[119]

---

[114] Erickson, Bindschadler, Theilig, and Vandermey, "Project Galileo: Surviving Io, Meeting Cassini," pp. 11–12.

[115] The terminator is the boundary between the part of a moon or a planet that is illuminated by sunlight and the part that is dark. *Encarta World English Dictionary* (North American Edition), 2001.

[116] Erickson, Bindschadler, Theilig, and Vandermey, "Project Galileo: Surviving Io, Meeting Cassini," pp. 11–12.

[117] Ibid., p. 12.

[118] "Europa; Another Water World?"; Clark R. Chapman, "What's Under Europa's Icy Crust?" Southwest Research Institute's *Technology Today* (Fall 1998), *http://www.swri.edu/3pubs/ttoday/fall98/europa.htm* (available in folder 18522, NASA Historical Reference Collection, Washington, DC).

[119] "Europa; Another Water World?"

Dr. Clark Chapman, a planetary scientist at the Southwest Research Institute in Boulder, Colorado, estimated the age of Europa's surface based on current data on the frequency at which solar system bodies are impacted and scarred by comet and meteoroid hits. Some of these data were developed by Gene and Carol Shoemaker, codiscoverers of the Shoemaker-Levy comet that hit Jupiter in July 1994. The Shoemakers led a telescopic survey of comets in near-Jupiter space, defining an impact rate. According to current data, a crater 20 kilometers in diameter or larger is created on Europa approximately every million years. Fewer than 10 such craters have been observed on Europa, a finding that indicates a surface averaging less than 10 million years old. Chapman believes that the sections of the surface located in nearly crater-free regions are far younger than this. "We're probably seeing areas a few million years old or less, which is about as young as we can measure on any planetary surface besides Earth."[120]

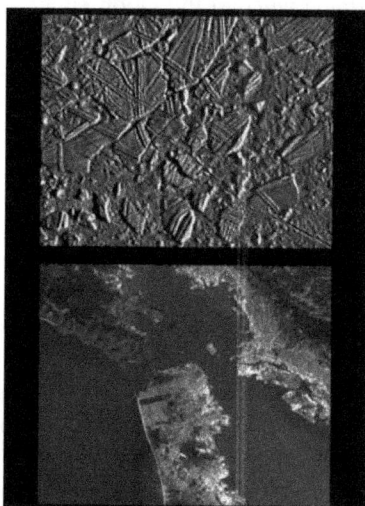

Figure 9.4. **A comparison of the size of Europa's iceberglike surface structures with features of Earth's San Francisco Bay area. Both images show areas of equal size, 34 by 42 kilometers (21 by 26 miles), and resolution, 54 meters (59 yards). North is to the top of the picture. Europa's crustal plates, ranging up to 13 kilometers (8 miles) across, have been broken apart and "rafted" into new positions, superficially resembling the disruption of pack ice on polar seas during spring thaws on Earth. The size and geometry of Europa's features suggest that motion was enabled by water underneath the ice rafts. (JPL image number PIA00597)**

Images from Galileo support the theory of active Europan tectonics that have wiped out old craters. The images depict blocks of crust resembling icebergs floating on an invisible ocean (see figure 9.4). Some of these blocks are tilted or rotated out of place, as if local stresses broke them free and turned them. Dark bands of ice and rock spread outward from central ridges. A mechanism that might account for these features is "tidal flexing," which occurs as Europa is pulled in different directions by the gravitational fields of Jupiter and the other Galilean moons. Europa gets stretched and compressed up to

---

[120]   Chapman,"What's Under Europa's Icy Crust?"; Donald Savage (NASA Headquarters) and Jane Platt (JPL), "New Images Hint at Wet and Wild History for Europa," NASA news release 97-66, 9 April 1997.

several tens of meters each day, fracturing its brittle ice crust and possibly leading to eruptions of ice volcanoes or geysers that shower the surface with material from below.[121]

Not all scientists agreed that Europa's surface was young. Dr. Michael Carr, a geologist with the U.S. Geological Survey and another Galileo imaging team member, put Europa's surface age at closer to one billion years. He compared the number of smaller craters on Ganymede with those on Earth's Moon and found indications that the cratering rate in the Jovian system was significantly less than in the Earth-Moon system. From this, he concluded that the relatively smooth surface of Europa was simply due to a lower rate of impacts with comets and meteoroids. As more observations of Europa are taken, this debate on the age of its surface may be resolved. Of special interest is evidence of current geological activity such as erupting geysers. If observations can establish that the satellite's surface has changed since the Voyager flyby in 1979, or during the years of Galileo flybys, then this evidence will support the hypothesis of a dynamic geology that can rapidly alter surface features.[122]

Evidence for a Subsurface Ocean. The results of many different research efforts, including surface observational data, laboratory experiments, magnetometer data, and gravity measurements, suggest that a liquid ocean may exist beneath Europa's icy crust.

The very young age of the Europan surface (see the discussion in the previous section) indicates that the surface is being rapidly renewed, possibly by the movements of a low-viscosity region below it.[123] The moon's surface topography also supports this possibility, especially the fractured chunks of ice that have been observed. "In some areas, the ice is broken up into large pieces that have shifted away from one another, but obviously fit together like a jigsaw puzzle," says Dr. Ronald Greeley, a geologist at Arizona State University and scientist on the Galileo imaging team. "This shows that the ice crust has been or still is lubricated from below by warm ice or maybe even liquid water."[124]

NIMS observations of Europa's surface added to the case for a subsurface ocean. In 1998, NASA reported that Europa's surface appeared to contain magnesium and sodium salts (sulfates and carbonates). The salt deposits were associated with surface regions that had experienced the most recent disruption. These were important findings because a likely source of such salt deposits is a brine ocean below the ice crust. If the water in this ocean is carbonated, it would explain how enough pressure could be generated to produce the sprays of debris that have been observed on the surface.[125]

Another piece of data supporting the existence of a brine ocean was provided by laboratory experiments involving salt deposits. Magnesium sulfate salts were formed from brines in experiments simulating the conditions that might be expected on a moon

---

[121] "Europa: Another Water World?"

[122] Savage and Platt, "New Images Hint."

[123] Krishan K. Khurana and Margaret G. Kivelson, "Potential for a Subsurface Ocean on Europa and Its Suitability for Life," abstract P21C-06 in *Eos Transactions*, AGU 82, no. 47, Fall Meeting Suppl., 2001; Chapman, "What's Under Europa's Icy Crust?"

[124] Douglas Isbell (NASA Headquarters) and Mary Beth Murrill (JPL), "Jupiter's Europa Harbors Possible 'Warm Ice' or Liquid Water," NASA news release 96-164, 13 August 1996.

[125] Mitchell et al., "Project Galileo: The Europa Mission," p. 13; Ronald Greeley, Christopher F. Chyba, James W. Head III, Thomas B. McCord, William B. McKinnon, Robert T. Pappalardo, and Patricio Figueredo, "Geology of Europa," in *Jupiter*, ed. Fran Bagenal, Tim Dowling, and Bill McKinnon (Port Melbourne, Australia: Cambridge University Press, 2004).

with a subsurface ocean that finds channels to the surface. The brines in the experiment were rapidly cooled on cold surfaces under a vacuum, and the resulting salt deposits exhibited reflectance spectra more similar to those for Europa than salts produced at room temperature.[126]

Galileo's magnetometer provided perhaps the strongest evidence for a liquid briny ocean on Europa. The magnetometer data showed that the direction of Europa's magnetic field changes in a manner suggesting that it is induced by Jupiter's field rather than generated independently of Jupiter through processes within the moon itself. The importance of an induced Europan magnetic field is that it can be explained by the existence of a spherical shell of electrically conducting material (such as salt water) just under Europa's surface. The Jovian field would induce electric currents in this conducting layer, and these currents would generate the observed Europan field.[127]

Data supporting an induced field were gathered by the team of Dr. Margaret Kivelson, a UCLA professor and principal investigator for Galileo's magnetometer experiment. The team had originally been looking for evidence of an internally generated Europan magnetic field. The team found, however, that the axis of the observed Europan magnetic field differed from the moon's rotation axis by approximately 90 degrees—difficult to explain for an internally generated field. Earth's magnetic field, for instance, is much more closely aligned with our planet's axis of rotation. The direction of the Europan field axis could be explained, however, if the field was induced by Jupiter's field.[128]

To settle the question of the type of field that Europa has, the Kivelson team asked JPL to design a Galileo flyby when the moon was passing through a region of the Jovian magnetic field in which the field was pointing opposite to the direction in which it had been pointing during previous flybys. If the Europan field was indeed induced by Jupiter's field, the opposite polarity of the Jovian field should cause the moon's field to flip its polarity. An internally generated Europan field, however, would not flip. A Europa flyby (E26) was planned for January 2000, when the spacecraft would fly through a reversed Jovian field region. The spacecraft magnetometer observations taken during the flyby conclusively demonstrated that Europa's field polarity reversed when Jupiter's field was pointing in the reverse direction, supporting the existence of an induced Europan field. Furthermore, a detailed analysis of data collected during several Europan flybys demonstrated that a global, salty Europan ocean approximately 6 to 100 kilometers (4 to 60 miles) in depth, with a salinity similar to that of Earth's ocean, could explain the field observed by the magnetometer.[129]

Gravity measurements revealed that Europa's surface layer or layers have a low density, similar to that of either liquid or frozen water. These results were consistent with the existence of an induced magnetic field and subsurface ocean, but only if one of the surface layers was liquid. Europa's induced magnetic field is believed to originate from currents in an electrically conducting layer near the moon's surface. Since ice is not a good conductor of electricity, it is a poor candidate for the composition of the layer responsible for the induced field. A layer of liquid water containing

[126] Greeley et al., "Geology of Europa."

[127] Mitchell et al., "Project Galileo: The Europa Mission," p. 9; Khurana and Kivelson, "Potential for a Subsurface Ocean"; Webster, "Galileo Evidence Points to Possible Water World."

[128] Krishan K. Khurana interview, telephone conversation, 8 October 2003.

[129] Khurana and Kivelson, "Potential for a Subsurface Ocean"; Khurana interview, 8 October 2003; Webster, "Galileo Evidence Points to Possible Water World."

dissolved salt ions has a much higher conductivity and can better explain Europa's induced magnetic field.[130]

If a liquid-water ocean does exist, it is not the Sun that keeps it in a liquid state. At Europa's great distance from the Sun, the intensity of solar radiation is less than 4 percent what it is on Earth's surface.[131] A mechanism that might generate enough heat to maintain a liquid Europan ocean is the moon's tidal flexing. Thermal energy produced by the satellite's expansion and contraction could be sufficient to melt part of its ice crust, thereby creating invisible seas below the surface. Heating from radioactive sources in the moon may add to the melting.[132]

A liquid-water sea cannot exist uncovered on the surface of Europa. Due to the moon's nearly nonexistent atmospheric pressure, water near the top of such a sea would quickly evaporate and be released into space or settle as snow or ice crystals on the moon's surface. In addition, the low temperature (-162°C or -260°F) would rapidly freeze a layer of water. These processes would quickly seal in a liquid-water sea, preventing further loss. Therefore, if liquid water does exist on Europa, it is under the moon's ice crust.[133]

Could a Europan Ocean Support Life?

Once the existence of an ocean beneath the ice crust of Europa was suspected, a most tantalizing question arose both within and outside the space science community: Might life exist on Europa? The three main factors that exobiologists seek when searching for extraterrestrial life are water, organic compounds, and adequate heat. The presence of a liquid-water ocean would satisfy the first criterion, and organic compounds are known to be prevalent in our solar system. As to the third criterion, imaging data of Europa demonstrate the existence, at least at one time and very likely at present, of enough heat to create convection currents that drove or drive ice flows on the surface of the moon. This circumstance does not prove that life actually exists—only that it might.[134]

Scientists consider Europa to be one of the handful of solar system locations that "possess an environment where primitive forms of life could possibly exist."[135] Mars and the Saturnian moon Titan are other such places. It cannot be stressed enough, however, that "could exist" does not by any means imply that the existence of life is even probable in those places. The search for extraterrestrial life-forms has to proceed slowly and meticulously, and all conclusions need to be based on good science. In the words of Daniel S. Goldin, former NASA Administrator:

---

[130] Khurana and Kivelson, "Potential for a Subsurface Ocean"; Mitchell et al., "Project Galileo: The Europa Mission," p. 9; Webster, "Galileo Evidence Points to Possible Water World."

[131] JPL, "Jupiter, Gas Giant," *Galileo: Journey to Jupiter,* 17 June 2002, *http://www.jpl.nasa.gov/galileo/jupiter/ jupiter.html.*

[132] "Europa: Another Water World?"; "Galileo Finds Evidence for an Ocean on Europa"; Kathy Sawyer, "Galileo Mission's Perseverance Extends Studies of Jupiter's Moons," *Washington Post* (26 December 1997): A3.

[133] LarryPlatonic, "An Ocean Discovered: Europa Surrenders Her Secrets," *Galileo Messenger* (May 1997); "Divining Water on Europa."

[134] Douglas Isbell (NASA Headquarters) and Franklin O'Donnell (JPL), "Ice Volcanoes Reshape Europa's Chaotic Surface," NASA news release 97-12, 17 January 1997; "Europa: Another Water World?"

[135] Douglas Isbell (NASA Headquarters) and Mary Beth Murrill (JPL), "Jupiter's Europa Harbors Possible 'Warm Ice' or Liquid Water," NASA news release 96-164, 13 August 1996.

A few days ago, I greeted the possibility of ancient microbial life on Mars with skeptical optimism, and invited further scientific examination and debate . . . . The potential for liquid water on Europa is an intriguing possibility, and another step in our quest to explore the solar system, the stars, and the answer to the great mystery of whether life exists anywhere else in the cosmos . . . . Once again, NASA will ask the scientific process to work.[136]

What Europan Energy Source Could Support Life? Europa has possibly had abundant water, rich in organics and minerals, from the time it was created. On Earth, life appeared within 700 million years of the planet's formation. Europa is as old as Earth, and the moon's oceans may have had adequate time for life to evolve in them.[137]

But what type of energy might support life on Europa? Most life on Earth depends, either directly or indirectly, on photosynthesis. According to Christopher Chyba of the Search for Extra-Terrestrial Intelligence (SETI) Institute, the first link in Earth's food chain "is chlorophyll's conversion of sunlight into chemically stored energy. But imagine an ocean on Europa, a huge, bottled-up body of water capped with miles of ice. Photosynthesis isn't going to work there." Nonetheless, according to Chyba, there are other ways to "make a metabolic living in those dark seas."[138]

According to Chyba and his colleague, Kevin Hand of Stanford University, the energy necessary to support life on Earth is usually obtained through oxidation-reduction reactions in which substances such as carbon and oxygen bond to share an electron and release energy during the reaction. In Earth's oceans, molecular oxygen, a product of photosynthesis, acts as the oxidizing agent in such reactions. But would molecular oxygen or other oxidizing agents be present in the depths of a Europan ocean? Chyba and Hand think perhaps yes. High-speed particles accelerated in Jupiter's magnetosphere routinely bombard Europa's ice crust, where they form oxidants such as hydrogen peroxide and molecular oxygen. The critical question is whether these molecules eventually get transported through Europa's ice crust—for instance, through geological upheavals where sections of the crust are overturned—and into an ocean below. "We can't be certain at this point whether the oxidants would actually make it into the water, even over geological timescales,"[139] said Chyba.

Even if such molecules do not get transported through the ice, other mechanisms exist that might supply molecular oxygen to the oceans. One of these is the radioactive decay of potassium isotope ^{40}K, which should be present in both the Europan ice crust and the liquid water below. This decay would split water molecules, generating oxygen that could conceivably support biological processes.[140]

Some scientists believe that geochemical energy could also help to support Europan life, although at significantly lower abundances than on Earth. This view is

---

[136] Laurie Boeder (NASA Headquarters contact), "Statement by Administrator Daniel S. Goldin on the Release of New Galileo Spacecraft Images of Europa," NASA news release 96-166, 13 August 1996.

[137] Larry Palkovic, "An Ocean Discovered: Europa Surrenders Her Secrets."

[138] SETI Institute, "Hidden Oceans Could Still Support Life," news release, Mountain View, CA, 14 June 2001, *http://www.seti-inst.edu/general/press_release/hiddenoceans.html.*

[139] Ibid.

[140] Ibid.

based in part on discoveries of Earth-based ecological communities on the ocean floor near volcanic vents and in hot springs, where life-forms employ chemical energy found in near-boiling water. Using models of geochemical activity from rock weathering combined with volcanic activity, Bruce Jakosky of the University of Colorado and Everett Shock of Washington University in St. Louis calculated the potential amount of life that geochemical energy could support on various planets. They found that sufficient chemical energy could be found on Mars for roughly 20 grams of organisms per square centimeter of surface to develop over 4 billion years. They estimated that a somewhat smaller amount of life could develop on Europa. For purposes of comparison, sufficient geochemical energy exists on Earth for 20 grams of organisms per square centimeter of surface to grow in only 1,000 years—4 million times faster than on Mars. Jakosky and Shock suggest that the most likely place to look for life on Europa is not in an ocean, but in the rocks under the ocean. Internal heat sources could possibly transfer life-sustaining energy to organisms living in the rocks.[141]

A Europan-Like Region on Earth. Can life exist in the frigid conditions of an icy solar system body such as Europa? The best way to find out is to send a spacecraft to Europa to conduct in situ analyses. But until such a mission is launched, studying similar regions that are a bit closer can help to determine how extreme an environment can be and still support life. One such extreme environment that resembles Europa in several ways is Lake Vostok in Antarctica.[142]

In 1974, an airborne research team passed over the Soviets' Vostok station in Antarctica. The team's sounding instruments detected a body of water that was located beneath 3,700 meters (12,000 feet) of ice and was about 50 by 220 kilometers (30 by 140 miles) in size—roughly as large as Lake Ontario. Scientists thought that the lake was kept liquid by geothermal energy, combined with the insulation provided by its nearly 4-kilometer-thick ice blanket.

In 1998, Richard Hoover of NASA Marshall Space Flight Center and Dr. S. S. Abyzov of the Russian Academy of Scientists used an electron microscope to examine ice cores from above Lake Vostok for evidence of microbial life. They found cyanobacteria, bacteria, fungi, spores, pollen grains, diatoms (a type of algae), and, according to Hoover, "some really bizarre things . . . not recognizable as anything we've ever seen before." For instance, some of the microbes were covered with a strange, fibrous structure made out of "nanofilaments" only 30 to 40 nanometers in diameter. An unusual feature of many of the cyanobacteria taken from ice at a depth of 1,200 meters (about a third of the way down to the water) was that they contained abnormal amounts of antimony, a toxic heavy metal.[143]

Richard Hoover explained that when the microorganisms froze, they went into an "anabiotic" state, which means that they shut down their functions and became

---

[141] "Limited Energy for Mars and Europa Life, Scientists Report," *SpaceViews—The Online Publication of Space Exploration* (26 August 1998), *http://www.spaceviews.com/1998/08/26a.html.*

[142] "The Frosty Plains of Europa," *Science@NASA*, 3 December 1998, *http://science.nasa.gov/newhome/headlines/ast03dec98_1.htm.*

[143] Ibid.; Marshall Space Flight Center, "Clues to Possible Life May Lie Buried in Antarctic Ice," 5 March 1988, *http://science.nasa.gov/newhome/headlines/ast05,ar98_1.html;* "Exotic-Looking Microbes Turn Up in Ancient Antarctic Ice," Science@NASA, 13 March 1998, *http://science.nasa.gov/newhome/headlines/ast12mar98_1.htm;* "Diatom," *Microsoft Encarta Online Encyclopedia*, 2001.

inactive, as if in suspended animation. Frozen microbes can often be revived, however. Russian scientists have revived and cultured bacteria, yeast, fungi, and other microbiota from ice cores.

The ice that harbored the Vostok finds varied in age up to 400,000 years. The ice was extracted using technology specially developed by scientists at the St. Petersburg Mining Institute in Russia for drilling ice cores without contaminating the samples. The Lake Vostok studies showed that life can exist under surprisingly harsh conditions. A subsurface Europan ocean may have a temperature profile, chemical composition, and ice cap resembling Lake Vostok's. Although such a correlation cannot prove the existence of life on Europa, it does make it easier to accept the possibility of life on the Jovian satellite.[144]

Europa's Atmosphere. In 1977, researchers suspected that an oxygen atmosphere might exist on icy bodies such as Ganymede and Europa. If so, water vapor would be present in at least small quantities. Solar radiation would slowly dissociate the water molecules into hydrogen and oxygen. The hydrogen, being lighter than oxygen, would escape the moons far more rapidly. Particles impacting the moons' surfaces would dislodge additional molecules of surface materials and send them into the moons' atmospheres.[145]

Scientists predicting a Europan atmosphere containing gaseous oxygen had to wait until the sensitive Hubble Space Telescope (HST) could confirm their theory. In February 1995, almost a year before Galileo reached Jupiter, a Johns Hopkins team using the HST reported that they had identified an "extremely tenuous atmosphere of molecular oxygen" around Europa. This was not an atmosphere that was likely to support life. Its surface pressure was estimated at one hundred-billionth that of Earth. Doyle Hall of Johns Hopkins, principal investigator of the team that detected that atmosphere, said that if all the oxygen estimated to be in the Europan atmosphere were compressed to the surface pressure of Earth's atmosphere, it would fill "only about a dozen Houston Astrodomes." The temperature of this thin atmosphere was -145°C (-230°F).[146]

During Galileo's December 1996 and February 1997 Europa encounters (E4 and E6), the spacecraft performed six occultation experiments in which Europa was positioned between the spacecraft and Earth. This configuration affected the radio signal sent from Galileo to Earth. Measurements of the signal by the Deep Space Network stations in Goldstone, California, and Canberra, Australia, showed that the radio beam from Galileo had been refracted by a layer of charged particles around Europa. According to Dr. Arvydas Kliore of JPL, the source of these ions was most likely water molecules from Europa's icy surface, dislodged by high-energy particles from Jupiter's magnetosphere to form a thin atmosphere. Water molecules from this atmosphere were then ionized,

---

[144] "Exotic-Looking Microbes Turn Up in Ancient Antarctic Ice"; "Frosty Plains of Europa."

[145] NASA Langley Research Center, *State of Knowledge of Europa*, taken from the *Report of the Europa Orbiter Science Definition Team*, June 1999, *http://centauri.larc.nasa.gov/outerplanets/Europa_SDT.pdf*. Available from the Outer Planets Program Library, an online library on the Langley Research Center Aerospace Systems, Concepts, and Analysis Web site. The library is accessible at *http://centauri.larc.nasa.gov/outerplanets/*. The Europa document is also available in Galileo—Meltzer Sources, folder 18522, NASA Historical Reference Collection, Washington, DC.

[146] Donald Savage (NASA Headquarters), Tammy Jones (Goddard Space Flight Center), and Ray Villard (Space Telescope Science Institute), "Hubble Finds Oxygen Atmosphere on Jupiter's Moon Europa," NASA news release 95-17, 23 February 1995.

which could have been done by ultraviolet radiation from the Sun or by collisions with additional energetic particles from the Jovian magnetosphere.[147]

Ganymede: Jupiter's Largest Moon Has a Ravaged Surface and a Magnetic Field

Ganymede, with a diameter of 5,270 kilometers (3,270 miles), is larger than Earth's Moon, larger than Pluto, and larger even than Mercury. It has been extensively bombarded by comets and meteorites and distorted by the same tectonic forces that make mountains and move continents on Earth. Ganymede has its own magnetosphere, a bubble-shaped region of charged particles surrounding the moon. This discovery was surprising to Galileo scientists because although many planets have magnetospheres, satellites generally do not.[148]

General Ganymede data are included in table 9.9.

Table 9.9. **Ganymede statistics.**[149]

Discovery	11 January 1610, by Galileo Galilei
Diameter	5,270 kilometers (3,270 miles)
Mass	1.48×10^{23} kilograms
Mass relative to that of Earth	0.0247 times the mass of Earth
Surface gravity (Earth = 1)	14.5 percent of Earth's gravity
Mean distance from Jupiter	1,070,000 kilometers (664,000 miles)
Mean distance from Jupiter in radii	15.1 R_J
Mean distance from the Sun	5.203 AU
Orbital period	7.154553 days
Rotational period	7.154553 days
Density	1.94 grams per cubic centimeter
Orbit speed	10.9 kilometers per second (24,400 mph)
Surface composition	Dirty ice

Ganymede Gravity-Assists Helped Make Critical Orbit Changes

Because of Galileo's trajectory as it approached the Jovian system, the spacecraft went into orbit around Jupiter inclined to the equatorial plane of the planet, as well as to the orbital planes of the Galilean satellites. This situation had to be changed. Orbital mechanics calculations showed that if Galileo, with its limited propellant, were to have

---

[147] Jane Platt (JPL), "Galileo Spacecraft Finds Europa Has an Atmosphere," JPL Public Information Office news release, 18 June 1997

[148] Galilean Satellite Sheet," *National Space Sciences Data Center (NSSDC) Planetary Sciences Home Page, http:// nssdc.gsfc.nasa.gov/planetary/factsheet/galileanfact_table.txt*; Mary Beth Murrill, "Galileo Spacecraft Makes New Discoveries at Ganymede," JPL Public Information Office news.

[149] Ganymede Fact Sheet," *JPL Galileo Home Page, http://www.jpl.nasa.gov/galileo/ganymede/fact.html* (accessed 6 June 2001).

close encounters with more than one of the Galilean satellites, the spacecraft's orbit would have to be altered so that it lay within the orbital plane of the satellites. Until a zero-inclination condition could be achieved, all of Galileo's close encounters had to be with the same moon. Ganymede was chosen as this moon because its large mass and orbital location were favorable for changing the spacecraft's orbital inclination, as well as its orbital period (the time the spacecraft took to complete an orbit around Jupiter), which was longer than desired. Galileo's initial orbital period was 210 days, or about seven months. A gravity-assist during G1 reduced the spacecraft's orbital period to 72 days, which allowed more orbits and thus more close encounters to be carried out each year. The G1 gravity-assist also slightly reduced the craft's orbital inclination. The second encounter, G2, was planned largely to put Galileo into an orbit that was coplanar with the satellites' orbits, thus enabling subsequent encounters with Europa, Io, and Callisto.[150]

Findings from Ganymede Flybys

Galileo's first close encounter of its Prime Mission satellite tour, G1, was with Ganymede on 27 June 1996 at an altitude of 835 kilometers (518 miles). The second Ganymede encounter, G2, took place on 6 September 1996 at an altitude of 260 kilometers (161 miles). G1 involved an equatorial-region flyby at closest approach, while G2 passed near a pole. Although Galileo's trajectory during these encounters was governed by the orbital mechanics factors described above, by good fortune it also provided an excellent opportunity for radio-science observations of different parts of the moon. These observations used the fact that any change in Galileo's velocity caused the frequency of the radio signal received on Earth to change. When the spacecraft passed close to Jupiter or to a moon, that body pulled on the craft, altering its velocity. The amount of velocity change depended on the mass of the body and on how that mass was internally distributed. Thus, by measuring the change in frequency of the radio signal received on Earth, the mass and internal structure of Jupiter or one of its satellites was estimated.[151]

The radio-science and other data received on Earth clearly indicated that Ganymede's interior was differentiated, probably into a core and a mantle, which were enclosed in an ice shell.[152] Besides revealing the moon's differentiated interior, the Ganymede encounters also provided important data supporting the existence of a self-generated magnetic field, a possible subsurface ocean, and past volcanism that might explain extensive surface features. A summary of Galileo's Ganymede flybys and observations is included in table 9.10.

---

[150] W. J. O'Neil, N. E. Ausman, J. A. Gleason, M. R. Landano, J. C. Marr, R. T. Mitchell, and R. J. Reichert, "Project Galileo at Jupiter" (paper number IAF-96-Q.2.01 of the International Astronautical Federation, Paris, France, presented at the 47th International Astronautical Congress, Beijing, China, 7–11 October 1996), p. 29.

[151] O'Neil et al., "Project Galileo at Jupiter"; "Radio Science," *Galileo's Science Instruments, JPL Galileo Home Page*, *http://www.jpl.nasa.gov/galileo/instruments/rs.html* (accessed 10 July 2001).

[152] Douglas Isbell (NASA Headquarters) and Mary Beth Murrill (JPL), "Big Icy Moon of Jupiter Found to Have a 'Voice' After All"; "Europa Flyby Next for Galileo," NASA news release 96-255, 12 December 1996, *http://www.jpl.nasa.gov/galileo/status961212.html*.

Table 9.10. **Ganymede encounters.**[153]

ORBIT	CLOSEST APPROACH IN KILOMETERS (MILES)	DATE	MISSION	COMMENTS
G1	835 (519)	27 June 1996	Prime	A gravity-assist during G1 reduced Galileo's orbital period from 210 to 72 days, which allowed more orbits and close encounters each year. The perijove of orbit (point of closest approach to Jupiter) was increased to keep the spacecraft out of the most intense radiation regions. A radio-science experiment analyzed Ganymede's gravitational field and internal structure. The instruments detected evidence of a self-generated magnetosphere around the moon.
G2	260 (161)	6 September 1996	Prime	A Ganymede gravity-assist put Galileo into coplanar orbit with other Galilean satellites, permitting subsequent encounters with them. A radio-science experiment analyzed Ganymede's gravitational field and internal structure. G1 and G2 radio-science and other data revealed that Ganymede had an interior that was probably differentiated into a core and a mantle. The plasma wave experiment and magnetometer data gave evidence of an internally generated magnetic field.
G7	3,102 (1,926)	5 April 1997	Prime	The spacecraft flew over the high latitudes of Ganymede and took high-resolution observations of high-energy impact regions, as well as Jupiter magnetosphere and aurora observations.
G8	1,603 (995)	7 May 1997	Prime	The spacecraft passed over the mid-latitudes of Ganymede, with closest approach longitudes 180° apart from those of the G7 encounter, allowing new terrain to be imaged.
G28	1,000 (600)	20 May 2000	GMM	Galileo's closest approach coincided with Cassini's. Joint Galileo-Cassini observations revealed solar wind effects and magnetospheric dynamics. High-resolution Ganymede images also were taken. Magnetometer data suggest that a salty water layer exists beneath the icy crust.

---

[153] O'Neil et al., "Project Galileo: Completing Its Primary Mission"; O'Neil et al., "Project Galileo at Jupiter"; Erickson et al., "Project Galileo Completing Europa, Preparing for Io"; Erickson et al., "Project Galileo: Surviving Io, Meeting Cassini."

Table 9.10. **Ganymede encounters.** (continued)

ORBIT	CLOSEST APPROACH IN KILOMETERS (MILES)	DATE	MISSION	COMMENTS
G29	2,321 (1,441)	28 December 2000	GMM	Real-time data were transmitted as Galileo flew from the inner magneto-sphere through the magnetopause and bow shock and into the solar wind. Remote sensing instruments targeted Jupiter, its rings, and the Galilean satellites.

Evidence of a Magnetosphere

Evidence from the G1 and G2 encounters indicates that Ganymede possesses a plan-etlike, self-generated magnetosphere, shielding the moon from the magnetic influence of its parent body, Jupiter. The Orbiter's plasma wave subsystem (PWS), which used a dipole antenna as a receiver, picked up electromagnetic wave activity characteristic of a magnetosphere and closely matching the electromagnetic signatures of Earth's, Saturn's, and Jupiter's magnetospheres. "The data we get back is in the form of a spectrogram, and reading it is kind of like looking at a musical score," said Dr. Donald Gurnett, a physicist at the University of Iowa and principal investigator of Galileo's PWS. "The instant I saw the spectrogram, I could tell we had passed through a magnetosphere at Ganymede." The approach to the moon was relatively quiet, and then "all of a sudden, there's a big burst of noise that signals entry into Ganymede's magnetosphere. Then, for about 50 minutes, we detected the kinds of noises that are typical of a passage through a magnetosphere. As we exited the magnetosphere, there was another big burst of noise."[154]

Abrupt changes in field intensity and direction also gave indications of a magneto-sphere. As the spacecraft approached the moon, the magnetometer measured, as expected, the field of Jupiter, which was spatially fairly uniform and pointed south. Then the field suddenly increased nearly five times in strength and swung around to point directly at Ganymede.

Ganymede provided the first example ever observed of a moon's magnetosphere contained within the magnetosphere of the parent planet. "We knew Ganymede was an interesting place," said Torrence Johnson, Galileo project scientist. "What we have just found makes it even more exciting."[155]

Source of Ganymede's Magnetism. UCLA's Dr. Margaret Kivelson, principal investiga-tor for Galileo's magnetometer experiment, confirmed the detection of a large magnetic field increase near Ganymede. Combined with the discovery of a self-generated magnetic field, the radio-science data indicated that Ganymede had a metallic core beginning at a depth of 400 to 1,300 kilometers (250 to 800 miles) below the surface. Depending on whether the core was pure iron or an iron/iron sulfide alloy, mission scientists estimated that it could account for as little as 1.4 percent or as much as one-third of Ganymede's total mass. Galileo scientists suspected that the moon's magnetic field is generated in a similar way to that of Earth's field, through a dynamo mechanism originating in the electrically conducting core.[156]

---

[154] Isbell and Murrill, "Big Icy Moon of Jupiter Found to Have a 'Voice' After All."

[155] Murrill, "Galileo Spacecraft Makes New Discoveries at Ganymede."

[156] Isbell and Murrill, "Big Icy Moon of Jupiter Found to Have a 'Voice' After All."

Surface Terrain

Ganymede's surface is a mixture of two types of terrain—40 percent highly cratered dark regions and 60 percent light, grooved terrain called "sulcus," meaning a groove or burrow. The dark regions on Ganymede are old and rough in texture. The dark, cratered terrain is believed to be the original crust of the satellite. Lighter regions are believed to be smoother and younger, having relatively few craters. The light-colored, grooved sulcus terrain may be formed by the release of water from beneath the surface. Galileo observed grooved ridges as high as 700 meters (2,000 feet) that ran for thousands of kilometers across Ganymede's surface.[157]

Ganymede's larger craters are nearly flat and lack the central depressions common to craters often seen on Earth's Moon. But our Moon has a rocky surface, whereas Ganymede has an icy one. Scientists think that the gradual flattening of Ganymede's large craters over millions of years is probably due to the softness of its icy surface. These large, flat "phantom craters" are called palimpsests, which is a term that originally referred to reused, ancient writing materials on which older writing was still visible underneath newer writing. Ganymede's palimpsest craters range from 50 to 400 kilometers (30 to 250 miles) in diameter, and they are surrounded by rays of ejecta.[158]

Does Ganymede Have a Hidden Ocean?

Ganymede has a strong, self-generated magnetic field and magnetosphere. In addition, it has a secondary field induced by Jupiter's magnetism. This makes Ganymede's magnetic data more complex to interpret than, for instance, data from Europa, which has only the secondary induced field. UCLA's Dr. Margaret Kivelson, principal investigator on Galileo's magnetometer experiment, believes that Ganymede's induced field has to originate in a material "more electrically conductive than solid ice." A liquid layer several kilometers thick, with a salt content approximately that of Earth's oceans and located within 200 kilometers (120 miles) of Ganymede's surface, could be responsible for the induced field.[159]

Infrared spectrometer analysis has identified sections of Ganymede's surface containing salt minerals. The salt-bearing materials may be mostly frozen magnesium sulfate brines that originally made their way to the surface through cracks, according to Dr. Thomas McCord, a University of Hawaii geophysicist. Image data show linear fractures on Ganymede that resemble such features on Europa. In addition, the salt minerals on Ganymede's surface are similar to minerals seen on Europa.

Galileo imaging data taken on the G28 encounter (May 2000) suggested further similarities between Ganymede and Europa. The Ganymede images show details of Arbela Sulcus, a bright, striated band of materials spread across an older, more heavily cratered landscape. According to Dr. Robert Pappalardo, a planetary scientist at

---

[157] Ganymede Fact Sheet," *JPL Galileo Home Page, http://www.jpl.nasa.ov/galileo/ganymede/fact.html* (accessed 6 June 2001).

[158] Ibid.

[159] Guy Webster, "Solar System's Largest Moon Likely has a Hidden Ocean," JPL Media Relations Office news release, 16 December 2000.

Brown University, "It is possible that Arbela Sulcus has formed by complete separation of Ganymede's icy crust, like bands on Europa."[160]

A subsurface ocean needs sufficient heat to maintain it in a liquid state. The tidal flexing that is thought to maintain Europa's ocean probably is not the mechanism responsible for one on Ganymede, whose nearly circular orbit would minimize such flexing. In order to explain the possible existence of a subsurface liquid-water layer on Ganymede, Galileo scientists made estimates of the thermal energy available from radioactivity in the satellite's rocky interior. According to Dr. Dave Stevenson, a planetary scientist at the California Institute of Technology in Pasadena, Ganymede has sufficient radioactive heat sources to maintain a stable layer of water at a depth of roughly 150 to 200 kilometers (90 to 120 miles) beneath the surface.[161]

A Mechanism To Explain Ganymede's Bright Bands of Frozen Water

Although tidal flexing may not be the energy source maintaining a Ganymede ocean, it likely contributed to the melting of the ice crust in the past and possibly to volcanism. A billion years ago, when Ganymede was moving into its present position relative to the other Galilean moons, it did not have nearly as circular an orbit as it does now. At different points in its old, elliptical orbit, the direction and magnitude of tidal forces on the moon would vary, slightly altering the shape of the moon. The expansion and contraction of different parts of the moon's interior would have resulted in heating, perhaps enough to have thinned its icy surface and enlarged a buried ocean, making it closer to the surface than now.[162]

During this time of tidal distortions, cracking of the crust and eruptions of slushy ice or water onto Ganymede's surface may have been more common than after the moon attained a nearly circular orbit. Such eruptions could explain Ganymede's extensive bright swaths of frozen water that overlay its darker, heavily cratered ice. Stereo images taken by Galileo show that the smoothest, brightest ice swaths lie in troughs 0.5 to 1 kilometer below the dark, heavily cratered areas and that volcanic, calderalike features occur at the edges of the swaths. The bright bands' low elevations and flatness could have resulted from water or slushy ice erupting and then flooding low-lying areas before refreezing. The eruptions might not have had enough energy to break through the additional ice of higher elevation surface areas.[163]

Louise M. Procter of the Johns Hopkins Applied Physics Laboratory said that the bright swaths' low elevation "is the most persuasive evidence that they could be volcanic" in origin. She raises the question, however, that if icy eruptions were common on Ganymede a billion or so years ago, then why is there not more visible evidence of them? According to Paul M. Schenk of Houston's Lunar and Planetary Institute, whose team created paired Voyager-Galileo images of areas on Ganymede, surface buckling and stretching since the eruptions could have obscured such visible

---

[160] Ibid.; "Hydrated Salt Minerals on Ganymede's Surface: Evidence of an Ocean Below," *JPL Galileo Home Page, http://www.jpl.nasa.gov/galileo/* (accessed 12 July 2001).

[161] Webster, "Solar System's Largest Moon Likely Has a Hidden Ocean."

[162] R. Cowen, "Images Suggest Icy Eruptions on Ganymede," *Science News* (3 March 2001): 133.

[163] Ibid.

evidence of icy eruptions in the past. Further analysis of Galileo and other spacecraft data may resolve the issue of how much volcanism occurred on the moon's surface in the past.[164]

Callisto: Why Is This Galilean Moon So Different from the Others?

The Scarred Face of Callisto

Callisto, the outermost of the four Galilean satellites, has the most pockmarked surface of any solar system body yet observed. It is the presence of so many craters that supports the theory of a very old, unchanging Callisto surface. Although scientists do not know the exact rate of impact-crater formation on Callisto, they estimate that it must have taken several billion years for the moon to accumulate its depressions. Active geological processes on the moon probably would have obliterated many of these craters. Since so many craters exist, scientists assume that Callisto has been geologically inactive for billions of years.[165]

Callisto's heavily cratered face is useful to scientists in providing a record of the bombardment history of the Jovian system. According to Dr. Torrence Johnson, Galileo project scientist, "The craters on Callisto are the visible record of what sizes of comets and other objects have pelted Jupiter and its moons and with what frequency . . . ."[166]

One puzzling feature regarding Callisto is that it appears to have notably few small craters. Galileo took high-resolution images in order to study more extensively the densities of small craters and how some of their features appear to be degraded or eroded. A dearth of small craters could reflect an active resurfacing mechanism that operates efficiently only on small scales. It could also indicate that relatively few small comets and meteoroids fly through the Jovian system and impact bodies within it.[167]

Callisto Encounters and General Data

Galileo made its first close pass of Callisto (C3) on 4 November 1996. The spacecraft flew to within 1,106 kilometers (686 miles) of the satellite's surface. Earth received the signals confirming the encounter 46 minutes later.[168] A summary of this and other Callisto encounters is included in table 9.11.

[164] Ibid.

[165] Douglas Isbell (NASA Headquarters) and Franklin O'Donnell (JPL), "Galileo Makes Close Pass By Callisto," NASA news release 96-226, 4 November 1996.

[166] Donald Savage (NASA Headquarters) and Guy Webster (JPL), "Galileo Gets One Last Close Encounter With Jupiter's Callisto," NASA news release 01-97, 22 May 2001.

[167] Southwest Research Institute, Boulder, CO, "October 1998 Update: Callisto's Perplexing Craters," Galileo Research (Imaging of the Galilean Satellites of Jupiter), 10 November 1998 (available in Galileo—Meltzer Sources, folder 18522, NASA Historical Reference Collection, Washington, DC); Savage and Webster, "Galileo Gets One Last Close Encounter With Jupiter's Callisto."

[168] Isbell and O'Donnell, "Galileo Makes Close Pass By Callisto."

Table 9.11. **Callisto encounters.**[169]

ORBIT	CLOSEST APPROACH IN KILOMETERS (MILES)	DATE	MISSION	OBSERVATIONS
C3	1,136 (705)	4 November 1996	Prime	Observations supported the theory that Callisto has a homogeneous internal structure, 60 percent rock and 40 percent ice.
C9	418 (260)	25 June 1997	Prime	The spacecraft passed through and studied the magnetotail region of the Jovian magnetosphere during the period between the C9 and C10 flybys. Analysis of the C3, C9, and C10 data suggests that Callisto may have a subsurface, salty ocean that is responsible for a variable magnetic field induced by Jupiter's field.
C10	539 (335)	17 September 1997	Prime	C10 data suggest that the internal structure of the moon is not homogeneous, but partially differentiated, with a higher percentage of rock than ice having settled toward the center of the satellite. Callisto is probably less differentiated than the other Galilean moons.
C20	1,315 (817)	5 May 1999	GEM	The perijove reduction campaign began; it involved incremental changes in the closest approach to Jupiter carried out over four Callisto encounters (C20–C23). The campaign was designed to set up flybys of Io, the Galilean moon closest to Jupiter.
C21	1,047 (650)	30 June 1999	GEM	NIMS studied the trailing edge of Callisto. The SSI camera observed dark surface material. The PPR studied equatorial region.
C22	2,296 (1,426)	14 August 1999	GEM	The spacecraft observed Callisto's ionosphere and measured the distribution of free electrons.
C23	1,057 (656)	16 September 1999	GEM	The spacecraft observed Callisto's ionosphere, measured the distribution of free electrons, and completed the perijove reduction campaign.
C30	138 (86)	25 May 2001	Post-GMM	The spacecraft observed the Asgard, Valhalla, and Bran craters in the closest flyby to date (in order to set up an Io encounter in August 2001). Camera problems were possibly due to continued radiation exposure that affected distant images taken of Io. Problems were corrected before the closest approach to Callisto.

---

[169] Guy Webster (JPL), "Galileo Succeeds in Its Closest Flyby of a Jovian Moon," JPL Media Relations Office news release, 25 May 2001; JPL, "Galileo Millennium Mission Status," JPL Media Relations Office news release, 24 May 2001; Erickson et al., "Project Galileo: Surviving Io, Meeting Cassini," p. 19.

General data regarding Callisto are included in table 9.12.

Table 9.12. **Callisto data.**[170]

Discovery	7 January 1610, by Galileo Galilei
Diameter	4,806 kilometers (2,980 miles)
Mass	1.077×10^{23} kilograms (about 2.4×10^{23} pounds)
Mass relative to that of Earth	0.01807
Surface gravity (Earth = 1)	12.7 percent of Earth's gravity
Mean distance from Jupiter	1,883,000 kilometers (1,167,000 miles)
Mean distance from Jupiter in radii	26.6 R_J
Mean distance from Sun	5.203 AU
Orbital period	16.68902 days
Rotational period	16.68902 days
Density	1.86 grams per cubic centimeter
Orbit speed	8.21 kilometers per second (18,400 mph)
Surface composition	Ice

Callisto's Internal Structure

Galileo's observations of Callisto reveal that it is a mixture of metallic rock and ice and that, unlike Ganymede, it has no identifiable central core. Data taken on Galileo's first Callisto encounter (C3) in November 1996 seemed to support the idea that Callisto had a homogeneous structure made up of 60 percent rock containing iron and iron sulfide and 40 percent compressed ice. Scientists believed that since Callisto was located further from Jupiter than the other Galilean moons, it was not subjected to the same gravitational pull and tidal forces and thus did not experience the heating that melted parts of the other moons. This heating allowed heavier materials on the three other Galilean moons to sink and lighter ones to rise, leading to the formation of differentiated layers. "Callisto had a much more sedate, predictable and peaceful history than the other Galilean moons," said Dr. John Anderson, a planetary scientist at JPL.[171]

Later data from Galileo's third Callisto encounter in September 1997 (C10) led scientists to modify their view of Callisto from a moon with a totally undifferentiated interior to one that is not completely uniform. According to Anderson, researchers have found signs that "interior materials, most likely compressed ice and rock, have partially settled, with the percentage of rock increasing toward the center of Callisto." The satellite is still unlike Io, Europa, and

---

[170] "Callisto, A Continuing Story of Discovery," *JPL Project Galileo Home Page, http://www.jpl.nasa.gov/galileo/callisto/ #overview* (accessed 8 July 2001).

[171] Douglas Isbell (NASA Headquarters) and Jane Platt (JPL), "Galileo Returns New Insights Into Callisto and Europa," NASA news release 97-110, 23 May 1997.

Ganymede, all of which have differentiated interiors with distinct layers. The gravitational tidal forces believed to have heated these other moons and led to their differentiation appears to have only "half-baked" Callisto and somewhat separated its constituents.[172]

According to Dr. Gerald Schubert, a UCLA planetary physics professor and Galileo gravity investigator, "Learning about the structure of these celestial bodies enhances our knowledge of how all planets and moons form and evolve, including our own Earth and Moon." Interior structure models of Jovian satellites were derived from radio Doppler data gathered by Galileo flybys. The moons' gravitational signatures, which are dictated by the densities and distributions of the materials inside them, affected the spacecraft's velocity and slightly altered the frequency of its radio signals. By analyzing those alterations, scientists were able to characterize the compositions and structures of the satellites.[173]

Magnetic Activity and a Possible Subsurface Sea

The initial analysis of plasma wave observations from Galileo's Callisto C3 encounter did not find evidence of a magnetic field or magnetosphere. Magnetometer studies of the satellite did not find such evidence either. The lack of a Callisto magnetosphere is consistent with its apparent lack of a metallic core. This is different from the situation at Ganymede, where a metallic core is likely responsible for that moon's self-generated magnetic field and magnetosphere.[174]

After evidence began to pour in supporting a subsurface ocean on Europa, data from Galileo's November 1996 (C3), June 1997 (C9), and September 1997 (C10) encounters were reexamined. Variations in Europa's Jupiter-induced magnetic field had been attributed to a possible brine ocean beneath its surface; researchers wondered if such a feature might also be found at Callisto. Dr. Margaret Kivelson, planetary science professor at UCLA and Galileo's magnetometer experiment principal investigator, and her colleagues found signs that Callisto did indeed have a variable magnetic field that could be explained by the presence of Jupiter-induced electrical currents flowing near the moon's surface. Kivelson's team set out to identify the source of the currents. Callisto was known to have an atmosphere, but it was extremely tenuous and lacking in charged particles that could carry an electric current. Scientists did not believe that it could generate the satellite's observed magnetic field. Callisto's ice crust could not be responsible for the field, either, for ice is too poor an electrical conductor. Because of the lack of other explanations for the observed magnetic field, a subsurface briny ocean on Callisto, such as likely exists on Europa, was suspected. Lending further credence to this theory, Galileo's observations showed that the electric currents induced in Callisto flowed in opposite directions at different times, a characteristic consistent with the existence of a brine ocean. "Until now, we thought Callisto was a dead and boring moon, just a hunk of rock and ice," said Kivelson. "The new data certainly suggests that something is hidden below Callisto's surface, and that something may well be a salty ocean."[175]

---

[172] Jane Platt (JPL), "Galileo Mission Finds Strange Interior of Jovian Moon," JPL Media Relations Office news release, 4 June 1998.

[173] Ibid.

[174] Isbell and Platt, "Galileo Returns New Insight Into Callisto and Europa."

[175] Douglas Isbell (NASA Headquarters) and Jane Platt (JPL), "Jupiter's Moon Callisto May Hide Salty Ocean," NASA news release 98-192, 21 October 1998.

Although the presence of a Europan ocean raises the possibility of life's past or present existence, scientists consider it far less likely that life might exist on Callisto. According to Dr. Torrence Johnson, Galileo project scientist, "The basic ingredients of life—what we call 'pre-biotic chemistry'—are abundant in many solar system objects such as comets, asteroids and icy moons. Biologists believe liquid water and energy are then needed to actually support life, so it's exciting to find another place where we might have liquid water. But, energy is another matter, and currently, Callisto's ocean is only being heated by radioactive elements, whereas Europa has tidal energy as well."[176]

Observations of Jupiter's Fields, Particles, and Atmosphere Made During the Satellite Tour

Some of the most interesting results reported by Galileo science teams during the satellite tour are related to the formation of the Jovian ring system. Jupiter has a main bright ring, a pair of faint "gossamer rings," and a cloudlike ring halo. The gossamer rings are located just outside the main ring; at its inner edge, the main ring merges gradually into the halo (see figure 9.5). Scientists believe that when interplanetary meteoroids and comet fragments collided with Jupiter's four small inner moons, Amalthea, Adrastea, Metis, and Thebe, they generated dust that formed the Jovian ring system. Galileo data also indicate that Amalthea's and Thebe's orbital characteristics are related to main- and gossamer-ring thicknesses.[177]

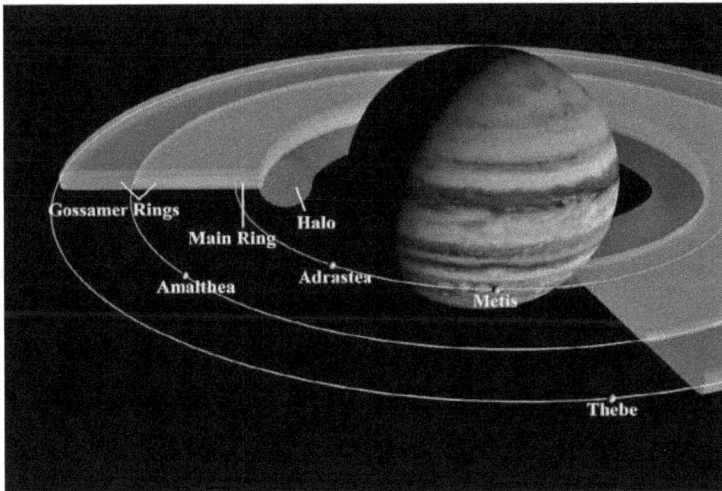

Figure 9.5. **Jupiter's inner satellites and ring components. (NASA image number PIA01627)**

[176] Ibid.

[177] Erickson et al., "Project Galileo: Completing Europa, Preparing for Io," p. 16; "Jupiter's Ring System," Rings Node Home Page, *http://pds-rings.seti.org/jupiter/jupiter.html* (accessed 7 September 2005) (available in folder 18522, NASA Historical Reference Collection, Washington, DC).

During GEM, Galileo was able to observe a unique Jupiter atmospheric event—the merging of two "white oval" storms. For half a century, three storms that appear as white ovals when seen from above have been observed in a band in Jupiter's equatorial region. At some time prior to the E17 orbit, two of the three ovals merged into one. Scientists could not determine the exact time that this occurred because Jupiter was in superior conjunction (i.e., behind the Sun from Earth's point of view). No viewing of the event was possible from Earth, and Galileo did not witness the event. Observations of the merged ovals were carried out during E17 by Galileo's near infrared mapping spectrometer, solid state imaging camera, and ultraviolet spectrometer. Prior to the merger, each oval had roughly two-thirds the diameter of Earth. The combined oval was as large as Earth and had characteristics that its parents did not. It was not visible at particular infrared frequencies, which indicated that it was not as deep as its parent storms. Those ovals had upwelling winds in their centers, with downwelling surrounding currents.[178]

Galileo also acquired the highest resolution pictures ever taken of Jupiter's auroral phenomena. This was accomplished during a campaign coordinated with the Hubble Space Telescope and instruments on Earth. The auroras seen in Galileo images were lower in the Jovian atmosphere than expected. Atmospheric scientists are generating new models to explain the auroral data supplied by Galileo and the other sources.[179]

Observations of Jupiter's Inner Moons

Four tiny moons orbit closer to Jupiter than the Galilean moons. Metis, the innermost, and Adrastea, the next closest, were discovered through analyses of Voyager data taken in 1979.[180] Amalthea, the third moon, was discovered by Edward Emerson Barnard in 1892. Thebe, like the two closest moons, was also discovered in 1979 through analysis of Voyager data. Table 9.13 lists characteristics of Jupiter's inner moons. The Jovian moons were named after women, boys, and nymphs who were associated with the Roman god Jupiter.[181]

Table 9.13. **Jupiter's inner moons.**[182]

SATELLITE	DISCOVERY	DISTANCE FROM JUPITER IN KILOMETERS (MILES)	DIAMETER OR DIMENSIONS IN KILOMETERS (MILES)
Metis	1979	128,000 (70,500)	40 (25)
Adrastea	1979	129,000 (80,000)	20 (12)
Amalthea	1892	181,300 (112,700)	270 x 166 x 150 (168 x 103 x 93)
Thebe	1979	222,000 (138,000)	100 x 90 (60 x 56)

---

[178] Ibid., pp. 8, 16.

[179] Mitchell et al., "Project Galileo: The Europa Mission," p. 13.

[180] Donald Savage (NASA Headquarters) and Guy Webster (JPL), "The Jupiter Millennium Mission," NASA press kit, October 2000, p. 14, *http://www.jpl.nasa.gov/jupiterflyby/documents/jupiterflyby.pdf*; Jane Platt, "New Views of Jupiter's Tiny Moons Are Online," JPL Media Relations Office news release, 24 April 2000.

[181] Savage and Webster, "Jupiter Millennium Mission," p. 14.

[182] "Inner Satellites and Rings," *JPL NASA Home Page, http://galileo.jpl.nasa.gov/moons/rings.html* (accessed 8 July 2001); "Jupiter's Inner Moons," from the University of Arizona's The Nine Planets, Students for the Exploration and Development of Space (SEDS) Web site, *http://seds.lpl.arizona.edu/nineplanets/nineplanets/amalthea.html* (accessed 8 July 2001).

There are dramatic differences in size between the Galilean moons (whose diameters range from 3,138 kilometers to 5,270 kilometers) and Jupiter's inner moons (the largest of which, irregularly shaped Amalthea, is only 270 kilometers long and 166 kilometers across).[183]

During the Galileo Europa Mission (GEM), project engineers adjusted the spacecraft's perijove so that Galileo would fly close by Io. This action also opened up the opportunity to study Jupiter's innermost moons in greater detail. Before the summer of 1999, each elliptical orbit of the spacecraft took it no closer to Jupiter's center than Europa's orbit, 1,100,000 kilometers (700,000 miles) distant. This path kept Galileo well away from the highest intensity magnetic fields and charged particles in the planet's radiation belts but also kept it quite far from the inner moons. The GEM perijove adjustment brought Galileo as close as about 400,000 kilometers (250,000 miles) from Jupiter's center. During August and November 1999, and again in January 2000, Galileo's solid state imaging camera obtained high-resolution pictures that revealed many features of three of Jupiter's innermost moons: Thebe, Amalthea, and Metis (see figure 9.6). These images are important because before Galileo, spacecraft and Earth-based telescopes were able to observe small moons such as Thebe and Metis only as specks of light.[184]

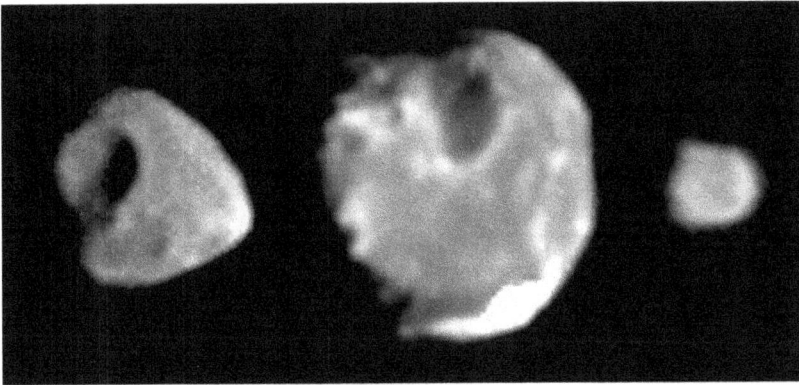

Figure 9.6. **Thebe, Amalthea, and Metis (left to right), taken in January 2000 by Galileo's solid state imaging camera. The images resolve surface features as small as 2 kilometers (1.2 miles). The prominent impact crater on Thebe is about 40 kilometers across. The large white region near the south pole of Amalthea is the brightest patch of surface material seen anywhere on these three moons. (JPL image number PIA02531)**

Previous spacecraft images had shown a round, bright spot on Amalthea. Galileo images taken during August and November 1999 revealed this "spot" to be a long, bright streak that was possibly formed by ejecta from a nearby meteoroid crater. It could also be the crest of a local ridge (see figure 9.7, where this streak appears in the upper left of each image).[185]

---

[183] Savage and Webster, "Jupiter Millennium Mission," p. 13.

[184] David Brand, "Galileo Takes Risky Trip to Dribble Back Data Revealing Best Images Yet of Jupiter's Cratered Inner Moons," Cornell University News Service, 24 April 2000, *http://www.news.cornell.edu/releases/April00/Simonelli.moons.deb.html.*

[185] Ibid.

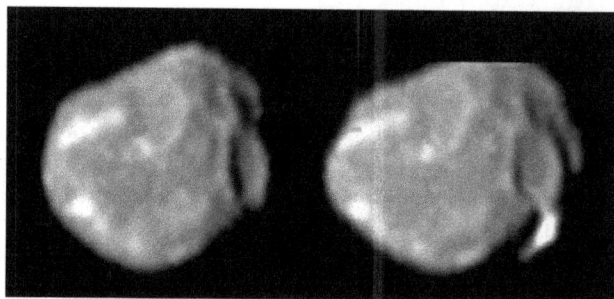

Figure 9.7. **Bright streak on Amalthea. Galileo's solid state imaging camera obtained the left image in August 1999 and the right in November 1999. The images show features as small as 3.8 kilometers (2.4 miles) across. The bright linear streak in the top left of the images is about 50 kilometers (30 miles) long. The large impact crater near the right-hand edge of the images is about 40 kilometers (25 miles) across. Two ridges, tall enough to cast shadows, extend from the top of the crater in a V shape resembling two rabbit ears. (JPL image number PIA02532)**

On 4 November 2002, on a trajectory called "Amalthea 34," Galileo flew closer than it ever had to Jupiter and Amalthea. Galileo's closest distance to the surface of Amalthea was 160 kilometers (99 miles). Scientists used data from the encounter in order to better determine the mass and density profile of Amalthea. Amalthea 34 observations, combined with previously determined shape and volume information, generated a bulk density (weight per volume) estimate near 1 gram per cubic centimeter, considerably lower than had been envisioned from the moon's dark albedo and its expected rocky composition. Scientists suggested a highly porous "rubble pile" of low-density rock or rock/ice mixtures, consistent with Amalthea's intense collisional past, as the most likely explanation for the low density.[186]

Just after the Amalthea encounter, Galileo's tape recorder failed to play back the data collected. This failure caused considerable grief for the project team because the Orbiter had not only collected potentially important Amalthea data, but it also had entered an extremely interesting region of Jupiter's magnetosphere. In this inner magnetosphere region, Jupiter resembles a star called a pulsar. The planet's magnetic field is so powerful in the inner magnetosphere that it imparts high energies to electrons trapped there and ejects them. Astronomers have detected pulsed radio signals indicative of this phenomenon. The Galileo spacecraft had just collected vital in situ data that could help our understanding of inner magnetospheric processes, but if the tape recorder could not be made to run, this information would never get to Earth.

After a detailed diagnostic investigation, the flight team concluded that the problem was not stuck tape, as had happened when Galileo was first approaching Jupiter in 1995, but was instead the result of radiation damage of one or more of the instrument's infrared light-emitting diodes (LEDs). The damage was believed to consist of the displacement of atoms in the LEDs' crystal lattices, which degraded the LED optical output to only

---

[186] T.V. Johnson and J. D. Anderson, "Galileo's Encounter with Amalthea" (abstract no. 7902 from the EGS-AGU-EUG (European Geophysical Society, American Geophysical Union, and European Union of Geosciences) Joint Assembly meeting held in Nice, France, 6–11 April 2003); Joseph A. Burns, Damon P. Simonelli, Mark R. Showalter, Douglas P. Hamilton, Carolyn C. Porco, Henry Throop, and Larry W. Esposito, "Jupiter's Ring-Moon System," in *Jupiter*, ed. Fran Bagenal, Tim Dowling, and Bill McKinnon (Port Melbourne, Australia: Cambridge University Press, 2004); JPL, "Galileo's Amalthea Flyby a 'Partial Success,'" Galileo Millennium Mission Status series, JPL Media Relations Office news release, 6 November 2002.

20 percent of its full power. Laboratory experiments suggested that for the tape recorder to run properly, the LED output had to be at least 50 percent.

The Galileo flight team conducted an exhaustive analysis of possible ways to work around the problem and developed a strategy that might partially repair the damaged lattices. JPL would send commands to the spacecraft to initiate electric currents passing through the LEDs. One member of the team in particular, Greg Levanas, had studied the tape-recorder design in considerable depth and believed that these currents would eventually cause some of the LEDs' atoms that had been physically dislodged by radiation to return to their original locations in their crystal lattices. Levanas's strategy did not immediately fix the LEDs, but after multiple applications of electric current, LED optical output increased to 60 percent, allowing the tape recorder to begin running again and download its stored data. Claudia Alexander, who later became the Galileo Millennium Mission Project Manager, called the recovery of the tape recorder after its being damaged by Jupiter's severe radiation environment "one of the finest technical accomplishments"[187] of the mission.[188]

The Demise of the Spacecraft

After fourteen years, Galileo is out of gas and it's going out with a bang!
—Gay Hill, "End of Mission" Webcast announcer[189]

Galileo's final orbit took it on an elongated loop away from Jupiter, from which it returned on 21 September 2003 to plow into the parent planet's 60,000-kilometer-thick atmosphere.[190] This demise was planned in order to avoid any chance that the spacecraft might strike and contaminate the moon Europa, where scientists believe that simple life-forms may exist. If such life-forms are discovered in future missions, scientists must be sure that they are not Earth organisms that were accidentally carried to Europa aboard Galileo.[191]

On 21 September 2003, hundreds of current and former Galileo project members and their families converged on JPL "for a celebration to bid the spacecraft goodbye."[192] Through speeches and toasts by project staff, the event attempted to put closure on the mission and to articulate its specialness. One of the memorable comments was by T. V. Johnson, project scientist, who said that "we haven't lost a spacecraft; we've gained a new stepping stone to the future." This echoed the sentiments of others who considered the mission's value to be

---

[187] Claudia Alexander interview, telephone conversation, 31 October 2003.

[188] Gregory C. Levanas, systems engineer, interview, Pasadena, CA, 21 September 2003; G. Levanas to Claudia Alexander and Tiffany Chiu, "Galileo OP133 Infrared Radiation Presentation," JPL Interoffice Memorandum 3483-014-2003, 25 August 2003 (provided to author by Claudia Alexander, Galileo Millennium Mission Project Manager); Gary M. Swift, Gregory C. Levanas, J. Martin Ratliff, and Allan H. Johnston, "In-Flight Annealing of Displacement Damage in GaAs LEDs," JPL internal presentation, 2003 (exact date not known) (provided to author by Claudia Alexander); Claudia Alexander e-mail, 11 September 2003; Alexander interview, 31 October 2003.

[189] Gay Hill, announcer on JPL's "End of Mission Webcast," originally Webcast on 21 September 2003, accessible from *http://www.jpl.nasa.gov/webcast/galileo/.*

[190] Webster, "Galileo Gets One Last Frequent-Flyer Upgrade"; Donald Savage (NASA Headquarters) and Guy Webster (JPL), "The Jupiter Millennium Mission," NASA press kit, October 2000, p. 13, *http://www.jpl.nasa.gov/jupiterflyby/documents/jupiterflyby.pdf*; Porter Anderson, "Eilene Theilig: Faith, by Jupiter," *CNN.Com/CAREER* Web site, 16 July 2001, *http://www.cnn.com/2001/CAREER/jobenvy/07/16/eilene.theilig/index.html.*

[191] "Galileo Spacecraft May Be Sent Crashing—But Not Soon," San Francisco Chronicle "SF Gate" online news service, Breaking News section, 2 March 2000, *http://www.sfgate.com/.*

[192] JPL, "Galileo End of Mission Status, September 21, 2003," news release, *http://www.jpl.nasa.gov/releases/2003/129.cfm.*

not only in its discoveries and engineering advances, but also in the path that Galileo blazed toward future exploration. For instance, NASA Administrator Sean O'Keefe commented in a phone call to those assembled at JPL that Galileo was a "marvelous chapter of NASA's exploration history" and that "we can utilize this great set of achievements to an even larger end." In particular, the probable discovery of a warm saltwater ocean under Europa's ice was such a compelling find that additional visits to the moon are already being planned.[193]

The spacecraft disintegrated in Jupiter's dense atmosphere at 11:57 a.m. PDT on 21 September. However, the Deep Space Network tracking station in Goldstone, California, did not receive the craft's last signal until 12:43 p.m. PDT, due to the time delay for the signal to travel to Earth. During the last seconds of Galileo's existence, the audience at JPL's Von Karman Auditorium counted down as if this were a launch, not the demise, of a spacecraft. Later in the day, past and present Galileo staff toasted the spacecraft (with soft drinks and water). One of the most memorable toasts was given by Claudia Alexander, Galileo's last Project Manager: "Following Balboa—who found an ocean—Galileo found another ocean. Here's to you and all who follow in your path."[194]

An aesthetic view of the spacecraft was given in a toast by Theodore Iskendarian, an engineer on the project. He said that "Galileo was a work of art—especially the dual spin, the antenna, the feathery umbrella antenna. Not a brick-like spacecraft."[195] Former Galileo Project Manager William O'Neil also expressed his view of Galileo's appearance in his Webcast comment, "It's without equal . . . and it's beautiful besides."[196]

While the spacecraft was most in the spotlight during the celebration (including in a musical revue), the Galileo team got many kudos as well. In the words of Greg Levanas, Galileo systems engineer, "We as a team would not take no for an answer. And the spacecraft just kept going."[197] Former Project Manager Eilene Theilig spoke in the same vein when she said, "I think this mission was special because of all the obstacles that arose. People drew together to surmount them."[198] And in a phone message to all those assembled in JPL's Von Karman Auditorium, NASA Administrator Sean O'Keefe called the mission "an extraordinary job well done . . . a testimonial to the persistence NASA demonstrates."[199]

Two former Galileo project managers summed up the mission very concisely. In his toast, Jim Erickson said, "Everything we did was on the backs of the guys and women who went before. Not only did we have a great mission, but we went out in style." And in Claudia Alexander's words, "We learned mind-boggling things. This mission was worth its weight in gold."[200]

---

[193] T. V. Johnson and Sean O'Keefe (NASA Administrator), speaking at the Galileo end-of-mission event at JPL, Pasadena, CA, as reported by author, 21 September 2003.

[194] Claudia Alexander, speaking at the Galileo end-of-mission event at JPL, Pasadena, CA, as reported by author, 21 September 2003.

[195] Theodore Iskendarian, cognizant engineer for Galileo's linear boom actuator, interview, JPL, 21 September 2003.

[196] William O'Neil, former Galileo Project Manager, speaking on the "End of Mission Webcast."

[197] Greg Levanas, systems engineer, speaking on the "End of Mission Webcast."

[198] Eilene Theilig, former Galileo Project Manager, interview, 21 September 2003.

[199] Sean O'Keefe, speaking at the Galileo end-of-mission event at JPL, Pasadena, CA, 21 September 2003.

[200] Jim Erickson and Claudia Alexander, speaking at the Galileo end-of-mission event at JPL, Pasadena, CA, 21 September 2003.

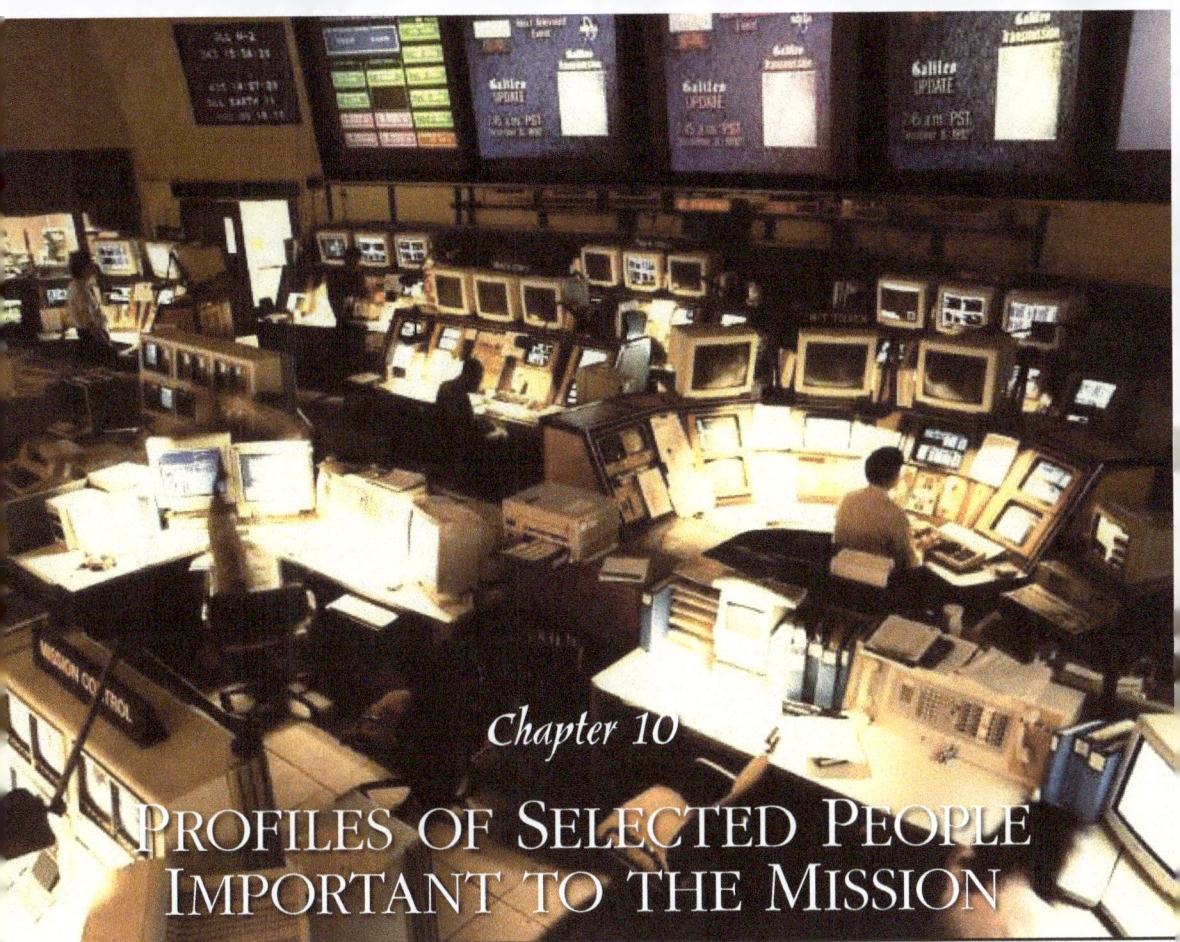

Chapter 10

PROFILES OF SELECTED PEOPLE
IMPORTANT TO THE MISSION

THE GALILEO PROJECT'S MANY ACHIEVEMENTS WOULD NOT HAVE been possible without an exceptional team that included a succession of notable project managers, scientists, engineers, and mission support personnel. The Galileo team was drawn from different NASA sites and a range of universities, research laboratories, and institutes. Earlier chapters, especially chapters 3 and 5, describe the significant contributions from this wide spectrum of organizations.

At the heart of the Galileo team were tremendously dedicated people who were key to the mission's success. It is beyond the scope of this book to document the individual contributions of all, or even a large fraction of the people responsible for Galileo's achievements. This chapter does, however, examine the experiences of several key members of the team who have articulated the mission experience and the importance of the work done by the Galileo team.

John R. Casani: Galileo Project Manager, 1977–88

I love the interaction with these people. Brilliant, creative people, with lots of challenges. Still trying to do stuff that's never been done before.

—John Casani in an interview with the author on 29 May 2001

When John Casani was a 20-year-old student at the University of Pennsylvania in 1952, he read a series of articles in Colliers' magazine that had a pivotal effect on his life. These were articles by Wernher von Braun, the German rocket engineer who came to the United States after World War II. Von Braun's articles promoted space exploration. He touted the construction of space stations and flying to the Moon, and his visions of what was possible made a deep impression on Casani. In 1956, a year after Casani graduated from the University of Pennsylvania with a bachelor's degree in electrical engineering, he hunted down a job at the Jet Propulsion Laboratory. Although he initially worked on inertial guidance systems for guided missiles, his position at JPL would give him the chance to help create some of the projects that von Braun had only imagined.[1]

At the time that Casani started work at JPL, NASA had not yet been formed. JPL was then a contractor to the Army's Ballistic Missile Agency in Huntsville, Alabama. After Congress created NASA, the new Agency gave various missions to its different Centers. Right from the beginning, JPL's mission was to develop interplanetary space travel, and Casani was deeply involved in this effort. Early Casani assignments at JPL included work on the Explorer, Pioneer, and Ranger missions. Explorer 1 was the first U.S. Earth satellite, which launched in January 1958 and discovered a radiation belt around Earth. Pioneer 3 discerned the two bands of radiation that encircled Earth. Pioneer 4 passed within 37,000 miles of the Moon before achieving a permanent solar orbit.

Even though his background was in electrical engineering, Casani soon transitioned into systems engineering, serving as the lead design engineer on the Ranger-class spacecraft and as the team lead on the Ranger 1 and 2 missions. Later, Casani was also the lead design engineer on the Mariner-class spacecraft and the team lead on the Mariner 3 and 4 missions. These were the first uses of the Mariner-class spacecraft. (The first Mariner missions, Mariner 1 and Mariner 2, actually used the older Ranger-class spacecraft design.)

Casani soon moved up the management chain into positions of increasing responsibility. On the 1969 Mariner Mars project, Casani was the Deputy Spacecraft System Manager; later, on the 1973 Mariner Venus-Mercury project, Casani served as Spacecraft System Manager. In April 1976, he assumed the role of Project Manager for the highly successful Voyager mission to the outer planets; the mission's two spacecraft launched from Cape Kennedy in August and September 1977.[2]

Casani's Management of the Galileo Mission

As Casani and his boss, Bob Parks, JPL's Assistant Laboratory Director for Flight Projects, were driving to the airport to return to Pasadena after the Voyager spacecraft launches, Parks offered Casani a new assignment. Would he manage what was then called the Jupiter Orbiter Probe, or JOP 77? Casani said yes. But one of the first things he did after assuming the Project Manager post was to solicit suggestions from his staff for a new mission title. Planetary missions were traditionally given names such as Mariner Venus 62 and Mariner Mars 64, and Casani thought that "all those letters and numbers lose their significance to anybody except people who work"[3] on the project. His staff

[1] John Casani interview, tape-recorded telephone conversation, 29 May 2001; "Biography of John C. [sic] Casani," John Casani Collection, Galileo Correspondence February 1978–March 1978, folder 42, box 5, JPL 14, JPL Archives.

[2] "Biography of John C. [sic] Casani"; "Early NASA Aeronautics and Astronautics Chronology (Pre-Apollo)," sections 1958 and 1959, and "Aeronautics and Astronautics Chronology," section 1961, all on the NASA History Division Web site, http://www.hq.nasa.gov/office/pao/History/timeline.html, excerpted from Eugene M. Emme, *Aeronautics and Astronautics: An American Chronology of Science and Technology in the Exploration of Space, 1915–1960*, (Washington, DC: NASA, 1961).

[3] Casani interview, 29 May 2001.

most frequently recommended the name "Galileo," which in Casani's view was especially appropriate because it was the astronomer Galileo who had discovered Jupiter's four largest moons and because the Galileo mission's primary objectives included "the study of Jupiter and the Galilean satellites."[4]

Table 10.1. **Galileo Project managers.**[5]

John Casani	October 1977–February 1988
Dick Spehalski	February 1988–March 1990
Bill O'Neil	March 1990–December 1997
Bob Mitchell	December 1997–June 1998
Jim Erickson	June 1998–January 2001
Eilene Theilig	January 2001–August 2003
Claudia Alexander	August 2003–September 2003

Other managers of the Galileo project (see table 10.1) considered Casani to be a strong leader who could command the respect of both the congressional and scientific communities. These skills were absolutely essential to Galileo, for the project was nearly killed several times by budget-conscious members of Congress and presidential staffs (see chapters 2 and 3). John Casani made many trips to Washington, DC, to present his views and garner support for the mission. He lauded the valuable science findings that Galileo would produce and stressed the importance of keeping planetary exploration alive. He made government officials in Washington aware that JPL had no other major in-house missions now that Voyager was on its way. If Congress did not approve the Galileo mission, the loss of the project would put JPL in danger of losing 400 very highly qualified engineers and scientists who were counting on participating in the project. Casani stressed that JPL was a critical national resource that would be severely damaged if Congress killed Galileo.[6]

Casani also played on the space program's appeal to the United States public. He reminded members of Congress of the excitement that Voyager's images had generated. When dealing with presidential staff, Casani adjusted his approach to the particular aims of the administration at the time. President Reagan, for instance, strongly supported Strategic Defense Initiative ("Star Wars") military defense technology. Casani capitalized on this by pointing out that Galileo was going to travel to Jupiter, over 400 million miles away, with an expected navigational accuracy that was absolutely remarkable. Casani

---

[4] Gia Scafidi, "John Casani Awarded Honorary Doctorate Degree," JPL Media Relations Office news release, 31 May 2000; Casani interview, 29 May 2001; "A Name for Project Galileo," JPL interoffice memo GLL-JRC-78-53, 6 February 1978, John Casani Collection, Galileo Correspondence 12/77–2/78, folder 43, box 5 of 6, JPL 14, JPL Archives. Quotation is from "A Name for Project Galileo."

[5] Guy Webster, JPL Press Office, telephone conversation, 29 August 2001; J. R. Casani to multiple addresses, "Revised Guidelines for Galileo Implementation," 30 January 1978, JPL Interoffice Memo GAL-JRC-78, folder 18522, NASA Historical Reference Collection, Washington, DC; Claudia Alexander, e-mail message, 11 September 2003.

[6] Casani interview, 29 May 2001.

impressed upon Reagan's administration that such a feat would, in a very nonthreatening way, convey to people around the world the level of technological capability that the United States had. Other countries would think that if the United States could achieve such an amazing feat with a civilian enterprise, imagine what it could do in a military situation. Galileo provided a way of demonstrating, without building military equipment, the technological strength of our country.[7]

The *Challenger* accident in 1986 resulted in millions of people around the nation and the world sharing a sense of sorrow and personal loss. As JPL employees, Casani and his staff felt this loss very intensely. But for them, there was also the possible loss of all they had been working toward for many years. "We had so much invested in the mission from 1977 through 1986,"[8] said Casani. After *Challenger's* explosion, NASA management delayed Galileo's launch from May 1986 until June 1987 and then, due to safety concerns, canceled the use of the Centaur liquid-fueled upper stage in the Space Shuttle. This was a low point for Galileo's staff, who feared that the spacecraft might never get off the ground. Casani's task was to keep his staff's morale up during this tremendously difficult time and focus its efforts on redesigning the mission and keeping it on track. The Centaur had to be replaced with a less powerful solid-fueled upper stage, but project engineers had not yet figured out how such an upper stage would be able to get Galileo to Jupiter.

The eventual discovery of the six-year VEEGA trajectory, in which Galileo first flew toward the Sun and then looped around Venus for a gravity-assist, buoyed Casani's spirits and those of his staff. "The most exciting time in my life," he said, began with "the realization that there was another way of doing [the mission] using the gravity assist of Venus."[9]

Casani and his staff successfully put the Galileo project back on the path toward a launch. By 1988, Casani decided that "in terms of the [Galileo] mission, everything was very, very secure"[10] and accepted an excellent opportunity to enter the upper management structure of JPL, eventually becoming its Assistant Laboratory Director for Flight Projects.[11]

The Robotic Versus Human Spaceflight Debate

An endless debate in the United States space science community focuses on why NASA puts the majority of its budget into human spaceflight, which is extremely expensive, when considerably more science per dollar can be achieved using robotic spacecraft. Although Casani participated in many robotic missions and managed several of them, he saw the need for both types of missions. He understood the less tangible benefits offered by a human mission, over and above the science findings. Casani's understanding of the value of human expeditions was illustrated in the following statement:

---

[7] Casani interview, 29 May 2001.

[8] Ibid.

[9] "From the Project Manager," *Galileo Messenger* (February 1986); "From the Project Manager," *Galileo Messenger* (July 1986); Casani interview, 29 May 2001.

[10] Casani interview, 29 May 2001.

[11] Dick Spehalski, e-mail message, March 2002.

To my mind, the debate is one of why do we have a NASA? The S in NASA doesn't stand for science. You can argue that everything should only be evaluated in terms of science. My own personal view is that NASA's a part of a president's armory for executing national and foreign policy and it can be used any way the president sees fit to use it. And many presidents have felt, and I agree, that it's a powerful mechanism for motivating and inspiring people. Young people and older people alike . . . . Kids and people like heroes. What if instead of Admiral Byrd going to the South Pole, it was just an automated sled? It might have been able to do more science than he could, but it wouldn't have inspired the kind of interest and public reaction that [Byrd] did. Having heroes is an important part of any enterprise like this.[12]

Our Space Program's Impact

One thing that gave Casani satisfaction over the years was the importance and interest that so many people in the world assigned to the U.S. space program. Once, on a street in Argentina, a man called out to him by name. The man, an Argentinian lawyer, followed our space program in great detail, and was very pleased to actually meet Casani and welcome him to Argentina. On another occasion, one of Casani's colleagues was on a wildlife safari in Africa, during which his guide commented on the similarities between their all-terrain vehicle and NASA's Mars Rover vehicles. These were not isolated incidents. Casani and his colleagues regularly met men and women in many different countries who were not involved in the U.S. space program but were well aware of what the program was doing. Sometimes, these people even knew intricate details about the particular equipment used on different space missions because those missions were of high interest to them.[13]

Casani believes that young people are inspired by our space program, just as he was inspired as a 20-year-old by Wernher von Braun's articles. "There are a lot of people in science and technology now because of the Apollo program," Casani says. "I'm a person who happens to believe that our whole world is technology-oriented. Anything we can do to increase people's understanding, interest, or thirst for technology is good."[14]

Future Plans

As to the future, Casani plans on working for JPL and the space program as long as he can "continue to contribute a little bit." He retired in 1999, but shortly afterwards, JPL management asked him to return to the Laboratory to lead the failure investigations of NASA's two ill-fated Mars missions. When that was done, NASA and JPL management asked him to advise on a range of other issues. Daniel Goldin, NASA Administrator at the time, invited him to sit on the Agency's prestigious Advisory Council. Casani says that if he can persist in remaining "relevant," he will keep on trying to help the space program in whatever way he can.[15]

---

[12] Casani interview, 29 May 2001.

[13] Ibid.

[14] Ibid.

[15] Ibid.

Torrence V. Johnson: Galileo Project Scientist Since the Beginning of the Mission

Galileo gave us a chance to explore the entire Jupiter system . . . looking at the interactions of all parts of the system and not just . . . one piece of the elephant.

—Torrence Johnson, Galileo project scientist, in an interview with the author on 31 July 2001

As Galileo's project scientist, Torrence "T. V." Johnson served as the mission's primary science advisor and the principal interface between project staff and investigators on science experiments. The mission's project scientist was also an ex officio member of all of Galileo's science teams. The criticality of Johnson's position was exemplified by the responsibilities that fell to him after Galileo's high-gain antenna failed to open. The Galileo team had to face the fact that the spacecraft was no longer capable of sending back all the data that scientists on Earth had been hungrily awaiting, some for as many as 20 years. As project scientist, Johnson needed to decide who got what. "The hardest thing I had to do as a project scientist was to figure out which objectives had the highest priority and how we were going to recover from not being able to send back all the data we wanted. I had to basically give the different science areas priority ratings, and give them allocations. Say, you're only going to get 30% of the data. See what you can do with it."[16]

Johnson's involvement with Galileo dated back to the mid-1970s, when he was a member of NASA's Mariner Jupiter Uranus planning group and the Jupiter Orbiter Probe (JOP) science working group. The JOP group developed the scientific rationale for the project that would later be called Galileo and was key in convincing Congress to fund the project.[17]

T. V. Johnson's excitement about space science, and especially about NASA interplanetary missions, is infectious. He has refused speaking engagements where he was given only 1 hour to talk about Galileo's most important accomplishments. He sums up how the space program makes our lives better by first saying that it is not because such inventions as Tang and Teflon came out of it. The value in exploring space is much grander than that. By studying Io's volcanoes or Jupiter's Great Red Spot or a black hole in a distant galaxy, "you are trying to use the universe as a laboratory to test out your theories which were developed here on Earth under very different conditions. That improves those theories . . . . It's not going to put more chicken in the kitchen but it is connective with the general human enterprise of trying to understand how the universe works."[18]

Space science, in Johnson's view, is simply an extension of our basic humanity. That is reason enough to conduct it. He is aware, however, of the tangible benefits that frequently surface from our quest to understand "how planets tick." For instance, a friend of his who pursued space science because of a fascination with Jupiter's clouds is now using the knowledge of atmospheric chemistry that he gained to do some of the world's leading research into ozone-depleting chemicals.[19]

---

[16] "Meet the Team," *Galileo Messenger* (December 1984): 4; Torrence V. Johnson interview, tape-recorded telephone conversation, 31 July 2001.

[17] "Meet the Team."

[18] Johnson interview, 31 July 2001.

[19] Ibid.

Dick Spehalski: Galileo Project Manager, 1988–90

If we had launched in 1986, if the Challenger had worked and we had been able to launch on schedule, we would have found out that [Galileo's] thrusters would fail. We would probably have lost the mission.

—Dick Spehalski, former Project Manager of Galileo

The most exciting moment of the Galileo mission for Dick Spehalski was the launch, or, more specifically, the countdown before the launch. "Spe" had been on the project since its official start date in 1977 and, in his managerial functions under John Casani, had helped work through one potentially mission-ending barrier after another. Now that the Galileo spacecraft was finally on the launchpad, tucked carefully inside the Space Shuttle, the countdown to takeoff seemed incredibly long. Spehalski's adrenaline got him through those final hours, he said. He had been waiting for 12 years for this day.

A year earlier, in February 1988, when Casani left the project, Spehalski had been chosen as his successor because JPL management viewed him as the best person to get Galileo off the ground and into space. Spehalski wanted to get the launch behind him. When the Shuttle Atlantis finally fired its rockets and carried Galileo aloft, then successfully deployed the spacecraft, Spehalski felt totally exhilarated but almost exhausted. "At the end of it," he said, "you feel like gee, I was just put through this wringer and I don't have any energy left." His work was not over, however. "Now we've got to fly this thing. It doesn't ease up."[20]

During John Casani's tenure, Dick Spehalski had been the Flight System Integration Manager. Spehalski had been responsible for mechanical engineering tasks in the design and construction of the Galileo Orbiter. It also had been his job to develop smooth interfaces between the Orbiter and its Atmospheric Probe, the Centaur upper stage launch vehicle (which was scrapped after the *Challenger* disaster), the spacecraft's propulsion system (used to maneuver the craft in interplanetary space and in the Jovian system), and the radioisotope thermal generator (RTG) electric power system. Spehalski had to work closely with industry contractors building Galileo's components; with the U.S. Department of Energy, which designed and fabricated the RTGs; and with the Federal Republic of Germany, which designed and built the spacecraft's propulsion system.[21]

Working with another country that was building a critical part of the spacecraft but was not under JPL's direct control was one of the most difficult challenges that Spehalski faced during the entire mission. He does not hold back in expressing strong opinions, either verbally or in his publications.[22] Spehalski criticized the decision to include a major international partner on Galileo in so vital a role. Although he recognized the additional capabilities that Germany brought to the mission, the cultural differences, along with varying ways of doing business, a lack of frequent face-to-face meetings, and the dissimilar motivations of the two governments led to serious problems.

---

[20] Dick Spehalski, interview, tape-recorded telephone conversation, 4 May 2001; Dick Spehalski, e-mail message, March 2002. Quotations are from the interview.

[21] "Meet the Team," *Galileo Messenger* (December 1982).

[22] R. J. Spehalski, "Galileo Spacecraft Integration: International Cooperation on a Planetary Mission in the Shuttle Era," *Earth-Orient. Applic. Space Technol.* 4, no. 3 (1984): 139–150.

The German-built propulsion system was an absolutely critical part of NASA's spacecraft, but its design and construction were not under NASA's direct control. NASA had made a cooperative agreement with Germany's Bundesministerium fuer Forschung und Technoligie (BMFT), which used a local industrial contractor, Messerschmitt-Boelkow-Blohm, to fabricate Galileo's propulsion module. The German organizations resisted NASA's and JPL's desires to set requirements for the propulsion system's design, guide its construction, and implement inspection schedules for verifying that the system was reliable for flight. In Spehalski's words, "the challenge is to get them to do what you want done with their money and resources . . . . This challenge leads to a lot of international harangue and argument . . . ."[23] A great deal of unplanned time was spent by both parties negotiating requirements and reviewing the design and test results, thereby causing significant cost and schedule impacts.[24]

Dick Spehalski's most difficult period on the Galileo mission began when he found out, one year before launch, that the thrusters Germany had built for the spacecraft had a serious problem. Discovering this was a strange benefit of the *Challenger* disaster and the delay in Galileo's launch resulting from it. If Galileo had been sent into space in 1986 as planned, NASA probably would have lost the mission.

Galileo's thrusters had a critical design flaw—the hot gases generated for propulsion burned holes through the walls of their combustion chambers. Germany had agreed to build Galileo's thrusters partly to develop a product that it could market for other applications. One of these applications was in communications satellites. It was after such a satellite was launched that the thruster problem was discovered. The satellite's thrusters overheated and failed, catastrophically, with the resultant loss of that mission. After that event, NASA's German partners tested Galileo's thrusters and found that they were susceptible to the same problem.[25]

Spehalski, his staff, and their German partners went through an extremely hectic period after the thruster problem was identified. The thrusters had to be made reliable in order for Galileo to be sent into space. Germany's BMFT, its industrial partner Messerschmitt-Boelkow-Blohm, and JPL staff had to find the root cause of the thruster failure and come up with a fix for the problem that satisfied JPL's requirements sufficiently to give the mission a green light to launch. And they had to do this in only one year, because management did not want to delay the launch yet again. But Spehalski added another task that increased the pressure on his staff even more. The team had to develop a backup design for a thruster system, in case Germany could not make their existing thrusters flight-ready in time for the launch. Spehalski explained that he "didn't want the mission held hostage to whether or not the Germans were going to be successful."[26] So he contracted with a supplier in southern California who made thrusters for the Air Force, and he negotiated to get a high priority within its production facilities. His staff quickly completed the backup thruster design during "a very stressful period. At JPL, designers were going full bore on the backup design while in Germany, the German team with JPL propulsion specialists' support were

---

[23] Spehalski interview, 4 May 2001; Spehalski e-mail message, March 2002.

[24] Spehalski, "Galileo Spacecraft Integration."

[25] Spehalski interview, 4 May 2001; Spehalski e-mail message, March 2002; Jim Erickson interview, tape-recorded telephone conversation, 8 May 2001

[26] Spehalski interview, 4 May 2001.

going full bore trying to solve the problem with the thrusters. Ultimately the Germans came through by finding a way to use the thrusters without overheating them. We flew the mission without incident."[27]

Bill O'Neil: Galileo Project Manager, 1990–97

Some of the senior engineers in the Probe area, when that signal came to say that everything worked, they actually were crying.

—Bill O'Neil, former Project Manager of Galileo

For Bill O'Neil, who was Galileo's Project Manager from 1990 until 1997, nothing on the mission was more exciting than Jupiter Arrival Day, especially the Probe's plunge into the Jovian atmosphere. "All the chips were on the table that day . . . . For five months we'd had no idea of the condition of the Probe." Due to Jupiter's enormous mass and gravity, it was extremely difficult to plan a successful entry into its thick atmosphere without the Probe falling so fast that it would burn up before taking all of its measurements. O'Neil believed that the entry maneuver was more difficult than it would have been at any other planet in the solar system. "The Probe had to survive this tremendous heating, and the entry loads that were estimated could go up as high as 350 g's. The parachute would have to work properly, and the separation of the heatshield would have to be proper." It was not until the Probe was on parachute and transmitting that scientists on Earth could confirm that the atmospheric entry had been successful. O'Neil remembers the pandemonium at JPL when this information was received. "Some of the senior engineers in the Probe area, when that signal came to say that everything worked, they actually were crying."[28]

O'Neil, who studied aeronautical engineering in college, was hired by JPL in 1963. He worked on the Surveyor Moon lander, the Mariner Mars 1971 orbiter, and the Viking Mars landers and orbiters prior to transferring to the Galileo project in 1980.[29] O'Neil brought key experience from his other missions that helped him on Galileo. On the Lunar soft-landing Surveyor mission, he learned intricate details of launch systems and trajectory design. His association with Hughes Space and Communications Company during the Lunar Surveyor project was useful because Hughes also built Galileo's Atmospheric Probe under contract to Ames Research Center, and O'Neil once again had to work closely with the company. O'Neil's tenure on the Viking mission was advantageous because the project involved a challenging atmospheric entry and an orbital insertion similar to those on the Galileo mission.[30]

When O'Neil joined the Galileo team in 1980, he assumed the position of Science and Mission Design Manager. In that capacity, he became a key figure in planning out details of the project. He was also responsible for interfacing with Galileo scientists and developing the science operations that the spacecraft would perform.

---

[27] Ibid.; Spehalski e-mail message, March 2002. Quotation is from the interview.

[28] Bill O'Neil interview, tape-recorded telephone conversation, 15 May 2001.

[29] "Meet the Team," *Galileo Messenger* (May 1982): 1.

[30] O'Neil interview, 15 May 2001.

In O'Neil's opinion, Galileo's premier accomplishment was the discovery of strong evidence for liquid water under Europa's ice crust. About this discovery, O'Neil said, "I find it absolutely ironic that the most significant thing that Galileo achieved has also given Galileo the death sentence. That it must be crashed into a body to make sure that it will never strike Europa." Galileo's observations of Europa raised the possibility that the moon harbored life. To make sure that Galileo would never contaminate the moon's ocean with microorganisms brought from Earth, in 2003 the spacecraft was sent at high speed into Jupiter's thick atmosphere, where it met a fiery demise.[31]

The most unpleasant task that O'Neil ever had to perform on Galileo occurred soon after its tape recorder jammed, a couple of months before arrival at Jupiter. Mission engineers got the recorder working again, although they did not know for how long. It was clear to O'Neil what he had to do to protect the tape recorder, but taking the appropriate actions was difficult because he knew that they would have hurtful consequences. He had to ensure that the tape recorder remained functional for its most important Arrival Day function—capturing data from the Atmospheric Probe's descent. That meant not risking its use on other extremely important operations.

"It was the worst decision that I ever had to make. I had to say there was no way we could do the approach imaging"[32] of Io and Europa that had been planned since the inception of the project. In fact, mission planners had picked 7 December 1995 as Arrival Day because of the excellent opportunity to view both moons before Jupiter Orbit Insertion took place, a maneuver that carried some risk of failure. These two satellite encounters had been precisely calculated and much anticipated, all the more so because no one knew whether Galileo would have a chance to view the moons closely again. The intense radiation around Io might preclude any more flybys in the future. "It fell to me to tell the scientists, without any debate—there was no point in debate as far as I was concerned"—that the images and other data for which they had been waiting could not be stored on the tape recorder and thus would be lost.[33]

A detailed analysis of the tape-recorder problem after Jupiter Arrival Day indicated that O'Neil's decision had been the right one. According to O'Neil, "After the fact, when we fully diagnosed [the tape recorder problem], we know that [the recorder] absolutely would have failed, had we tried to do that imaging." Such a failure would have prevented most of the Probe's descent data from ever being sent to Earth.[34]

One of the tasks that O'Neil was thankful he did not have to perform was declaring the mission a failure. "Can you imagine the circumstance that we had the whole world, at least a lot of the nations and Europe, watching our day of arrival, and we had the auditorium at JPL all full of VIPs and we had the senior, the top management of NASA. Imagine if I had to go to the podium at the end of the day . . . and tell them that the mission had failed! That would clearly have been the most difficult task."[35]

---

[31] Ibid.

[32] Ibid.

[33] Ibid.

[34] O'Neil interview, 15 May 2001.

[35] O'Neil interview, 15 May 2001.

Bob Mitchell: Galileo Project Manager, 1997–98

When the Challenger explosion accident occurred, we found ourselves in a situation . . . where we didn't have any way to get to Jupiter . . . . Now we just took this on as kind of our own personal challenge.

—Bob Mitchell, former Galileo Project Manager

After the *Challenger* tragedy took place in January 1986 and after NASA management decided, a few months later, that it was too dangerous to fly the Centaur upper stage in the Shuttle, Galileo mission staff found themselves in trouble. The solid-fueled alternative upper stages to the Centaur did not have enough thrust to propel Galileo on a direct trajectory to the Jovian system.[36]

Bob Mitchell was Galileo's Mission Design Manager and Navigation Team Chief at the time. His job was to figure out a way for the spacecraft to reach Jupiter. Given the constraints imposed, Mitchell's task appeared to be extremely difficult, if not impossible. In his view, "most people had pretty much accepted that the Galileo project, the spacecraft, was going to make a one-way trip to the Smithsonian."[37] In other words, the mission was on the verge of getting scrapped.[38]

Galileo management desperately needed its staff to find a spacecraft trajectory that would allow the mission to be completed using a solid-fueled upper stage. To meet this goal, management brought together a group of four people with some of the best trajectory mechanics knowledge at JPL: Dennis Byrnes, Louis D'Amario, Roger Diehl, and their leader, Bob Mitchell. They made it their personal mission to design a trajectory with sufficiently low energy requirements that a solid-fueled booster could drive Galileo to Jupiter. The team conducted numerous brainstorming sessions in which they developed "really far-out, unconstrained ideas."[39] They envisioned various gravity-assists that might give the spacecraft the extra energy it needed. But the team had to abide by one constraint on the process that proved to be very frustrating. They had been told by their managers that Galileo could not fly closer to the Sun than just inside Earth's orbit. Galileo designers worried that if the spacecraft flew any closer to the Sun, it would receive more thermal input than its instruments could withstand.[40]

After a month of "pretty serious brainstorming and head-scratching," Mitchell's team concluded that there was no trajectory that was going to work, given the constraints imposed. So one day, he said to the group, "Forget about this inside-of-the-Earth constraint. If you need to go inside Earth's orbit, let's do that." They all went off and did some more calculations, and one morning, Roger Diehl came into Mitchell's office with a new idea. The team had been considering gravity-assists by Earth, Venus, and other bodies, but what Diehl envisioned was a Venus gravity-assist followed by two Earth assists—in other words, a "VEEGA" trajectory. The thing that was new about Diehl's scheme, and the

---

[36] Bob Mitchell interview, tape-recorded telephone conversation, 10 May 2001.

[37] Ibid.

[38] Dick Spehalski commented that statements about the spacecraft ending up in the Smithsonian Institution were more "figurative than literal" in that if Galileo had not been launched, he did not think it would actually have been displayed in the Smithsonian. In an e-mail message to the author (March 2002), Spehalski wrote that although "people made such comments," if Galileo had not been launched, "the hardware most likely would have been used on other missions."

[39] Mitchell interview, 10 May 2001.

[40] "Meet the Team," *Galileo Messenger* (August 1982): 4; Mitchell interview, 10 May 2001.

reason Mitchell thinks that nobody had thought of it before, was that the second Earth encounter would not give Galileo any more energy. But it turned out that the spacecraft did not need additional energy; what it needed was to have its trajectory bent in a different direction. The second Earth encounter gave Galileo just the bending required to send it in the right direction.[41]

Mitchell called Bill O'Neil, who was then the science and Mission Design Manager, and relayed Diehl's idea to him. O'Neil and Casani, Galileo's Project Manager, agreed that a VEEGA trajectory could work. At that point, team members Louis D'Amario and Dennis Byrnes took Diehl's initial trajectory concept and refined it to a much greater level of accuracy. D'Amario's and Byrnes's contributions were critical because Diehl's work was based on design tools that did not have the accuracy to demonstrate that the VEEGA trajectory would actually be successful.[42]

Working out the VEEGA trajectory was the high point of the Galileo mission for Mitchell. He says about his experience during the effort: "Even though Roger was the one that made the discovery, and Lou and Dennis were the ones that designed it to where it was really workable and flyable, I was the leader of the team, and the involvement, the effort, the result of that was very exciting and very rewarding. We took a situation where we almost surely had the spacecraft going to the Smithsonian and turned it around."[43]

Mitchell believes that the most significant result of the Galileo mission was the detailed understanding that we attained of the Jovian system. The importance of the spacecraft's discoveries was underlined in a meeting, which Mitchell attended, of the mission's entire science group. One of the scientists asserted that as a result of the data collected by Galileo at Jupiter, they were going to "have to throw away all the textbooks and start to rewrite 'em on Jupiter." And all the other scientists in the room agreed.[44]

Jim Erickson: Galileo Project Manager, 1998–2001

Immediately after the high-gain antenna had failed . . . [I had to go] into a meeting to explain what we had just seen in the telemetry, and what it meant to the team, and what it meant to the project . . . [and] to keep everybody from getting into the mode of, "Gee, the mission has failed." . . . We were a bunch of smart people, and we would find a way to solve this problem.

—Jim Erickson, former Galileo Project Manager

Jim Erickson was Galileo's sequence team chief at the time that the high-gain antenna failed to open on Galileo's interplanetary trajectory to Jupiter. Erickson's job as sequence team chief was to run the team of 60 people who took all of the project staff's "science experiment desires" and "engineering desires" and created sequences of commands necessary for the spacecraft to achieve those desires. Transmission of these sequences to Galileo depended on the spacecraft's having an operable antenna. The high-gain antenna's failure jeopardized the transmissions and threatened to shut down the mission.

---

[41] Mitchell interview, 10 May 2001.

[42] Ibid.

[43] Ibid.

[44] Ibid.

Immediately following the antenna failure, nobody on Erickson's team had a clue how to solve the problem. Erickson said that the most difficult thing he ever had to do on the mission was "facing the 60 people who worked for me, trying to have them not despair of Galileo working."[45] He had to maintain their morale and convince them that they would discover a means of making the mission successful. Ultimately, they did.[46]

Erickson was attracted to the Galileo mission soon after its inception in 1977 because it "seemed to be the most challenging mission on the horizon. The hardest job." Galileo management liked his work and kept giving him new jobs to do, including serving as deputy manager of the Galileo Engineering Office, manager of the Galileo Science and Sequence Office, Deputy Project Manager, and finally, in 1998, Galileo Project Manager.[47]

Eilene Theilig: Galileo Project Manager, 2001–03

I really like watching people presented with a problem, and using their creativity to overcome it. And we've had lots of examples of that . . . . The people who work here are really dedicated. It's not just a job.[48]

—Eilene Theilig, former Galileo Project Manager

What Eilene Theilig enjoyed about working on the mission was that it changed character every few years. The spacecraft always presented her and the rest of the team with challenges, as did the evolving complexion of the mission—the six-year interplanetary journey, the two-year Primary Mission observations of Jupiter and its moons, and the extended missions activities that had to be accomplished with greatly reduced staffs and budgets.[49]

Theilig was tremendously impressed with the design and construction of a spacecraft "that is so robust and is still going . . . . It's done far more than people thought it would. With the troubles that we've overcome, we still made numerous high-level, high-priority scientific observations at Jupiter." Observations that the spacecraft made have implications that reach far beyond the space science community. For instance, the possible identification of water on Europa is driving many Earth studies of "extremophiles"—life-forms that exist in areas we did not think possible, like in Earth's very deep oceans or under kilometers of Antarctic ice.[50]

Galileo's most important accomplishment, in Theilig's mind, was to observe lava flowing on the surface of Io, in an environment similar to what existed on Earth two billion

---

[45] Jim Erickson interview, telephone conversation, 8 May 2001.

[46] Ibid.

[47] Ibid.; Jane Platt and Mary Beth Murrill, "New Managers Appointed for Cassini and Galileo Missions," JPL Media Relations Office news release, 4 June 1998.

[48] Eilene Theilig interview, tape-recorded telephone conversation, 8 May 2001.

[49] Porter Anderson, "Eilene Theilig: Faith, by Jupiter," CNN.com/CAREER Web site, 16 July 2001, *http://www.cnn.com/2001/CAREER/jobenvy/07/16/eilene.theilig/index.html* (available in folder 18522, NASA Historical Reference Collection, Washington, DC).

[50] Theilig interview, 8 May 2001.

years ago. As a planetary geologist specializing in lava flows, she was thrilled by the Io pictures, in part because they provided us with windows into our own planet's distant past.[51]

Theilig's love of her work came across very strongly. Her enthusiasm extended to both the people on the project and the science achieved. But she tried hard to balance her never-ending Galileo tasks with nonscientific pursuits. She recognized that hers was "a career type that could eat you up." Although she spent many hours after the workday and on weekends at JPL, she was also deeply committed to her faith community, in which she was an elder. "Our society is so mobile that very few of us live in an area with the large extended family that we used to have," Theilig said. "I think faith communities can help fill that gap."[52]

Theilig found the Galileo project very welcoming to women. She noted the high number of women working on the mission, particularly in the science and design areas. Theilig's assignments on the mission were largely engineering-oriented, and she found it taxing to be a scientist trying to fit into engineering niches. "The terminology and the way engineers think was more of a challenge for me [than being a woman on the project]."[53]

Of the many technical challenges she has faced, the most difficult was redesigning the mission after the high-gain antenna failed to open and figuring out how the team would operate the spacecraft at Jupiter, given the new constraints. She and the rest of the flight team had to create a very different mission quickly, one that did not depend on real-time transmission of most of its data. Galileo staff had to modify the spacecraft's on-board programming to consolidate, compress, and store as much of the science data as possible in order to send it to Earth slowly through the spacecraft's low-gain antenna. That whole reprogramming had to be done while the team was also very busy flying the spacecraft and collecting data, such as from the Shoemaker-Levy comet. Meeting all of these objectives at the same time was extremely difficult. But it was also exhilarating. In Theilig's words, "I really like watching people presented with a problem and using their creativity to overcome it. And we've had lots of examples of that."[54]

Mission Support Personnel

If you cut us, we bleed Galileo.[55]

—Nagin Cox, quoting Shadan Ardalan, Galileo engineer

Galileo's managers fought the political and budget battles and made the high-level technical decisions that gave the mission life and kept it alive during crises. But it was the mission support staff who provided hands-on expertise, quickly interpreting data streaming onto their monitors from the spacecraft, determining whether the craft was doing what it was supposed to, and fixing it from 450 million miles away when it ran into problems. Successful Jovian moon flybys would not have been achieved without the vital activities performed by mission support engineers and technicians, sometimes under extreme

[51] Ibid.

[52] Porter Anderson, "Eilene Theilig: Faith, by Jupiter."

[53] Theilig interview, 8 May 2001.

[54] Ibid.

[55] Nagin Cox interview, tape-recorded telephone conversation, 15 May 2001. Cox was quoting a statement made by Shadan Ardalan, Galileo Attitude and Articulation Control Subsystem Engineer.

pressure. The Galileo spacecraft could not be stopped while support crew analyzed a malfunction. If a problem was not fixed "on the fly," before an encounter was completed, the opportunity might be lost to collect key data that a science team had been waiting years to receive.

One of many critical mission support personnel was Nagin Cox. Her experiences on various Galileo teams mirror those of many of her colleagues and illustrate the vital responsibilities that these people had. During most of Galileo's Prime Mission, Nagin Cox served on the Orbiter engineering team, focusing on "system fault protection." Galileo was so far away from Earth, and it took such a long time to communicate with the craft, that it needed to "take care of itself for a while" when problems arose. Cox helped to maintain and apply the on-board algorithms that protected Galileo and put it into a safe mode or told it how to respond if something went wrong. These fault protection algorithms were complex enough that a separate team was formed to work on them and ensure that Galileo was always configured properly to take care of itself.[56]

Cox was appointed deputy team chief for the spacecraft and sequence team on the Galileo Europa Mission, which ran from 1997 through 1999. Her Galileo experiences were most intense on GEM, after project staff had been greatly reduced from Prime Mission levels of about 200 people to only 60. The spacecraft was beginning to exhibit the cumulative effects of radiation exposure, and new problems in its operation were surfacing. Project staff became a very tight-knit squad as they dealt with these issues that often surfaced only hours before an encounter. A breakneck pace of work ensued for two years and included many satellite encounters and long days and nights keeping the spacecraft operational. Just a week following GEM's kickoff on 8 December 1997, or, in Cox's words, "seven days after we were handed the keys," radiation-related malfunctions in the ship's gyros arose. The gyros were used to point Galileo's scan platform, on which sat the SSI camera and other remote sensing instruments, and also to help ascertain spacecraft orientation. The team determined that a damaged field-effect transistor switch was responsible for the gyro problem. The switch was supposed to respond to the spacecraft's degree of rotation but had begun to over-react, putting out more pulses than it should have for every degree of movement. For example, the spacecraft might only have rotated 10 degrees, but the signal that the gyros sent would say that it had rotated perhaps 50 degrees. Cox's team had to write an algorithm to account for the gyros' altered behavior.[57]

When Cox was hired by JPL, it was the fulfillment of a desire that she'd had since she was 14. She said that "working at JPL is all I've ever wanted to do." She was more attracted to JPL's robotic space exploration program than to the human missions conducted by other NASA sites because robotic spacecraft could travel the furthest into the unknown, uncharted regions of space. When the chance came to join the Galileo mission, it was particularly enticing. In her words, "What could be more cool than the first spacecraft to go into orbit around [an outer] planet? The first probe to take in situ measurements of a gas giant? For some engineers, . . . once it's launched, they're fine with that, and they tend to go on and build something else. Or once it's in orbit around Jupiter, they're fine with that, and they'd like to go do the hard engineering work getting something else into orbit. For me, I was totally thrilled and entranced to actually be doing

---

[56] Cox interview, 15 May 2001.

[57] Ibid.

the day-to-day operations. I'd walk outside at night and look up at Jupiter, and think, our ship's up there."[58]

As Galileo approached Io prior to its I24 encounter, the responsibility of Cox's position as deputy team chief weighed heavily on her. She wondered whether she had just spent the last guilt-free night of her life. If anything went wrong on the encounter, if her team didn't get the data from Io that the science teams were counting on, she would feel partly responsible.

One of Cox's new duties would be to lead the mission's anomaly response team. For two years, she would serve as the project's "first call," which meant that whenever a problem occurred with the spacecraft, she would be contacted first by the mission controller on duty. She would stay on call 24 hours a day, 7 days a week, unless she made prior arrangements. It turned out to be a challenging two years, but Cox gained a deep satisfaction from being "so connected to the ship," as well as to her team.[59]

Although mission support activities were not as widely publicized as those of Galileo's managers, mission support personnel's actions were no less vital to the completion of the Galileo mission. It was mission support's charge to keep the spacecraft running and provide the key link between the dreams and hopes of Galileo's scientific teams and the achievement of those desires.

This chapter has highlighted only some of the personnel who made Galileo a success. The Galileo team members brought many different skill sets to the project, but they all shared a high degree of dedication to the mission and a determination to do whatever it took to solve the many problems that arose, some of which could have ended the mission. For all of these people, Galileo was more than just a job.

---

[58] Ibid.

[59] Ibid.

Chapter 11

CONCLUSION

THIS BOOK HAS ATTEMPTED TO CAPTURE THE IMPORTANCE OF THE Galileo mission to our country and to the world. The Galileo spacecraft explored new, exciting territory, using instruments that were, in essence, the eyes, ears, and fingertips of humankind. As the Galileo Probe buried itself in Jupiter's atmosphere, as the Orbiter skimmed over Io's fire fountains and Europa's ice and possible buried ocean, it was *we* who were exploring uncharted frontiers.

These were not merely geographic frontiers being studied, but technological ones as well. To even get the Galileo spacecraft to Jupiter, NASA and the rest of the U.S. space science community had to push the envelope of what was possible. First, they had to create a vision for the mission that was so compelling that it would inspire a skeptical Congress and several recalcitrant White House administrations to fund the project. Then, they had to create a space vehicle robust enough to stay alive through all kinds of unexpected occurrences. And the challenge did not end with the spacecraft's launch. When systems aboard Galileo malfunctioned and threatened to shut down our eyes and ears, mission staff had to invent new ways of operating and then reprogram the spacecraft through a thin lifeline of impaired radio communication from a distance of hundreds of millions of miles.

The mission logged so many notable achievements. One of the most difficult was simply getting the spacecraft off the ground. Years of battles with OMB and Congress, accompanied by one redesign of the spacecraft after another (including a major redesign after the *Challenger* disaster), made the eventual launch of Galileo a triumph of perseverance and willpower. Overcoming postlaunch mechanical difficulties with the spacecraft was also a premier achievement. For instance, the ingenuity that was required to keep the mission going after its main antenna refused to open can only be described as awesome.

No less impressive were the years of discoveries within the Jovian system. In situ profiling of Jupiter's upper atmosphere, observations of Io's extreme volcanism, and the discovery that an internal Europan ocean might harbor life were products of the Galileo effort that will long be remembered. So will the studies of the Shoemaker-Levy comet and the massive interplanetary dust storm. What may be less remembered, but was no less critical, were the decades of united effort by thousands of engineers, scientists, technicians, and politicians to create and carry out one of the greatest and most successful voyages of discovery ever attempted.

Figure 11.1. **The demise of Galileo.**

Acronyms and Abbreviations

^{40}K	potassium-40, an isotope of potassium
ABMA	Army Ballistic Mission Agency
AGU	American Geophysical Union
AIAA	American Institute of Aeronautics and Astronautics
ASC	Astro Sciences Center; part of IITRI
ASI	atmospheric structure instrument
AU	Astronomical Unit (distance from Earth to the Sun)
BMFT	Bundesministerium fuer Forschung und Technoligie
Caltech	California Institute of Technology
CB	Citizens Band
CCD	charge-coupled device
CFC	chlorofluorocarbon
CH_4	methane
COMPLEX	Committee on Planetary and Lunar Exploration
DC	direct current
DDS	dust detection system
DOD	Department of Defense
DOE	Department of Energy
DPS	Division for Planetary Sciences
DSN	Deep Space Network
EDT	eastern daylight time
EGA	Earth gravity-assist
EGS	European Geophysical Society
EPD	energetic particles detector
EPI	energetic particles instrument
ESRO	European Space Research Organization
ESTEC	European Space Research and Technology Centre
EUG	European Union of Geosciences
EUV	extreme ultraviolet
FEIS	final environmental impact statement
FSAR	final safety analysis report
FY	Fiscal Year
GAO	General Accounting Office
GC	gas chromatograph
GC/MS	gas chromatograph–mass spectrometer
GEM	Galileo Europa Mission
GMM	Galileo Millennium Mission
GMT	Greenwich mean time

GPO	Government Printing Office
H_2	molecular hydrogen
H_2O	water
H_3	hydrogen isotope
HAD	helium abundance detector
HEOS	Highly Eccentric Orbit Satellite
HGA	high-gain antenna
HST	Hubble Space Telescope
HUD	Housing and Urban Development
IITRI	Illinois Institute of Research Technology Research Institute
INSRP	Interagency Nuclear Safety Review Panel
IRAS	Infrared Astronomy Satellite
IRD	Interface Requirements Document
IUS	Interim Upper Stage; later, Inertial Upper Stage
JOI	Jupiter Orbit Insertion
JOP	Jupiter Orbiter Probe
JOPSWG	JOP science working group
JPL	Jet Propulsion Laboratory
JSC	Johnson Space Center
keV	thousand (kilo) electron volts
KSC	Kennedy Space Center
LED	light-emitting diodes
LEMMS	low-energy magnetospheric measurements system
LGA	low-gain antenna
LRD	lightning and radio emissions detector
MeV	million electron volts
mph	miles per hour
N_2O	nitrous oxide
NASA	National Aeronautics and Space Administration
NEP	nephelometer
NEPA	National Environmental Policy Act
NFR	net flux radiometer
NH_3	ammonia
NH_4SH	ammonium hydrosulfide
NIMS	near infrared mapping spectrometer
NMS	neutral mass spectrometer
NO_3	oceanic nitrate
NOAA	National Oceanic and Atmospheric Administration
NSSDC	National Space Science Data Center
OMB	Office of Management and Budget
PDT	Pacific daylight time
PLS	plasma subsystem

PPR	photopolarimeter radiometer
psi	pounds per square inch
PST	Pacific standard time
PWS	plasma wave subsystem
RF $^{\bullet}$	radio frequency
RFP	Request for Proposal
R_J	Jupiter radii
RPM	retropropulsion module (part of the Jupiter Orbiter)
rpm	revolutions per minute
RTG	radioisotope thermal generator
S/L-9	Comet Shoemaker-Levy 9
SAF	Spacecraft Assembly Facility (JPL)
SAG	Science Advisory Group
SAR	safety analysis report
SBA	spin bearing assembly
SEC	Sun-Earth-Craft
SEDS	Students for the Exploration and Development of Space
SETI	Search for Extraterrestrial Intelligence
SITURNS	spacecraft inertial turns, also called science turns
SOE	sequence of events
SOPE	Strategy for Outer Space Exploration
SRM	solid rocket motor
SSI	solid state imaging
Star Wars	nickname for the Strategic Defense Initiative
TCM	trajectory correction maneuver
TDM	time division multiplex
TDRS	Tracking and Data Relay Satellite
TOPS	Thermoelectric Outer Planet Spacecraft
U.S.	United States
UCAR	University Corporation for Atmospheric Research
UCLA	University of California, Los Angeles
UT	universal time
UV	ultraviolet
UVS	ultraviolet spectrometer
VEEGA	Venus-Earth-Earth gravity-assist
VLF	Very Low Frequency
VOIR	Venus Orbiter Imaging Radar

Index

Note: Terms such as "NASA," "Jet Propulsion Laboratory (JPL)," "Orbiter," and "Probe" are not included because of their high frequency.

NASA History Series

Reference Works, NASA SP-4000

Grimwood, James M. *Project Mercury: A Chronology*. NASA SP-4001, 1963.

Grimwood, James M., and C. Barton Hacker, with Peter J. Vorzimmer. *Project Gemini Technology and Operations: A Chronology*. NASA SP-4002, 1969.

Link, Mae Mills. *Space Medicine in Project Mercury*. NASA SP-4003, 1965.

Astronautics and Aeronautics, 1963: Chronology of Science, Technology, and Policy. NASA SP-4004, 1964.

Astronautics and Aeronautics, 1964: Chronology of Science, Technology, and Policy. NASA SP-4005, 1965.

Astronautics and Aeronautics, 1965: Chronology of Science, Technology, and Policy. NASA SP-4006, 1966.

Astronautics and Aeronautics, 1966: Chronology of Science, Technology, and Policy. NASA SP-4007, 1967.

Astronautics and Aeronautics, 1967: Chronology of Science, Technology, and Policy. NASA SP-4008, 1968.

Ertel, Ivan D., and Mary Louise Morse. *The Apollo Spacecraft: A Chronology, Volume I, Through November 7, 1962*. NASA SP-4009, 1969.

Morse, Mary Louise, and Jean Kernahan Bays. *The Apollo Spacecraft: A Chronology, Volume II, November 8, 1962–September 30, 1964*. NASA SP-4009, 1973.

Brooks, Courtney G., and Ivan D. Ertel. *The Apollo Spacecraft: A Chronology, Volume III, October 1, 1964–January 20, 1966*. NASA SP-4009, 1973.

Ertel, Ivan D., and Roland W. Newkirk, with Courtney G. Brooks. *The Apollo Spacecraft: A Chronology, Volume IV, January 21, 1966–July 13, 1974*. NASA SP-4009, 1978.

Astronautics and Aeronautics, 1968: Chronology of Science, Technology, and Policy. NASA SP-4010, 1969.

Newkirk, Roland W., and Ivan D. Ertel, with Courtney G. Brooks. *Skylab: A Chronology*. NASA SP-4011, 1977.

Van Nimmen, Jane, and Leonard C. Bruno, with Robert L. Rosholt. *NASA Historical Data Book, Volume I: NASA Resources, 1958–1968*. NASA SP-4012, 1976, rep. ed. 1988.

Ezell, Linda Neuman. *NASA Historical Data Book, Volume II: Programs and Projects, 1958–1968*. NASA SP-4012, 1988.

Ezell, Linda Neuman. *NASA Historical Data Book, Volume III: Programs and Projects, 1969–1978*. NASA SP-4012, 1988.

Gawdiak, Ihor Y., with Helen Fedor, compilers. *NASA Historical Data Book, Volume IV: NASA Resources, 1969–1978*. NASA SP-4012, 1994.

Rumerman, Judy A., compiler. *NASA Historical Data Book, 1979–1988: Volume V, NASA Launch Systems, Space Transportation, Human Spaceflight, and Space Science*. NASA SP-4012, 1999.

Rumerman, Judy A., compiler. *NASA Historical Data Book, Volume VI: NASA Space Applications, Aeronautics and Space Research and Technology, Tracking and Data Acquisition/Space Operations, Commercial Programs, and Resources, 1979–1988*. NASA SP-2000-4012, 2000.

Astronautics and Aeronautics, 1969: Chronology of Science, Technology, and Policy. NASA SP-4014, 1970.

Astronautics and Aeronautics, 1970: Chronology of Science, Technology, and Policy. NASA SP-4015, 1972.

Astronautics and Aeronautics, 1971: Chronology of Science, Technology, and Policy. NASA SP-4016, 1972.

Astronautics and Aeronautics, 1972: Chronology of Science, Technology, and Policy. NASA SP-4017, 1974.

Astronautics and Aeronautics, 1973: Chronology of Science, Technology, and Policy. NASA SP-4018, 1975.

Astronautics and Aeronautics, 1974: Chronology of Science, Technology, and Policy. NASA SP-4019, 1977.

Astronautics and Aeronautics, 1975: Chronology of Science, Technology, and Policy. NASA SP-4020, 1979.

Astronautics and Aeronautics, 1976: Chronology of Science, Technology, and Policy. NASA SP-4021, 1984.

Astronautics and Aeronautics, 1977: Chronology of Science, Technology, and Policy. NASA SP-4022, 1986.

Astronautics and Aeronautics, 1978: Chronology of Science, Technology, and Policy. NASA SP-4023, 1986.

Astronautics and Aeronautics, 1979–1984: Chronology of Science, Technology, and Policy. NASA SP-4024, 1988.

Astronautics and Aeronautics, 1985: Chronology of Science, Technology, and Policy. NASA SP-4025, 1990.

Noordung, Hermann. *The Problem of Space Travel: The Rocket Motor.* Edited by Ernst Stuhlinger and J. D. Hunley, with Jennifer Garland. NASA SP-4026, 1995.

Gawdiak, Ihor Y., Ramon J. Miro, and Sam Stueland, comps. *Astronautics and Aeronautics, 1986–1990: A Chronology.* NASA SP-4027, 1997.

Gawdiak, Ihor Y. and Shetland, Charles. *Astronautics and Aeronautics, 1990–1995: A Chronology.* NASA SP-2000-4028, 2000.

Orloff, Richard W. *Apollo by the Numbers: A Statistical Reference.* NASA SP-2000-4029, 2000.

Management Histories, NASA SP-4100

Rosholt, Robert L. *An Administrative History of NASA, 1958–1963.* NASA SP-4101, 1966.

Levine, Arnold S. *Managing NASA in the Apollo Era.* NASA SP-4102, 1982.

Roland, Alex. *Model Research: The National Advisory Committee for Aeronautics, 1915–1958.* NASA SP-4103, 1985.

Fries, Sylvia D. *NASA Engineers and the Age of Apollo.* NASA SP-4104, 1992.

Glennan, T. Keith. *The Birth of NASA: The Diary of T. Keith Glennan*. J. D. Hunley, editor. NASA SP-4105, 1993.

Seamans, Robert C., Jr. *Aiming at Targets: The Autobiography of Robert C. Seamans, Jr.* NASA SP-4106, 1996.

Garber, Stephen J., editor. *Looking Backward, Looking Forward: Forty Years of U.S. Human Spaceflight Symposium*. NASA SP-2002-4107, 2002.

Mallick, Donald L. with Peter W. Merlin. *The Smell of Kerosene: A Test Pilot's Odyssey*. NASA SP-4108, 2003.

Iliff, Kenneth W. and Curtis L. Peebles. *From Runway to Orbit: Reflections of a NASA Engineer*. NASA SP-2004-4109, 2004.

Chertok, Boris. *Rockets and People, Volume 1*. NASA SP-2005-4110, 2005.

Chertok, Boris. *Rockets and People: Creating a Rocket Industry, Volume II*. NASA SP-2006-4110, 2006.

Laufer, Alexander, Todd Post, and Edward Hoffman. *Shared Voyage: Learning and Unlearning from Remarkable Projects*. NASA SP-2005-4111, 2005.

Dawson, Virginia P. and Mark D. Bowles. *Realizing the Dream of Flight: Biographical Essays in Honor of the Centennial of Flight, 1903–2003*. NASA SP-2005-4112, 2005.

Project Histories, NASA SP-4200:

Swenson, Loyd S., Jr., James M. Grimwood, and Charles C. Alexander. *This New Ocean: A History of Project Mercury*. NASA SP-4201, 1966; rep. ed. 1998.

Green, Constance McLaughlin, and Milton Lomask. *Vanguard: A History*. NASA SP-4202, 1970; rep. ed. Smithsonian Institution Press, 1971.

Hacker, Barton C., and James M. Grimwood. *On the Shoulders of Titans: A History of Project Gemini*. NASA SP-4203, 1977.

Benson, Charles D., and William Barnaby Faherty. *Moonport: A History of Apollo Launch Facilities and Operations*. NASA SP-4204, 1978.

Brooks, Courtney G., James M. Grimwood, and Loyd S. Swenson, Jr. *Chariots for Apollo: A History of Manned Lunar Spacecraft*. NASA SP-4205, 1979.

Bilstein, Roger E. *Stages to Saturn: A Technological History of the Apollo/Saturn Launch Vehicles*. NASA SP-4206, 1980, rep. ed. 1997.

SP-4207 not published.

Compton, W. David, and Charles D. Benson. *Living and Working in Space: A History of Skylab*. NASA SP-4208, 1983.

Ezell, Edward Clinton, and Linda Neuman Ezell. *The Partnership: A History of the Apollo–Soyuz Test Project*. NASA SP-4209, 1978.

Hall, R. Cargill. *Lunar Impact: A History of Project Ranger*. NASA SP-4210, 1977.

Newell, Homer E. B*eyond the Atmosphere: Early Years of Space Science*. NASA SP-4211, 1980.

Ezell, Edward Clinton, and Linda Neuman Ezell. *On Mars: Exploration of the Red Planet, 1958–1978*. NASA SP-4212, 1984.

Pitts, John A. *The Human Factor: Biomedicine in the Manned Space Program to 1980.* NASA SP-4213, 1985.

Compton, W. David. *Where No Man Has Gone Before: A History of Apollo Lunar Exploration Missions.* NASA SP-4214, 1989.

Naugle, John E. *First Among Equals: The Selection of NASA Space Science Experiments.* NASA SP-4215, 1991.

Wallace, Lane E. *Airborne Trailblazer: Two Decades with NASA Langley's Boeing 737 Flying Laboratory.* NASA SP-4216, 1994.

Butrica, Andrew J., editor. *Beyond the Ionosphere: Fifty Years of Satellite Communication.* NASA SP-4217, 1997.

Butrica, Andrew J. *To See the Unseen: A History of Planetary Radar Astronomy.* NASA SP-4218, 1996.

Mack, Pamela E., editor. *From Engineering Science to Big Science: The NACA and NASA Collier Trophy Research Project Winners.* NASA SP-4219, 1998.

Reed, R. Dale, with Darlene Lister. *Wingless Flight: The Lifting Body Story.* NASA SP-4220, 1997.

Heppenheimer, T. A. *The Space Shuttle Decision: NASA's Search for a Reusable Space Vehicle.* NASA SP-4221, 1999.

Hunley, J. D., editor. *Toward Mach 2: The Douglas D-558 Program.* NASA SP-4222, 1999.

Swanson, Glen E., editor. *"Before this Decade Is Out . . .": Personal Reflections on the Apollo Program.* NASA SP-4223, 1999.

Tomayko, James E. *Computers Take Flight: A History of NASA's Pioneering Digital Fly-by-Wire Project.* NASA SP-2000-4224, 2000.

Morgan, Clay. *Shuttle-Mir: The U.S. and Russia Share History's Highest Stage.* NASA SP-2001-4225, 2001.

Leary, William M. *"We Freeze to Please": A History of NASA's Icing Research Tunnel and the Quest for Flight Safety.* NASA SP-2002-4226, 2002.

Mudgway, Douglas J. *Uplink-Downlink: A History of the Deep Space Network 1957–1997.* NASA SP-2001-4227, 2001.

Dawson, Virginia P. and Mark D. Bowles. *Taming Liquid Hydrogen: The Centaur Upper Stage Rocket, 1958–2002.* NASA SP-2004-4230, 2004.

Center Histories, NASA SP-4300

Rosenthal, Alfred. *Venture into Space: Early Years of Goddard Space Flight Center.* NASA SP-4301, 1985.

Hartman, Edwin P. *Adventures in Research: A History of Ames Research Center, 1940–1965.* NASA SP-4302, 1970.

Hallion, Richard P. *On the Frontier: Flight Research at Dryden, 1946–1981.* NASA SP-4303, 1984.

Muenger, Elizabeth A. *Searching the Horizon: A History of Ames Research Center, 1940–1976.* NASA SP-4304, 1985.

Hansen, James R. *Engineer in Charge: A History of the Langley Aeronautical Laboratory, 1917-1958*. NASA SP-4305, 1987.

Dawson, Virginia P. *Engines and Innovation: Lewis Laboratory and American Propulsion Technology*. NASA SP-4306, 1991.

Dethloff, Henry C. *"Suddenly Tomorrow Came . . .": A History of the Johnson Space Center*. NASA SP-4307, 1993.

Hansen, James R. *Spaceflight Revolution: NASA Langley Research Center from Sputnik to Apollo*. NASA SP-4308, 1995.

Wallace, Lane E. *Flights of Discovery: 50 Years at the NASA Dryden Flight Research Center*. NASA SP-4309, 1996.

Herring, Mack R. *Way Station to Space: A History of the John C. Stennis Space Center*. NASA SP-4310, 1997.

Wallace, Harold D., Jr. *Wallops Station and the Creation of the American Space Program*. NASA SP-4311, 1997.

Wallace, Lane E. *Dreams, Hopes, Realities: NASA's Goddard Space Flight Center, The First Forty Years*. NASA SP-4312, 1999.

Dunar, Andrew J., and Stephen P. Waring. *Power to Explore: A History of the Marshall Space Flight Center*. NASA SP-4313, 1999.

Bugos, Glenn E. *Atmosphere of Freedom: Sixty Years at the NASA Ames Research Center*. NASA SP-2000-4314, 2000.

Schultz, James. *Crafting Flight: Aircraft Pioneers and the Contributions of the Men and Women of NASA Langley Research Center*. NASA SP-2003-4316, 2003.

Bowles, Mark D. *Science in Flux: NASA's Nuclear Program at Plum Brook Station, 1955–2005*. NASA SP-2006–4317, 2006.

General Histories, NASA SP-4400

Corliss, William R. *NASA Sounding Rockets, 1958–1968: A Historical Summary*. NASA SP-4401, 1971.

Wells, Helen T., Susan H. Whiteley, and Carrie Karegeannes. *Origins of NASA Names*. NASA SP-4402, 1976.

Anderson, Frank W., Jr. *Orders of Magnitude: A History of NACA and NASA, 1915–1980*. NASA SP-4403, 1981.

Sloop, John L. *Liquid Hydrogen as a Propulsion Fuel, 1945–1959*. NASA SP-4404, 1978.

Roland, Alex. *A Spacefaring People: Perspectives on Early Spaceflight*. NASA SP-4405, 1985.

Bilstein, Roger E. *Orders of Magnitude: A History of the NACA and NASA, 1915–1990*. NASA SP-4406, 1989.

Logsdon, John M., editor, with Linda J. Lear, Jannelle Warren–Findley, Ray A. Williamson, and Dwayne A. Day. *Exploring the Unknown: Selected Documents in the History of the U.S. Civil Space Program, Volume I, Organizing for Exploration*. NASA SP-4407, 1995.

Logsdon, John M., editor, with Dwayne A. Day and Roger D. Launius. *Exploring the Unknown: Selected Documents in the History of the U.S. Civil Space Program, Volume II, Relations with Other Organizations*. NASA SP-4407, 1996.

Logsdon, John M., editor, with Roger D. Launius, David H. Onkst, and Stephen J. Garber. *Exploring the Unknown: Selected Documents in the History of the U.S. Civil Space Program, Volume III, Using Space.* NASA SP-4407, 1998.

Logsdon, John M., general editor, with Ray A. Williamson, Roger D. Launius, Russell J. Acker, Stephen J. Garber, and Jonathan L. Friedman. *Exploring the Unknown: Selected Documents in the History of the U.S. Civil Space Program, Volume IV, Accessing Space.* NASA SP-4407, 1999.

Logsdon, John M., general editor, with Amy Paige Snyder, Roger D. Launius, Stephen J. Garber, and Regan Anne Newport. *Exploring the Unknown: Selected Documents in the History of the U.S. Civil Space Program, Volume V, Exploring the Cosmos.* NASA SP-2001-4407, 2001.

Siddiqi, Asif A. *Challenge to Apollo: The Soviet Union and the Space Race, 1945–1974.* NASA SP-2000-4408, 2000.

Hansen, James R., editor. *The Wind and Beyond: Journey into the History of Aerodynamics in America, Volume 1, The Ascent of the Airplane.* NASA SP-2003-4409, 2003.

Monographs in Aerospace History, NASA SP-4500

Launius, Roger D. and Aaron K. Gillette, compilers, *Toward a History of the Space Shuttle: An Annotated Bibliography.* Monograph in Aerospace History, No. 1, 1992.

Launius, Roger D., and J. D. Hunley, compilers, *An Annotated Bibliography of the Apollo Program.* Monograph in Aerospace History, No. 2, 1994.

Launius, Roger D. *Apollo: A Retrospective Analysis.* Monograph in Aerospace History, No. 3, 1994.

Hansen, James R. *Enchanted Rendezvous: John C. Houbolt and the Genesis of the Lunar-Orbit Rendezvous Concept.* Monograph in Aerospace History, No. 4, 1995.

Gorn, Michael H. *Hugh L. Dryden's Career in Aviation and Space.* Monograph in Aerospace History, No. 5, 1996.

Powers, Sheryll Goecke. *Women in Flight Research at NASA Dryden Flight Research Center, from 1946 to 1995.* Monograph in Aerospace History, No. 6, 1997.

Portree, David S. F. and Robert C. Trevino. *Walking to Olympus: An EVA Chronology.* Monograph in Aerospace History, No. 7, 1997.

Logsdon, John M., moderator. *Legislative Origins of the National Aeronautics and Space Act of 1958: Proceedings of an Oral History Workshop.* Monograph in Aerospace History, No. 8, 1998.

Rumerman, Judy A., compiler, *U.S. Human Spaceflight, A Record of Achievement 1961–1998.* Monograph in Aerospace History, No. 9, 1998.

Portree, David S. F. *NASA's Origins and the Dawn of the Space Age.* Monograph in Aerospace History, No. 10, 1998.

Logsdon, John M. *Together in Orbit: The Origins of International Cooperation in the Space Station.* Monograph in Aerospace History, No. 11, 1998.

Phillips, W. Hewitt. *Journey in Aeronautical Research: A Career at NASA Langley Research Center.* Monograph in Aerospace History, No. 12, 1998.

Braslow, Albert L. *A History of Suction-Type Laminar-Flow Control with Emphasis on Flight Research.* Monograph in Aerospace History, No. 13, 1999.

Logsdon, John M., moderator. *Managing the Moon Program: Lessons Learned From Apollo*. Monograph in Aerospace History, No. 14, 1999.

Perminov, V. G. *The Difficult Road to Mars: A Brief History of Mars Exploration in the Soviet Union*. Monograph in Aerospace History, No. 15, 1999.

Tucker, Tom. *Touchdown: The Development of Propulsion Controlled Aircraft at NASA Dryden*. Monograph in Aerospace History, No. 16, 1999.

Maisel, Martin D., Demo J. Giulianetti, and Daniel C. Dugan. *The History of the XV-15 Tilt Rotor Research Aircraft: From Concept to Flight*. NASA SP-2000-4517, 2000.

Jenkins, Dennis R. *Hypersonics Before the Shuttle: A Concise History of the X-15 Research Airplane*. NASA SP-2000-4518, 2000.

Chambers, Joseph R. *Partners in Freedom: Contributions of the Langley Research Center to U.S. Military Aircraft in the 1990s*. NASA SP-2000-4519, 2000.

Waltman, Gene L. *Black Magic and Gremlins: Analog Flight Simulations at NASA's Flight Research Center*. NASA SP-2000-4520, 2000.

Portree, David S. F. *Humans to Mars: Fifty Years of Mission Planning, 1950–2000*. NASA SP-2001-4521, 2001.

Thompson, Milton O., with J. D. Hunley. *Flight Research: Problems Encountered and What They Should Teach Us*. NASA SP-2000-4522, 2000.

Tucker, Tom. *The Eclipse Project*. NASA SP-2000-4523, 2000.

Siddiqi, Asif A. *Deep Space Chronicle: A Chronology of Deep Space and Planetary Probes, 1958–2000*. NASA SP-2002-4524, 2002.

Merlin, Peter W. *Mach 3+: NASA/USAF YF-12 Flight Research, 1969–1979*. NASA SP-2001-4525, 2001.

Anderson, Seth B. *Memoirs of an Aeronautical Engineer-Flight Tests at Ames Research Center: 1940–1970*. NASA SP-2002-4526, 2002.

Renstrom, Arthur G. *Wilbur and Orville Wright: A Bibliography Commemorating the One–Hundredth Anniversary of the First Powered Flight on December 17, 1903*. NASA SP-2002-4527, 2002.

No monograph 28.

Chambers, Joseph R. *Concept to Reality: Contributions of the NASA Langley Research Center to U.S. Civil Aircraft of the 1990s*. NASA SP-2003-4529, 2003.

Peebles, Curtis, editor. *The Spoken Word: Recollections of Dryden History, The Early Years*. NASA SP-2003-4530, 2003.

Jenkins, Dennis R., Tony Landis, and Jay Miller. *American X-Vehicles: An Inventory-X-1 to X-50*. NASA SP-2003-4531, 2003.

Renstrom, Arthur G. *Wilbur and Orville Wright: A Chronology Commemorating the One-Hundredth Anniversary of the First Powered Flight on December 17, 1903*. NASA SP-2003-4532, 2002.

Bowles, Mark D. and Robert S. Arrighi. *NASA's Nuclear Frontier: The Plum Brook Research Reactor*. NASA SP-2004-4533, 2003.

Matranga, Gene J., Wayne C. Ottinger, Calvin R. Jarvis, and Christian D. Gelzer. *Unconventional, Contrary, and Ugly: The Lunar Landing Research Vehicle*. NASA SP-2006-4535, 2006.

McCurdy, Howard E. *Low Cost Innovation in Spaceflight: The History of the Near Earth Asteroid Rendezvous (NEAR) Mission.* NASA SP-2005-4536, 2005.

Seamans, Robert C. Jr. *Project Apollo: The Tough Decisions.* NASA SP-2005-4537, 2005.

Lambright, W. Henry. *NASA and the Environment: The Case of Ozone Depletion.* NASA SP-2005-4538, 2005.

Chambers, Joseph R. *Innovation in Flight: Research of the NASA Langley Research Center on Revolutionary Advanced Concepts for Aeronautics.* NASA SP-2005-4539, 2005.

Phillips, W. Hewitt. *Journey Into Space Research: Continuation of a Career at NASA Langley Research Center.* NASA SP-2005-4540, 2005.

Electronic Media, SP-4600 Series

Remembering Apollo 11: The 30th Anniversary Data Archive CD-ROM. NASA SP-4601, 1999.

The Mission Transcript Collection: U.S. Human Spaceflight Missions from Mercury Redstone 3 to Apollo 17. NASA SP-2000-4602, 2001, CD-ROM.

Shuttle–Mir: The United States and Russia Share History's Highest Stage. NASA SP-2001-4603, 2002, CD-ROM.

U.S. Centennial of Flight Commission presents Born of Dreams-Inspired by Freedom. NASA SP-2004-4604, 2004, DVD data disk.

Of Ashes and Atoms: A Documentary on the NASA Plum Brook Reactor Facility. NASA SP-2005-4605, DVD video.

Taming Liquid Hydrogen: The Centaur Upper Stage Rocket Interactive CD-ROM. NASA SP-2004-4606, 2004.

Fueling Space Exploration: The History of NASA's Rocket Engine Test Facility DVD. NASA SP-2005-4607, DVD video.

Conference Proceedings, SP-4700 Series

Dick, Steven J. and Keith L. Cowing, ed. *Risk and Exploration: Earth, Sea and the Stars.* NASA SP-2005-4701.

Dick, Steven J. and Roger D. Launius. *Critical Issues in the History of Spaceflight.* NASA 20064702.

www.ingramcontent.com/pod-product-compliance
Lightning Source LLC
Chambersburg PA
CBHW082137210326
41599CB00031B/6012